Tyrannosaurid Paleobiology

Life of the Past   James O. Farlow, editor

Indiana University Press   Bloomington & Indianapolis

Tyrannosaurid Paleobiology

Life of the Past   James O. Farlow, editor

Indiana University Press   Bloomington & Indianapolis

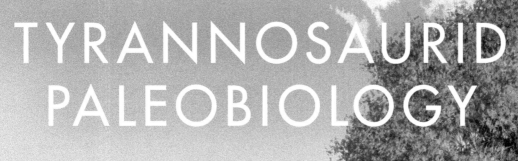

TYRANNOSAURID
PALEOBIOLOGY

EDITED BY **J. MICHAEL PARRISH,**
RALPH E. MOLNAR, PHILIP J. CURRIE,
AND **EVA B. KOPPELHUS**

This book is a publication of

Indiana University Press
Office of Scholarly Publishing
Herman B Wells Library 350
1320 East 10th Street
Bloomington, Indiana 47405 USA

iupress.indiana.edu

Telephone orders   800-842-6796
Fax orders   812-855-7931

*Manufactured in the
United States of America*

*Library of Congress
Cataloging-in-Publication Data*

Tyrannosaurid paleobiology / edited by
J. Michael Parrish, Ralph E. Molnar, Philip J.
Currie, and Eva B. Koppelhus.
    pages cm.–(Life of the past)
    "This volume had its genesis in a
conference held in Rockford, Illinois,
on September 16-18, 2005, titled 'The
Origin, Systematics, and Paleobiology of
Tyrannosauridae,' and jointly sponsored by
the Burpee Museum of Natural History and
Northern Illinois University"–Introduction.
    Includes index.
    ISBN 978-0-253-00930-2 (cl : alk.
paper)–ISBN 978-0-253-00947-0 (eb)
1. Tyrannosauridae. 2. Paleobiology.
3. Paleontology–Cretaceous. I. Parrish,
J. Michael, 1953- editor of compilation.
II. Molnar, Ralph E., editor of compilation.
III. Currie, Philip J., editor of compilation.
IV. Koppelhus, Eva B. (Eva Bundgaard),
editor of compilation.
    QE862.S3T96 2013
    567.912'9–dc23
                                                2013002879

1 2 3 4 5   18 17 16 15 14 13

Contents

C

Contents

Contributors

William L. Abler
Department of Geology
Field Museum of
Natural History
1400 South Lake Shore Drive
Chicago, IL 60605

Stephen L. Brusatte
Division of Paleontology
American Museum
of Natural History
Central Park West
at 79th Street
New York, NY 10024;
and Department of Earth
and Environmental Sciences
Columbia University
New York, NY
sbrusatte@amnh.org

Kenneth Carpenter
USU-CEU Prehistoric Museum
155 East Main
Price, UT 84501
Ken.Carpenter@usu.edu

Ralph E. Chapman
295 Bryce Avenue
Los Alamos, NM 87544

Philip J. Currie
Room Z 418
Department of
Biological Sciences
University of Alberta
Edmonton, Alberta
Canada T6G 2E9

James O. Farlow
Department of Geosciences
Indiana University-
Purdue University at Fort
Wayne, IN 46805

William F. Harrison
Northern Illinois University
DeKalb, IL 60115
aboramushi@gmail.com

Michael D. Henderson
Burpee Museum of
Natural History
737 North Main Street
Rockford, IL 61103; and
Northern Illinois University
DeKalb, IL 60115
fossilcat@aol.com

Thomas R. Holtz, Jr.
Department of Geology
University of Maryland
College Park, MD 20742

David W. E. Hone
Institute of Vertebrate
Paleontology and
Paleoanthropology
Chinese Academy of Sciences
PO Box 643
Beijing 100044
People's Republic of China

Grant R. Hurlburt
Department of Natural History
Royal Ontario Museum
Toronto, ON M5S 2C6
Canada
ghurlburt70@yahoo.com

Tyler Keillor
Department of Organismal
Biology and Anatomy
Fossil Laboratory
University of Chicago
5620 S. Ellis Ave.
Chicago, IL 60637
tylerkeillor@gmail.com

Eva B. Koppelhus
Room CW-405
Department of
Biological Sciences
University of Alberta
Edmonton, Alberta
Canada T6G 2E9

David A. Krauss
Science Department
Borough of Manhattan
Community College
City University of New York
199 Chambers St.
New York, NY 10007
dkrauss@bmcc.cuny.edu

Peter Larson
Black Hills Institute of
Geological Research, Inc.
PO Box 643
Hill City, SD 57745

Ralph E. Molnar
Museum of Paleontology
1101 Valley Life
Sciences Building
University of California
Berkeley, CA 94720

Nate L. Murphy
Judith River Dinosaur Institute
PO Box 51177
Billings, MT 59105
jrdi@bresnan.net

†Douglas J. Nichols
U.S. Geological Survey
Denver, CO 80225

J. Michael Parrish
College of Science
San Jose State University
San Jose, CA 95192
mparrish@science.sjsu,edu

Ryan C. Ridgely
Department of
Biomedical Sciences
Ohio University College of
Osteopathic Medicine
Athens, OH 45701
ridgely@ohio.edu

†John M. Robinson
Department of Physics
Indiana University-
Purdue University at
Fort Wayne, IN 46805

Bruce M. Rothschild
Northeast Ohio
Medical University
PO Box 95
Rootstown, Ohio
www.bmr@neomed.edu

Tanya Samman
formerly Department of
Geology and Geophysics
University of Calgary
Calgary, AB T2N 1N4
Canada
tanya_samman@yahoo.com

Reed P. Scherer
Northern Illinois University
DeKalb, IL 60115
reed@niu.edu

Walter W. Stein
PaleoAdventures
3082 Sikeston Ave.
North Port, FL 34286
stein151@verizon.net;
www.paleoadventures.com

David Trexler
Two Medicine
Dinosaur Center
PO Box 794
Bynum, MT 59419
dinoguy10@yahoo.org

Michael Triebold
Triebold Paleontology, Inc.;
and the Rocky Mountain
Dinosaur Resource Center
211 Fairview Ave.
Woodland Park, CO 80863
mike@rmdrc.com;
www.rmdrc.com

Christopher P. Vittore, MD
University of Illinois
College of Medicine and
Rockford Health System
Department of Radiology
2400 N. Rockton Avenue
Rockford, IL 61103

Lawrence M. Witmer
Department of
Biomedical Sciences
Ohio University College of
Osteopathic Medicine
Athens, OH 45701
witmer@oucom.ohiou.edu

Trevor H. Worthy
School of Biological
Earth and Environmental
Sciences
University of New
South Wales
Sydney, NSW 2052
Australia

Xu Xing
Institute of Vertebrate
Paleontology and
Paleoanthropology
Chinese Academy of Sciences
PO Box 643
Beijing 100044
People's Republic of China

Introduction

J. Michael Parrish and Ralph E. Molnar

Tyrannosaurus rex is assuredly the dinosaur with the greatest public visibility, and it has been cast as a heavy in countless films dating back to Harry Hoyt's (1925) adaptation of Sir Arthur Conan Doyle's (1912) *Lost World*. However, as of 1980, only seven specimens of the dinosaur were known (Larson 2008). In the last three decades, this number has swelled at least sevenfold (Larson 2008), and our knowledge of the relationships, anatomy, and biology of *T. rex* and its close relatives has expanded dramatically both through new specimens coming to light and through a plethora of analytical studies. This volume had its genesis in a conference held in Rockford, Illinois, on September 16–18, 2005, titled "The Origin, Systematics, and Paleobiology of Tyrannosauridae," and jointly sponsored by the Burpee Museum of Natural History and Northern Illinois University. The symposium was held in conjunction with the development of the Burpee's new dinosaur hall, the centerpiece of which was a skeletal reconstruction of "Jane" (BMR P2002.4.1), a relatively complete and very well preserved specimen of a juvenile tyrannosaur recovered by the Burpee Museum in 2002 from Carter County, Montana, and now mounted on display at the museum.

This was one of two tyrannosaur symposia that year, the other held at the Black Hills Natural History Museum in Hill City, South Dakota. The proceedings of that meeting have already been published by Indiana University Press as Tyrannosaurus rex, *the Tyrant King* (Larson and Carpenter 2008).

The initial motivation for the Burpee meeting was the relevance of "Jane" on the status of *Nanotyrannus lancensis* as either a valid taxon or a juvenile specimen of *Tyrannosaurus rex*. The ambit of the symposium, however, was broader and also included other issues of tyrannosaur paleobiology. Of the 30 presentations given, 8 concentrated on tyrannosaur ontogeny, 21 on other aspects of tyrannosaur paleobiology, and 1 each about dating ("Jane") and about Barnum Brown. A few contributions to this volume did not appear at the meeting and were included afterward. The results of some of the presentations given at the meeting have already appeared elsewhere (Erickson et al. 2004, 2006; Schweitzer et al. 2005a, 2005b; Snively and Russell 2007a, 2007b, 2007c; Sereno and Brusatte 2009; Witmer and Ridgley 2010).

"Jane" (BMR P2002.4.1) has been identified as a juvenile *Tyrannosaurus rex* and may bear on the question of whether the type skull of *Nanotyrannus* is also a juvenile. *Nanotyrannus lancensis* was originally described as species of *Gorgosaurus* by Gilmore (posthumously) in 1946.

Rozhdestvenskii, in 1965, published results of his work on the ontogenies of Mongolian dinosaurs, concluding that *Tarbosaurus bataar*, *Tarbosaurus efremovi*, and *G. lancinator* were ontogenetic stages of a single taxon, *T. bataar*. At about this time, *G. lancensis* was first proposed as a juvenile of *T. rex* in an unpublished report by Alan Tabrum, based on its co-occurrence with *T. rex* in the Hell Creek Formation of Montana. *Gorgosaurus lancensis* was then described as a valid taxon, in the new genus *Nanotyrannus*, by Bakker, Williams, and Currie in 1988, although they recognized that the holotype was from an immature animal. This taxon was later referred to *T. rex* as a juvenile by Carr in 1999. Thus the question is not whether the holotype skull of *N. lancensis* is from an immature individual, but whether it is from an immature *T. rex*.

Lawson (1978) described an isolated maxilla from Big Bend National Park in Texas he believed to derive from a juvenile *Tyrannosaurus rex*. The taxonomic status of this specimen is still unresolved, but it is generally believed not to represent *T. rex*. Its substantial differences in form, proportion, and the position of the fenestra maxillaris from the maxillae of *Nanotyrannus* suggest that these two specimens pertain to different taxa.

The whole issue of the identity of *Nanotyrannus* hinges on the question of how to distinguish and identify juvenile specimens in the fossil record, not just as immature, but as pertaining to taxa known from adult material. It is often believed that workers of the nineteenth and early twentieth centuries did not recognize juveniles but instead referred small forms to new taxa. This is not exactly correct, as shown, for example, by Lull (1933), Gilmore (1937), and Sternberg (1955). For modern forms the situation is, in principle, straightforward, for one can watch juvenile animals grow up or analyze DNA for evidence of relationships. These techniques are not generally available for fossils, with which one must rely upon three criteria: (1) similarity in geographic and stratigraphic range, (2) change in form consistent with changes seen in modern relatives, and/or (3) a large series of minimally (morphologically) different specimens, so that difference between any two "adjacent" forms is trivial but that the whole sequence shows a consistently changing form from obvious juveniles to obvious adults. Ideally, of course, one would wish to have all three.

The first criterion is not always reliable: in the case of *Dryosaurus* (Horner et al. 2009), for example, adults are not known and so presumably did not occupy (or were not preserved in) the same range as the juveniles. The second depends on a (sometimes subjective) choice of relatives and the assumption that the fossil forms did not deviate substantially in their growth trajectories from related modern taxa. Because related forms— such as *Tyrannosaurus rex* and *Nanotyrannus lancensis*—derive from a common ancestor, the degree of difference between a juvenile *Nanotyrannus* (assuming it is a distinct taxon) and a juvenile *Tyrannosaurus* is expected to be minor, possibly so much as to make them difficult to distinguish. The third criterion is obviously the best. But there are still

problems, basically those of recognizing paleospecies in general. Such problems include recognizing different forms that result from sexual dimorphism, polymorphism, and sibling species. The latter two factors open the possibility of errors resulting from either mistaking conspecifics for different taxa or mistaking different taxa for conspecifics. Such considerations also lead into problems of whether all alleged *T. rex* specimens derive from a single monomorphic species, a single sexually dimorphic species, two (or more) monomorphic species, or two (or more) dimorphic species and whether *N. lancensis* might actually be a valid species but a second species of *Tyrannosaurus*, rather than a separate genus.

Other issues that require further attention are what kinds and degrees of change can be plausibly attributed to growth in tyrannosaurs (treated by Rozhdestvenskii [1965] and Carr [1999]), and how one can distinguish persistently plesiomorphic tyrannosaur taxa from juveniles of contemporaneous advanced forms. In the last case, one hopes one could find both juvenile and adult specimens of the plesiomorphic taxa. Conclusions regarding the classification of a specimen as a juvenile or valid taxon, like many paleontological results, should be treated as hypotheses subject to further verification.

Larson (2008) records 45 specimens of *Tyrannosaurus rex*, of which 38 have been collected since 1980.

The chapters in this volume fall into three broad categories: (1) systematic studies and descriptions of new material, (2) projects incorporating functional morphology or life reconstruction, and (3) contributions focusing on paleoecology, taphonomy, and paleopathology.

The systematic and descriptive studies include a chapter by Brusatte and colleagues assessing the phylogenetic status of the Chinese tyrannosauroid *Chingkankousaurus fragilis*, Larson's argument for the generic status of *Nanotyrannus*, and Stein and Triebold's preliminary description of the tyrannosaurid "Sir William."

Functional studies include Abler's analysis of tooth serrations, Farlow et al.'s extensive analysis of pedal proportions in large theropods, a chapter by Hurlburt and colleagues assessing relative brain size in alligators and tyrannosaurs, and Samman's paper on tyrannosaurid craniocervical function. Chapters dealing more directly with soft-tissue reconstruction are Molnar's study of large theropod jaw musculature and Keillor's account of the process of flesh reconstruction of "Jane."

The third section of the volume includes paleopathology studies by Rothschild (focusing on tyrannosaurid claws) and by Vittore and Henderson (describing an apparent Brodie abscess in the Burpee tyrannosaurid). Harrison and colleagues provide a study of the palynology, leaf taphonomy, and paleomagnetism of the site that produced the Burpee theropod. Krauss posits *"Triceratops* tipping" as a hunting strategy for *Tyrannosaurus rex*, Carpenter weighs the ecological role of *T. rex* feeding, and Murphy and colleagues provide physical evidence of predation in a large theropod.

These contributions emphasize the far-ranging and vital state of the field of tyrannosaurid dinosaur studies. This is the golden age not only for discovery of new tyrannosaurid specimens but also of groundbreaking, interdisciplinary studies of their relationships, functional anatomy, and life histories.

Literature Cited

Bakker, R. T., M. Williams, and P. Currie. 1988. *Nanotyrannus,* a new genus of pygmy tyrannosaur, from the latest Cretaceous of Montana. *Hunteria* 1(5):1–30.

Carr, T. D. 1999. Craniofacial ontogeny in Tyrannosauridae (Dinosauria, Coelurosauria). *Journal of Vertebrate Paleontology* 19:497–520.

Doyle, A. C. 1912. *The Lost World.* London: Hodder & Sloughton.

Erickson, G. M., P. J. Currie, B. D. Inouye, and A. A. Winn. 2006. Tyrannosaur life tables: an example of nonavian dinosaur population biology. *Science* 313:213–217.

Erickson, G. M., P. J. Makovicky, P. J. Currie, M. A. Norell, S. A. Yerby, and C. A. Brochu. 2004. Gigantism and comparative life-history parameters of tyrannosaurid dinosaurs. *Nature* 430:772–775.

Gilmore, C. W. 1937. On the detailed skull structure of a crested hadrosaurian dinosaur. *Proceedings of the United States National Museum* 84:481–491.

Gilmore, C. W. 1946. A new carnivorous dinosaur from the Lance Formation of Montana. *Smithsonian Miscellaneous Collections* 106:1–19.

Horner, J. R., A. de Ricqles, K. Padian, and R. D. Scheetz. 2009. Comparative long bone histology and growth of the "hypsilophodontid" dinosaurs *Orodromeus makelai, Dryosaurus altus,* and *Tenontosaurus tillettii* [sic] [Ornithischia: Euornithopoda]. *Journal of Vertebrate Paleontology* 29:734–747.

Hoyt, H. O. (director). 1925. *The Lost World.* Screenplay by M. Fairfax; special effects by W. O'Brien. First National Pictures, Burbank, California.

Larson, N. L. 2008. One hundred years of *Tyrannosaurus rex:* the skeletons; pp. 1–55 in P. Larson and K. Carpenter (eds.), Tyrannosaurus rex, *the Tyrant King.* Indiana University Press, Bloomington.

Larson, P., and K. Carpenter. 2008. Tyrannosaurus rex, *the Tyrant King.* Indiana University Press, Bloomington.

Lawson, D. A. 1978. *Tyrannosaurus* and *Torosaurus,* Maestrichtian dinosaurs from Trans-Pecos, Texas. *Journal of Paleontology* 50:158–164.

Lull, R. S. 1933. A revision of the Ceratopsia or horned dinosaurs. *Memoirs of the Peabody Museum of Natural History* 3(3):1–135.

Rozhdestvenskii, A. K. 1965. Vozrastnaia izmenchivosti i nekotorie voprosi sistematiki dinozavrov azii. *Paleontologicheskii Zhurnal* 1965:95–109.

Schweitzer, M. H., J. L. Wittmeyer, J. R. Horner, and J. B. Toporski. 2005a. Soft tissue, vessels and cellular preservation in *Tyrannosaurus rex. Science* 307:1952–1955.

Schweitzer, M. H., J. L. Wittmeyer, J. R. Horner, and J. B. Toporski. 2005b. Gender-specific reproductive tissue in ratites and *Tyrannosaurus rex. Science* 308:1456–1460.

Sereno, P. C., and S. L. Brusatte. 2009. Comparative assessment of tyrannosaurid interrelationships. *Journal of Systematic Palaeontology* 7:455–470.

Snively, E., and A. P. Russell. 2007a. Craniocervical feeding dynamics of *Tyrannosaurus rex. Paleobiology* 33:610–638.

Snively, E., and A. P. Russell. 2007b. Functional variation of neck muscles and their relation to feeding style in Tyrannosauridae and other large theropods. *Anatomical Record* 290:934–957.

Snively, E., and A. P. Russell. 2007c. Functional morphology of neck musculature in the Tyrannosauridae (Dinosauria, Theropoda) as determined via a hierarchical inferential approach. *Zoological Journal of the Linnean Society* 151:759–808.

Sternberg, C. M. 1955. A juvenile hadrosaur from the Oldman Formation of Alberta. *National Museum of Canada Bulletin* 136:120–122.

Witmer, L. M., and R. C. Ridgely. 2010. The Cleveland tyrannosaur skull (*Nanotyrannus* or *Tyrannosaurus*): new findings based on CT scanning, with special reference to the braincase. *Kirtlandia* 57:61–81.

Systematics and Descriptions

1

1.1. Photographs and line drawings of the holotype of *Chingkankousaurus fragilis* Young, 1958 (IVPP V 836, right scapula). A) Photograph in lateral view (dorsal to top). B) Photograph in medial view (dorsal to bottom). C) Line drawing in medial view (dorsal to bottom). D) Cross sections from the three indicated areas (lateral to top). Abbreviations: mr, medial ridge; rug, ruosities on posterior expansion of blade. Top scale bar equals 10 cm; bottom scale bar (for cross sections) equals 2 cm.

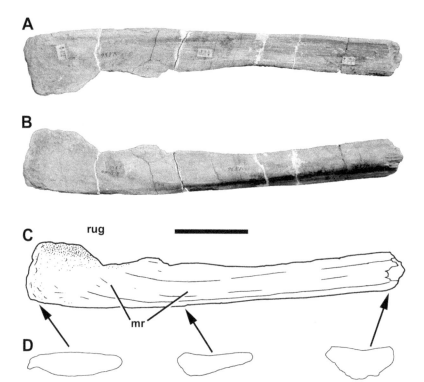

Phylogenetic Revision of *Chingkankousaurus fragilis,* a Forgotten Tyrannosauroid from the Late Cretaceous of China

1

Stephen L. Brusatte, David W. E. Hone, and Xu Xing

Recent discoveries, especially the feathered theropods of the Jehol Biota, have placed China at the forefront of contemporary dinosaur research (e.g., Chen et al. 1998; Xu et al. 2003; Norell and Xu 2005; Xu and Norell 2006). However, vertebrate paleontology has a long history in China, and the country's rich dinosaur fossil record has been explored for over a century. Much of the pioneering work on China's dinosaurs was led by C. C. Young (Yang Zhongjian), the "father of Chinese vertebrate paleontology." For over 40 years, from the early 1930s until his death in 1979, Young spearheaded expeditions across China and discovered many of the country's most recognizable dinosaurs, such as the colossal sauropod *Mamenchisaurus* and the prosauropods *Lufengosaurus* and *Yunnanosaurus* (Dong 1992).

In 1958, Young described a single fragmentary bone from the Late Cretaceous (Campanian-?Maastrichtian; see Weishampel et al. 2004; Zhao et al. 2008) Wangshi Series of Shandong Province as a new genus and species of giant theropod, *Chingkankousaurus fragilis.* This specimen, the posterior region of a large right scapula (IVPP V 836), has long been ignored because of its fragmentary condition. However, those authors who have considered this specimen have often disagreed about its phylogenetic affinities. Young himself (1958) noted similarities with *Allosaurus,* and much later Dong (1992) formally assigned the specimen to Allosauridae. Steel (1970) and Dong (1979) placed the specimen within Megalosauridae, a wastebasket assemblage of large theropods that are now regarded as basal tetanurans (Benson 2010; Benson et al. 2010). Finally, Molnar et al. (1990:199) referred IVPP V 836 to Tyrannosauridae "on the basis of its very slender scapular blade." This referral was taken one step further by Holtz (2004), who synonymized *Chingkankousaurus* with the common Asian Late Cretaceous tyrannosaurid *Tarbosaurus.* Unfortunately, most of these referrals have been based on vague criteria and were often simply asserted instead of supported by explicit discussion of characters and measurements. This was often unavoidable at the time, but an influx of new theropod discoveries from Asia and elsewhere over the past two decades now allows a firm basis for comparison.

In this chapter, we reassess IVPP V 836 based on firsthand examination of the specimen, compare it with the scapulae of other theropods,

Introduction

and use this information to comment on the taxonomy and phylogenetic placement of *Chingkankousaurus fragilis*. Although a systematic revision of a fragmentary specimen may seem trivial, it is important to establish the phylogenetic affinities of IVPP V 836 because this specimen has been referred to many disparate theropod groups and comes from an area (Shandong) where the theropod fauna has been more poorly sampled than in many other regions in China. If it truly does represent an allosauroid or megalosaurid, then this specimen would be among the last surviving members of these groups, would greatly expand their stratigraphic ranges in Asia, and would indicate that more basal theropods persisted alongside tyrannosaurids in the large predator niche of Late Cretaceous Asia (contrary to Brusatte et al. 2009b). However, if IVPP V 836 represents a tyrannosaurid or a closely related form, it is further evidence that that these enormous theropods were the sole large predators during the waning years of the Cretaceous in Laurasia.

Phylogenetic Definitions and Phylogenetic Framework

In this chapter we use the phylogenetic definitions of Sereno et al. (2005) for Tyrannosauroidea and Tyrannosauridae. Tyrannosauroidea is defined as the most inclusive clade containing *Tyrannosaurus rex* but not *Ornithomimus edmontonicus*, *Troodon formosus*, or *Velociraptor mongoliensis*. The more derived Tyrannosauridae is defined as the least inclusive clade containing *T. rex*, *Gorgosaurus libratus*, and *Albertosaurus sarcophagus*. In our discussion of tyrannosauroid phylogeny, we follow the phylogenetic analysis and cladogram presented by Brusatte et al. (2010). This cladogram is depicted in Figure 1.5, and major clades are denoted.

Identification

Although fragmentary, IVPP V 836 (Fig. 1.1) can be identified as a partial right scapula owing to its shape and features of its morphology. This bone was originally described as a scapula by Young (1958), an identification that has been followed by subsequent authors (e.g., Molnar et al. 1990). However, Chure (2000) questioned this identification, noting that the symmetrical cross section figured by Young (1958) is unusual for a scapula. Although Young (1958) describes the cross section as symmetrical, in fact the medial surface is convex, and the lateral surface is flat to slightly concave, as is usual for theropod scapulae (Fig. 1.2). This results in a triangular cross section at mid-shaft and a semi-ovoid cross section anteriorly at the broken edge (Fig. 1.1D). The medial convexity is due to a pronounced ridge, described below, which is a normal feature for tyrannosaurid (e.g., Brochu 2003:fig. 80) and other theropod scapulae (Fig. 1.2A–B). Other features of the bone, such as the slightly concave lateral surface and weakly rugose distal end, are also present in theropod scapulae (Fig. 1.2C–D).

Other possible identifications for the bone, including the possibility that it is part of a dorsal rib or a gastral element, are untenable. The specimen is straight along its entire length, whereas theropod dorsal ribs

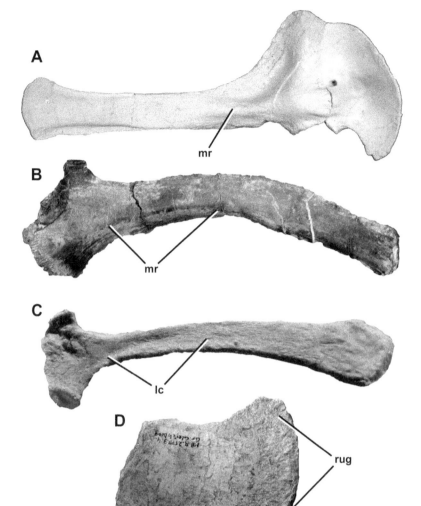

A

mr

B

mr

C

lc

D

rug

are strongly curved, and only very small fragments would appear straight if observed in isolation (e.g., Madsen 1976:pl. 40; Brochu 2003:fig. 64). Additionally, the dorsal ribs of large theropods often bear a thick ridge on their anterior surface, which is paralleled by a depressed groove (e.g., *Daspletosaurus*: AMNH 5468). The posterior surface is often corrugated, with a deep groove corresponding to the ridge on the lateral surface. This morphology is not present in IVPP V 836, which has a single ridge on one surface and a flat to slightly concave opposing surface. Although the distal ends of anterior dorsal ribs are sometimes expanded to articulate with the sternum, these expansions are usually slight and rarely, if ever, more than twice mid-shaft depth, as is the case in IVPP V 836 (e.g., Lambe 1917:figs. 6, 7; Brochu 2003:fig. 64).

Similarly, gastral elements of the largest theropods, such as *Tyrannosaurus*, are smaller than IVPP V 836, and their detailed morphology differs (e.g., Brochu 2003:fig. 70). In particular, although the medial ends of the medial gastral elements may be expanded relative to the mid-shaft, these expansions are usually irregular in shape (not spatulate as in IVPP V 836), extremely rugose, and often fused to the opposing medial gastral element. Additionally, IVPP V 836 is extremely large for a theropod gastral element.

Description

IVPP V 836 is the posterior end of a right scapula, measuring 520 mm long anteroposteriorly as preserved (Fig. 1.1). It is 47 mm deep dorsoventrally at its broken proximal end, and it maintains a relatively constant depth for most of the length of the shaft. However, it thins slightly to 43 mm in depth before expanding distally into a spatulate shape. As preserved, this expansion is 83 mm deep, but both its dorsal and ventral margins are eroded. When the preserved dorsal and ventral margins of the more proximal shaft are extended distally, filling in some of the missing regions, it appears as if the distal expansion was at least 94 mm deep. It is likely, however, that it was even deeper in life, as both the dorsal and ventral edges of the expansion are still quite thick, whereas they usually taper to a thin crest in most large theropod scapulae.

The lateral surface of the scapula is flat to slightly concave (Fig. 1.1A). The concavity is deepest dorsally, where it is overhung by a thickened ridge that parallels the dorsal margin of the blade. The ridge is thickest at the midpoint of the preserved fragment and thins out both proximally and distally. Ventrally the lateral concavity becomes progressively weaker until the lateral surface flattens out. This flat region, which corresponds to a flat surface on the medial surface of the blade, occupies approximately one half of the blade height.

The medial surface of the scapula is generally convex, due to the presence of a medial ridge (Fig. 1.1B–C). The ridge is strongest proximally: here it is most convex medially and also most extensive dorsoventrally, as it comprises the entire medial surface of the blade. Distally the ridge becomes weaker, as it becomes less convex and offset and thins into a more discrete crest that sweeps dorsally to parallel the dorsal margin of the blade. The ridge eventually funnels out into a broad triangular shape, which smoothly merges with the flat medial surface of the distal expansion.

Both the lateral and medial surfaces of the spatulate distal expansion are rugose (Fig. 1.1). This rugosity is most pronounced on the medial surface and takes the form of a mottled array of pits and raised mounds. A similar pattern of rugosity is seen on well-preserved theropod scapulae and corresponds to a number of muscle attachment sites (Brochu 2003:fig. 81). Additionally, in some tyrannosaurids the medial surface of the scapular expansion is more rugose than the lateral surface (e.g, *Albertosaurus:* AMNH 5458).

The ventral margin of the scapula, as seen in lateral and medial views, is straight for a short region proximally but describes a broad, concave arch distally. The dorsal margin, in contrast, is straight for most of its

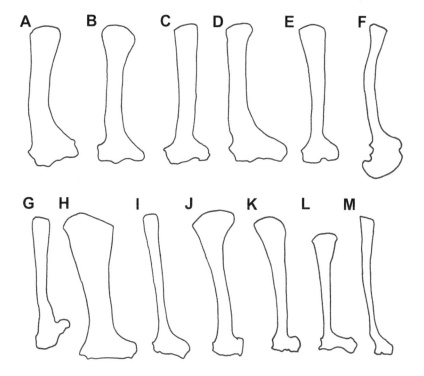

1.3. Comparative figure showing general outlines of several theropod scapulae: A–F) non-tyrannosauroids; G–M) tyrannosauroids. A) *Ceratosaurus* (Madsen and Welles 2000). B) *Piatnitzkysaurus* (Bonaparte 1986) C) *Sinraptor* (Currie and Zhao 1993). D) *Aerosteon* (MCNA-PV-3137). E) *Allosaurus* (Madsen 1976). F) *Acrocanthosaurus* (Currie and Carpenter 2000). G) *Guanlong* (IVPP V 14532). H) *Dilong* (Xu et al. 2004). I) *Raptorex* (LH PV18). J) *Albertosaurus* (Parks 1928). K) *Gorgosaurus* (Romer 1956). L) *Tyrannosaurus* (Brochu 2003). M) *Tarbosaurus* (MPC-D107/05). F shows the orientation of the drawings (ventral to left, dorsal to right). Images have been reflected where necessary from drawings of left scapulae to provide a better comparison. F and G show both the scapula and coracoid.

length. There is a small region distally that appears to be convex, but this may be an artifact of erosion. However, there is a slightly convex, raised margin in this region in some tyrannosaurid scapulae (Brochu 2003:fig. 80), suggesting that it may be a real feature.

Comparisons and Phylogenetic Affinity

Despite being a fragment of a single bone, IVPP V 836 exhibits a number of features that can be compared with those of other theropods (Fig. 1.3), allowing for a reasonable discussion and determination of its phylogenetic affinities. Importantly, the fact that the minimum shaft depth is preserved allows for the estimation of two important ratios that quantify scapula gracility and the relative size of the distal expansion (Fig. 1.4).

Although complete measurements are not possible, IVPP V 836 is clearly an elongate, gracile, and strap-like scapula. The length of the bone is at least 12 times greater than its minimum dorsoventral height, which is known with certainty (Table 1.1). A blade that is more than 10 times longer than deep has been used as a phylogenetic character in tyrannosauroid cladistic analyses and is optimized as a synapomorphy of Tyrannosauridae or slightly more or less inclusive clades (Sereno et al. 2009:character 69; Brusatte et al. 2010:character 234). As shown in Table 1.1, all tyrannosauroids except *Dilong* and *Guanlong* possess this character, although the latter taxon approaches this condition, whereas only a few non-tyrannosauroid theropods exhibit such strap-like scapulae.

Additionally, although complete measurements are again impossible, the distal expansion of IVPP V 836 is extensive compared to depth of

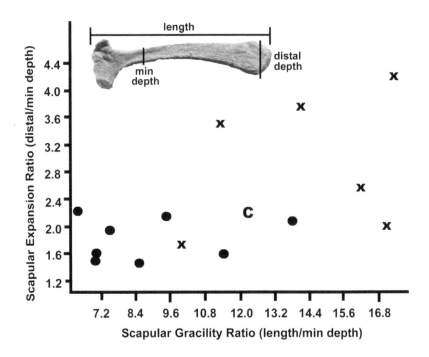

1.4. Bivariate plot of scapular expansion ratio vs. scapular gracility ratio (see the plot axes, as well as Tables 1.1 and 1.2, for definitions of these ratios). Tyrannosauroid theropods are represented by an x, non-tyrannosauroid large-bodied theropods by a filled circle, and IVPP V 836 by "C." Tyrannosauroids fall into the upper right-hand corner of the plot, and IVPP V 836 falls on the edge of this cluster. However, measurements of IVPP V 836 are incomplete due to breakage, and the more complete specimen could only migrate further into the upper right-hand corner. This supports the tyrannosauroid affinities of the specimen. Note that *Dilong,* which has an abnormally short and stout scapula among tyrannosauroids, is not figured in this plot, but its measurements are included in Tables 1.1 and 1.2.

the blade itself (Table 1.2). The ratio of the expansion depth to the minimum depth of the blade is at least 2.2 and was probably much greater in life. An expansion that is more than twice the minimum blade depth has been used as a phylogenetic character in tyrannosauroid cladistic analyses and also optimizes as a synapomorphy of Tyrannosauroidea or proximate clades (Holtz 2001:character 82; Holtz 2004:character 386; Sereno et al. 2009:character 70). As shown in Table 1.2, all tyrannosauroids except *Guanlong* possess this character, whereas the scapulae of other large theropods have relatively less expanded distal ends. There is also phylogenetically informative variation within tyrannosauroids, as all taxa more derived than *Raptorex* possess an expansion that is more than 2.5 times minimum blade depth (see also Brusatte et al. 2010:character 235).

When these two ratios are plotted against each other in a simple bivariate plot, most tyrannosauroids are seen to occupy the upper right-hand corner of the graph, whereas other theropods fall into the lower left-hand corner (Fig. 1.4). IVPP V 836 falls on the edge of the tyrannosauroid cluster but could only shift deeper within the tyrannosauroid region of the plot if more complete measurements were possible (because, for IVPP V 836, the scapular gracility ratio must have been greater than the plotted 12.23, and the scapular expansion ratio must have been greater than the plotted 2.2). In other words, even though IVPP V 836 is incomplete and likely missing much of its proximal region and distal end, the fragmentary preserved remains are themselves enough to quantitatively document similarities with tyrannosauroids to the exclusion of other theropods. The complete scapula could only be more strap-like, with a relatively larger distal expansion. In short, it could only be more tyrannosauroid-like.

Taxon	Ratio	Source
"Chingkankousaurus"	>12.23	IVPP V 836
Tyrannosauroidea:		
Tyrannosaurus	17.27	Brochu (2003)
Gorgosaurus	14.00	Lambe (1917)
Albertosaurus	11.25	Parks (1928)
Tarbosaurus	16.15	MPC-D107/05
Raptorex	17.00	LH PV18
Dilong	6.09	IVPP V 14242
Guanlong	9.95	IVPP V 14531
Other Large Theropods:		
Acrocanthosaurus	11.5	Currie and Carpenter (2000)
Aerosteon	7.00	MCNA-PV-3137
Allosaurus	13.8	Madsen (1976)
Ceratosaurus	7.03	Madsen and Welles (2000)
Dilophosaurus	6.40	Welles (1984)
Neovenator	9.50	Brusatte et al. (2008)
Piatnitzkysaurus	7.5	Bonaparte (1986)
Sinraptor	8.54	Currie and Zhao (1993)

Table 1.1. Scapular gracility ratio (ratio of anteroposterior length to minimum dorsoventral depth) in *Chingkankousaurus,* tyrannosauroids, and other large theropods. For a visual description of this ratio, see Figure 1.4.

Taxon	Ratio	Source
"Chingkankousaurus"	>2.22	IVPP V 836
Tyrannosauroidea:		
Tyrannosaurus	4.20	Brochu (2003)
Gorgosaurus	3.75	Lambe (1917)
Albertosaurus	3.50	Parks (1928)
Tarbosaurus	2.61	MPC-D107/05
Raptorex	2.00	LH PV18
Dilong	2.45	IVPP V 14242
Guanlong	1.70	IVPP V 14531
Other Large Theropods:		
Acrocanthosaurus	1.58	Currie and Carpenter (2000)
Aerosteon	1.58	MCNA-PV-3137
Allosaurus	2.07	Madsen (1976)
Ceratosaurus	1.53	Madsen and Welles (2000)
Dilophosaurus	2.20	Welles (1984)
Neovenator	~2.13	Brusatte et al. (2008)
Piatnitzkysaurus	1.94	Bonaparte (1986)
Sinraptor	1.45	Currie and Zhao (1993)

Table 1.2. Scapular expansion ratio (ratio of the dorsoventral depth of the distal expansion to the minimum depth of the blade) in *Chingkankousaurus,* tyrannosauroids, and other large theropods. For a visual description of this ratio, see Figure 1.4.

Other features of the scapula are shared with tyrannosauroids as well. The straight dorsal margin and concave ventral margin are seen in *Albertosaurus* (Parks 1928), *Dilong* (IVPP V 14242), *Eotyrannus* (MIWG 1997 550), *Gorgosaurus* (Lambe 1917), *Guanlong* (Xu et al. 2006), *Tarbosaurus* (Maleev 1974), and *Tyrannosaurus* (Brochu 2003). Other large theropods exhibit different morphologies (Fig. 1.3). For instance, in most allosauroids, both margins are straight (*Aerosteon:* MCNA-PV-3137; *Allosaurus:* UMNH UUVP 4423; *Neovenator:* Brusatte et al. 2008; *Sinraptor:*

Gao 1999). In *Acrocanthosaurus* (Currie and Carpenter 2000), as well as *Ceratosaurus* (Madsen and Welles 2000) and *Piatnitzkysaurus* (Bonaparte 1986), the dorsal margin is concave, and the ventral margin is straight or convex. Finally, it is also possible that the pronounced rugosity on the medial surface of the distal end, seen in IVPP V 836 and *Albertosaurus* (AMNH 5458), may be a synapomorphy of tyrannosauroids or a less inclusive clade, but it is only apparent on well-preserved specimens. Only additional material can clarify this feature.

Systematic and Phylogenetic Placement

As shown, IVPP V 836 shares features with tyrannosauroids that are otherwise unknown, or rare, in other large theropods. Additionally, it comes from a time (Late Cretaceous) and place (Asia) in which tyrannosaurids were common animals and likely the sole apex predators in most terrestrial ecosystems (Currie 2000; Brusatte et al. 2009b). Therefore, we assign IVPP V 836 to Tyrannosauroidea. Within Tyrannosauroidea, IVPP V 836 is more derived than the basal taxa *Guanlong* and *Dilong* in both of the scapular ratio characters considered above (Tables 1.1 and 1.2), and therefore it can be assigned to the unnamed tyrannosauroid clade that includes *Eotyrannus, Stokesosaurus, Xiongguanlong, Raptorex, Bistahieversor, Dryptosaurus, Appalachiosaurus,* and Tyrannosauridae (see Brusatte et al. 2010). This phylogenetic position is visually shown in the cladogram in Figure 1.5.

It is tempting to assign IVPP V 836 to even less inclusive clades, such as Tyrannosauridae or even *Tarbosaurus*. Indeed, Holtz (2004) formally assigned IVPP V 836 to *Tarbosaurus* and sunk *Chingkankousaurus fragilis,* which he considered a nomen dubium, into the genus *Tarbosaurus*. We agree that *C. fragilis* is a nomen dubium – there are no clearly autapomorphic features on IVPP V 836, nor a unique combination of characters that can diagnose it relative to other tyrannosauroids. However, we hesitate to refer the specimen to a less inclusive clade than the *Eotyrannus* + *Stokesosaurus* + more derived tyrannosauroid clade.

Referring IVPP V 836 to *Tarbosaurus* is problematic for two reasons. First, *Tarbosaurus* does not possess any clearly autapomorphic features of the scapula, and we prefer synapomorphy-based assessments (sensu Nesbitt and Stocker 2008) when referring fragmentary fossils to established taxa. Second, there are at least two other large tyrannosauroids that lived during the Late Cretaceous of Asia, *Alioramus* (Kurzanov 1976; Brusatte et al. 2009a) and *Alectrosaurus* (Gilmore 1933; Mader and Bradley 1989). Scapulae are unknown for both of these genera, thus precluding any comparison with IVPP V 836.

In a similar vein, we hesitate to refer IVPP V 836 to Tyrannosauridae, as the various phylogenetically informative features discussed above characterize the more inclusive clade Tyrannosauroidea (or, more precisely, the *Eotyrannus* + *Stokesosaurus* + more derived tyrannosauroid node). Indeed, there are few unequivocal features of the scapula, and certainly no features on the region of the scapula preserved in IVPP V

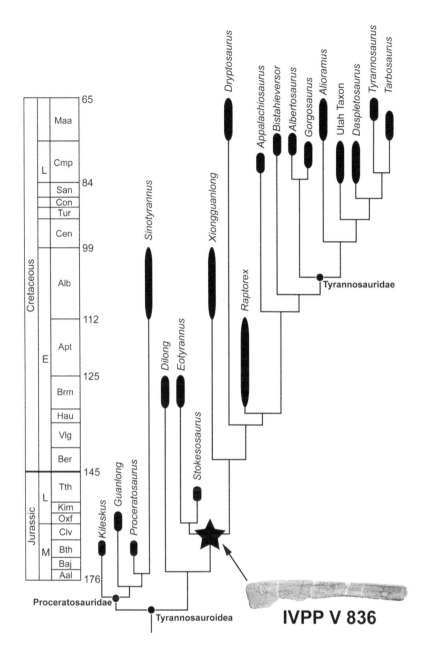

1.5. A phylogeny of tyrannosauroids indicating the most resolved phylogenetic position of IVPP V 836. The phylogeny, which is scaled against the geological time scale, is based on that of Brusatte et al. (2010). The large star indicates the tyrannosauroid ingroup clade to which IVPP V 836 can be assigned. IVPP V 836 is more derived than proceratosaurids (e.g., *Guanlong*) and *Dilong* based on its possession of two more derived scapular ratio characters (see text), but a more precise phylogenetic placement is not possible based on the fragmentary nature of the specimen. In other words, IVPP V 836 does not possess any unequivocal derived characters that allow its assignment to Tyrannosauridae or other more derived clades. Thick black bars on the figure represent the finest age resolution of each taxon, not actual duration.

836, that diagnose Tyrannosauridae relative to more basal tyrannosauroids (Brusatte et al. 2010). One characteristic synapomorphy of Tyrannosauridae, a scapular expansion ratio of greater than 2.5 (Brusatte et al. 2010), cannot be assessed in IVPP V 836 because of breakage. Although many basal tyrannosauroids are small animals, there are several large-bodied non-tyrannosaurid tyrannosauroids that were present in the Late Cretaceous of North America (*Dryptosaurus*: Carpenter et al. 1997; *Appalachiosaurus*: Carr et al. 2005; and *Bistahieversor*: Carr and Williamson 2010). These animals were of sufficient size to possess a scapula as large as IVPP V 836.

Conclusions

The species *Chingkankousaurus fragilis* is a nomen dubium as it is not diagnostic relative to other theropod genera and species. However, several features show that it is a tyrannosauroid, as has often been suggested but never conclusively demonstrated. Therefore, it does not represent a late-surviving allosauroid or megalosaurid, as some previous authors have suggested. It is possible that the type and only known specimen of "*C. fragilis*," IVPP V 836, belongs to *Tarbosaurus*, as asserted by Holtz (2004), but this cannot be shown with certainty. As a result, it is not possible to determine whether IVPP V 836 belongs to a previously known Asian tyrannosauroid or represents a new taxon, which is possible since it may come from the Campanian, which is more poorly sampled relative to the Maastrichtian in Asia. Unfortunately, this enigma is unlikely to be resolved, since "*C. fragilis*" is based on a single, isolated, fragmentary specimen. Although it was normal procedure to name such scrappy fossils during Young's (1958) time, taxonomic quandaries such as those generated by the fragmentary type of "*C. fragilis*" are a prime example of why such practice should be discontinued.

Acknowledgments

We are grateful to Ralph E. Molnar and Lu Junchang for their reviews. We thank Sean Gallagher for photos of IVPP V 836. Roger Benson provided the photo in Figure 1.2A, and Philip Currie provided photos of *Tarbosaurus* scapulae for comparison. We thank numerous curators for access to specimens in their care, most importantly Magdalena Borsuk-Białynicka, Steve Hutt, Randall Irmis, Mark Norell, Carl Mehling, Scott Williams, and Paul Sereno. SLB is supported by a National Science Foundation Graduate Research Fellowship and by the American Museum of Natural History. DWEH and XX are supported by the Chinese Academy of Sciences, and XX is supported by the Natural Science Foundation of China.

Literature Cited

Benson, R. B. J. 2010. A description of *Megalosaurus bucklandii* (Dinosauria: Theropoda) from the Bathonian of the United Kingdom and the relationships of Middle Jurassic theropods. *Zoological Journal of the Linnean Society* 158:882–935.

Benson, R. B. J., M. T. Carrano, and S. L. Brusatte. 2010. A new clade of archaic large-bodied predatory dinosaurs (Theropoda: Allosauroidea) that survived to the latest Mesozoic. *Naturwissenschaften* 97:71–78.

Bonaparte, J. F. 1986. Les dinosaurs (carnosaures, allosauridés, sauropodes, cétiosauridés) du Jurassic moyen de Cerro Cóndor (Chubut, Argentina). *Annales de Paléontologie* 72:247–289.

Brochu, C. A. 2003. Osteology of *Tyrannosaurus rex*: insights from a nearly complete skeleton and high-resolution computed tomographic analysis of the skull. *Society of Vertebrate Paleontology Memoir* 7:1–138.

Brusatte, S. L., R. B. J. Benson, and S. Hutt. 2008. The osteology of *Neovenator salerii* (Dinosauria: Theropoda) from the Wealden Group (Barremian) of the Isle of Wight. *Monograph of the Palaeontographical Society* 162(631):1–166.

Brusatte, S. L., T. D. Carr, G. M. Erickson, G. S. Bever, and M. A. Norell. 2009a. A long-snouted, multihorned tyrannosaurid from the Late Cretaceous of Mongolia. *Proceedings of the National Academy of Sciences (USA)* 106:17261–17266.

Brusatte, S. L., R. B. J. Benson, D. J. Chure, X. Xu, C. Sullivan, and D. E. W. Hone. 2009b. The first definitive carcharodontosaurid (Dinosauria: Theropoda) from Asia and the delayed ascent of tyrannosaurids. *Naturwissenschaften* 96:1051–1058.

Brusatte, S. L., M. A. Norell, T. D. Carr, G. M. Erickson, J. R. Hutchinson, A. M. Balanoff, G. S. Bever, J. N. Choiniere, P. J.

Makovicky, and X. Xu. 2010. Tyrannosaur paleobiology: new research on ancient exemplar organisms. *Science* 329:1481–1485.

Carpenter, K., D. Russell, D. Baird, and R. Denton. 1997. Redescription of the holotype of *Dryptosaurus aquilunguis* (Dinosauria: Theropoda) from the Upper Cretaceous of New Jersey. *Journal of Vertebrate Paleontology* 17:561–573.

Carr, T. D., and T. E. Williamson. 2010. *Bistahieversor sealeyi,* gen. et sp. nov., a new tyrannosauroid from New Mexico and the origin of deep snouts in Tyrannosauroidea. *Journal of Vertebrate Paleontology* 30:1–16.

Carr, T. D., T. E. Williamson, and D. R. Schwimmer. 2005. A new genus and species of tyrannosauroid from the Late Cretaceous (Middle Campanian) Demopolis Formation of Alabama. *Journal of Vertebrate Paleontology* 25:119–143.

Chen, P.-J., Z.-M. Dong, and S.-N. Zhen. 1998. An exceptionally well-preserved Theropod dinosaur from the Yixian Formation of China. *Nature* 391:147–152.

Chure, D. J. 2000. A new species of *Allosaurus* from the Morrison Formation of Dinosaur National Monument (UT–CO) and a revision of the theropod family Allosauridae. Unpublished Ph.D. dissertation, Columbia University, New York.

Currie, P. J. 2000. Theropods from the Cretaceous of Mongolia; pp. 135–144 in M. J. Benton, M. A. Shishkin, D. M. Unwin, and E. N. Kurochkin (eds.), *The Age of Dinosaurs in Russia and Mongolia.* Cambridge University Press, Cambridge.

Currie, P. J., and K. Carpenter. 2000. A new specimen of *Acrocanthosaurus atokensis* (Theropoda, Dinosauria) from the Lower Cretaceous Antlers Formation (Lower Cretaceous, Aptian) of Oklahoma, USA. *Geodiversitas* 22:207–246.

Currie, P. J., and X.-J. Zhao. 1993. A new large theropod (Dinosauria, Theropoda) from the Jurassic of Xinjiang, People's Republic of China. *Canadian Journal of Earth Sciences* 30:2037–2081.

Dong, Z.-M. 1979. Reptilia; pp. 88–249 in Z.-M. Dong, T. Qi, and Y. You (eds.), *Handbook of Chinese Fossil Vertebrates.* Science Press, Beijing.

Dong, Z.-M. 1992. *Dinosaurian Faunas of China.* China Ocean Press, Beijing.

Gao, Y. 1999. *A Complete Carnosaur Skeleton from Zigong, Sichuan:* Yangchuanosaurus hepingensis. Sichuan Science and Technology Press, Chengdu, 1–100 pp.

Gilmore, C. W. 1933. On the dinosaurian fauna of the Iren Dabasu Formation. *Bulletin of the American Museum of Natural History* 67:23–78.

Holtz, T. R., Jr. 2001. The phylogeny and taxonomy of the Tyrannosauridae; pp. 64–83 in K. Carpenter and D. Tanke (eds.), *Mesozoic Vertebrate Life.* Indiana University Press, Bloomington.

Holtz, T. R., Jr. 2004. Tyrannosauroidea; pp. 111–136 in D. B. Weishampel, P. Dodson, and H. Osmólska (eds.), *The Dinosauria,* 2nd ed. University of California Press, Berkeley.

Kurzanov, S. M. 1976. A new carnosaur from the Late Cretaceous of Nogon-Tsav, Mongolia. *Joint Soviet-Mongolian Paleontological Expedition Transactions* 3:93–104.

Lambe, L. B. 1917. The Cretaceous theropodous dinosaur *Gorgosaurus. Memoirs of the Geological Survey of Canada,* 100 pp.

Mader, B. J., and R. L. Bradley. 1989. A redescription and revised diagnosis of the syntypes of the Mongolian tyrannosaur *Alectrosaurus olseni. Journal of Vertebrate Paleontology* 9:41–55.

Madsen, J. H. 1976. *Allosaurus fragilis:* a revised osteology. *Utah Geological Survey Bulletin* 109:1–163.

Madsen, J. H., and S. P. Welles. 2000. *Ceratosaurus* (Dinosauria, Theropoda): a revised osteology. Miscellaneous Publication no. 00–2. Utah Geological Survey, Salt Lake City, 80 pp.

Maleev, E. A. 1974. Gigantic carnosaurs of the family Tyrannosauridae. *Sovmestnaia Sovestsko-Mongol'skaia Paleontologicheskaia Ekspeditsiia Trudy* 1:132–191.

Molnar, R., S. Kurzanov, and Z.-M. Dong. 1990. Carnosauria; pp. 169–209 in D. B. Weishampel, P. Dodson, and H. Osmólska (eds), *The Dinosauria.* University of California Press, Berkeley.

Nesbitt, S. J., and M. R. Stocker. 2008. The vertebrate assemblage of the Late Triassic Canjilon Quarry (Northern New Mexico, USA), and the importance of apomorphy-based assemblage comparisons. *Journal of Vertebrate Paleontology* 28:1063–1072.

Norell, M. A., and X. Xu. 2005. Feathered dinosaurs. *Annual Review of Earth and Planetary Sciences* 33:277–299.

Parks, W. A. 1928. *Albertosaurus arctunguis,* a new species of therapodous dinosaur from the Edmonton Formation of Alberta. *University of Toronto Studies, Geological Series* 25:1–42.

Romer, A. S. 1956. *Osteology of the Reptiles.* University of Chicago Press, Chicago.

Sereno, P. C., S. McAllister, and S. L. Brusatte. 2005. TaxonSearch: a relational database for suprageneric taxa and phylogenetic definitions. *PhyloInformatics* 8:1–21.

Sereno, P. C., L. Tan, S. L. Brusatte, H. J. Kriegstein, X.-J. Zhao, and K. Cloward. 2009. Tyrannosaurid skeletal design first evolved at small body size. *Science* 326:418–422.

Steel, R. 1970. Saurischia. *Handbuch der Paleoherpetologie,* part 14. Gustav Fischer Verlag, Stuttgart, 87 pp.

Weishampel, D. B., P. M. Barrett, R. A. Coria, J. Le Loeuff, X. Xu, X. Zhao, A. Sahni, E. Gomani, and C. R. Noto. 2004. Dinosaur distribution; pp. 517–606 in D. B. Weishampel, P. Dodson, and H. Osmólska (eds.), *The Dinosauria,* 2nd ed. University of California Press, Berkeley.

Welles, S. P. 1984. *Dilophosaurus wetherilli* (Dinosauria, Theropoda) osteology and comparisons. *Palaeontographica Abteilung A: Palaeozoologie–Stratigraphie* 185:85–180.

Xu, X., and M. A. Norell. 2006. Non-avian dinosaur fossils from the Lower Cretaceous Jehol Group of western Liaoning, China. *Geological Journal* 41:419–437.

Xu, X., M. A. Norell, X. Kuang, X. Wang, Q, Zhao, and C. Jia. 2004. Basal tyrannosauroids from China and evidence for protofeathers in tyrannosauroids. *Nature* 431:680–684.

Xu, X., Z. Zhou, X. Wang, X. Kuang, F. Zhang, and X. Du. 2003. Four-winged dinosaurs from China. *Nature* 421:335–340.

Xu, X., J. M. Clark, C. A. Forster, M. A. Norell, G. M. Erickson, D. A. Eberth, C. Jia, and Q. Zhao. 2006. A basal tyrannosauroid dinosaur from the Late Jurassic of China. *Nature* 439:715–718.

Young, C. C. 1958. The dinosaurian remains of Laiyang, Shantung. *Palaeontologia Sinica,* n.s., C 42:1–138.

Zhao, Q., X. Xu, C. Jia, and Z. Dong. 2008. Order Saurischia; pp. 279–335 in J. Li, X. Wu, and F. Zhang (eds.), *The Chinese Fossil Reptiles and Their Kin.* Science Press, Beijing.

2.1. Holotype of *Nanotyrannus lancensis,* CMNH 7541.

The Case for *Nanotyrannus*

2

Peter Larson

The genus *Nanotyrannus* was erected in 1988 by Bakker, Williams, and Currie, redescribing a skull (CMNH 7541) from the Maastrichtian (Lancian) Hell Creek Formation of Montana, first described as *Gorgosaurus lancensis* by Gilmore (1946). In part due to the absence of additional specimens, the validity of *Nanotyrannus* came under question by various researchers, culminating in 1999 when Carr assigned the specimen to *Tyrannosaurus rex*. Carr presented a compelling argument that CMNH 7541 was a juvenile and that characters separating *Nanotyrannus* from *Tyrannosaurus* were ontogenetic.

In 2001 a second specimen was located that compared very well with the type of *Nanotyrannus*. This new specimen (BMR P2002.4.1), nicknamed "Jane," consists of a beautifully preserved partial skull and skeleton. Although some researchers are convinced that BMR P2002.4.1 confirms Carr's juvenile *Tyrannosaurus rex* hypothesis, this paper questions that conclusion.

Fusion of the scapula-coracoid, fusion of the pelvis, and fusion and partial fusion of the centra to the dorsal spines throughout the represented vertebral column indicate cessation or near cessation of growth. A ninefold increase in size for BMR P2002.4.1 to reach the adult weight of FMNH PR 2081 ("Sue") seems a "stretch." BMR P2002.4.1 and the holotype have 15 or 16 tooth positions in their maxillae; all specimens unquestioningly ascribed to *Tyrannosaurus rex* have 11 or 12. BMR P2002.4.1's dentaries have 17 tooth positions; *T. rex* has 13 or 14. BMR P2002.4.1 and the type possess an incisiform and small first maxillary tooth, a character shared with *Gorgosaurus* and *Albertosaurus* but not with *T. rex*. A score of cranial and several post-cranial characters present in BMR P2002.4.1 and the type of *Nanotyrannus lancensis* are absent in *T. rex*. This leads to the conclusion that *Nanotyrannus* is a valid taxon.

Abstract

Nanotyrannus lancensis was named by Bakker, Williams, and Currie in 1988 for a well-preserved and uncrushed skull and lower jaws (CMNH 7541; Fig. 2.1) collected from the Hell Creek Formation of Carter County, Montana. The type specimen was originally described as *Gorgosaurus lancensis* by Gilmore in 1946. Bakker et al. (1988) argued that certain derived characters (including the construction of the basicranium, the angle of the occipital condyle, the maxillary tooth count, overall tooth morphology, the relative narrowness of the snout, and expansion of the

Introduction

temporal region of the skull) were sufficient to erect a new tyrannosaurid genus. Bakker et al. (1988) suggested that *Nanotyrannus* was, in fact, more closely aligned with *Tyrannosaurus* than with *Gorgosaurus*. They went on to state that CMNH 7541 was clearly an adult with closed cranial sutures and had reached "maximum size" for that individual (1988:17).

In a review of the tyrannosaurids, Carpenter (1992) noted that fusion of cranial bones is variable in dinosaurs and that the oval shape of the orbit may well be a juvenile character. He concluded that *Nanotyrannus lancensis* could be a juvenile *Tyrannosaurus rex*. This idea had been first proposed by Rozhdestvenskii in 1965 after synonymizing a number of described species of tyrannosaurids as different growth stages of *Tarbosaurus bataar*.

In 1999, Thomas Carr examined 17 specimens referred to *Albertosaurus libratus* and described an ontogenetic series and growth stages for this species. Following Carpenter (1992) and Rozhdestvenskii (1965), and based upon bone texture, lack of fusion, shape of the orbit, and overall skull morphology, Carr considered *Nanotyrannus lancensis* to be a juvenile *Tyrannosaurus rex*, placing CMNH 7541 into the youngest of his growth stages. In later arguments, Carr and Williamson (2004) established a growth series for *T. rex* and a sequence of changes from a small juvenile, LACM 28471 (stage 1), followed by the type of *Nanotyrannus* (stage 2), through sub-adults LACM 23845 (stage 3) and AMNH 5027 (stage 4), to the full adults LACM 23844 and FMNH PR 2081 (stage 5).

Although Carr's arguments have received wide acceptance (Brochu 2003; Holtz 2004), not all paleontologists concur with Carr's assessment. Currie (2003:223) pointed out that "most of the characters . . . used to demonstrate that *Nanotyrannus* and *Tyrannosaurus* are synonymous are also characters of *Tarbosaurus* and *Daspletosaurus*." To add to this, as Jørn Hurum (pers. comm., June 2005) pointed out, because the growth series is rooted in the argument that *Nanotyrannus* is a juvenile *T. rex*, if Carr is wrong, his concept of ontogenetic change and ontogenetic stages in *Tyrannosaurus rex* is in question. Carr's 1999 paper kindled a debate that has grown hotter by the year.

One of the problems with resolving the question of the validity of the genus *Nanotyrannus* has been the lack of specimens. CMNH 7541 is a skull and lower jaws (still articulated, obscuring much detail), with no post crania. A second referable specimen is a small, poorly preserved, and fragmentary skull (LACM 28471) described by Molnar (1978) as the "Jordan Theropod" from the Hell Creek Formation of Garfield County, Montana. This specimen was later referred to *Aublysodon* by Molnar and Carpenter (1989) and synonymized with *Tyrannosaurus rex* by Carr and Williamson (2004). A third specimen (BHI 6235), an isolated left lachrymal comparable in size and morphology to the type (CMNH 7541), was found associated with FMNH PR 2081 in the Hell Creek Formation of Dewey County, South Dakota, and identified as a juvenile *T. rex* by Larson (1997a). As of 2001 there were a total of three specimens, only one that provided much information, and none had associated post crania.

2.2. "Jane," BMR P2002.4.1.

In the summer of 2002 a field crew from the Burpee Museum of Rockford, Illinois, led by Mike Henderson, excavated an additional specimen that should have ended the debate. "Jane," as she was nicknamed (BMR P2002.4.1; see Fig. 2.2), is clearly referable to *Nanotyrannus lancensis*, if the taxon is indeed valid. In addition to many uncrushed, well-preserved, and mostly disarticulated skull elements with a nearly complete dentition, BMR P2002.4.1 also retains much of the post-cranial skeleton. This well-preserved and well-prepared fourth specimen – and the unlimited access granted to researchers by the staff of the Burpee Museum – make this reevaluation of the status of *Nanotyrannus lancensis* possible.

Ontogenetic Stage

In 2004, Erickson et al. presented findings of ontogenetic age for a number of tyrannosaurid dinosaur specimens. This was accomplished by thin sectioning tyrannosaurid pubes, fibulae, ribs, gastralia, and post-orbitals and by counting annual growth rings (along with those of other reptiles of known ages). Utilizing femoral lengths and circumferences, Erickson et al. estimated body mass for each individual. Combining this information, a graph was generated showing the logistic growth curves for *Tyrannosaurus*, *Daspletosaurus*, *Gorgosaurus* and *Albertosaurus* (see Fig. 2.3).

The three specimens referable to *Nanotyrannus lancensis* (LACM 28471, BHI 6235, BMR P2002.4.1), because of character affinities, along with the type (CMNH 7541), represent a growth series. LACM 28471 is the smallest individual, whose skull, were it complete, would measure

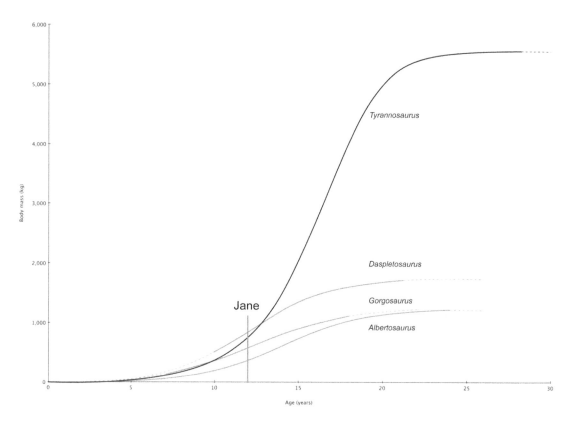

2.3. Logistic growth of tyrannosaurs, after Erickson et al. (2004).

approximately 450 mm. The type (CMNH 28471) is 580 mm in length, and the isolated lachrymal (BHI 6235) represents an animal of approximately the same size. BMR P2002.4.1 is by far the largest specimen, with a skull length of 720 mm (an adult *Tyrannosaurus rex* skull can be in excess of 1400 mm in length).

BMR P2002.4.1's ontogenetic age was determined to be 12 years (Greg Erickson, pers. comm., October 2005). With a femur length of 720 mm and a circumference of 250 mm, BMR P2002.4.1's live weight (calculated from the formula $W = 0.16Cf\,2.73$, after Anderson et al. 1985) would have been 560 kg. If BMR P2002.4.1 is a *Tyrannosaurus rex*, an age of 12 would place this specimen low on the logistic growth curve with a potential increase in mass of ten fold (FMNH PR 2081, the largest *T. rex* in the sample, weighed 5600 kg). If, however, *Nanotyrannus* is valid, and its logistic growth curve was more like other tyrannosaurs (*Gorgosaurus*, *Albertosaurus*, and *Daspletosaurus*), at 12 years of age, BMR P2002.4.1 would have achieved half her adult weight, and her skeleton would be nearing cessation of growth (see Fig. 2.3, after Erickson et al. 2004). Also worth considering is the possibility that, if *Nanotyrannus* is valid, it could follow a logistic growth curve that is unique, unlike those of the four genera examined by Erickson et al.

Another possible way to assess BMR P2002.4.1's ontogenetic stage is to observe post-cranial suture closure. Suture closure has been used to evaluate ontogenetic stage in sauropods. Ikejiri et al. (2005) used the terms "open" (separated), "closed" (together but still visible), and "fused" (no longer visible) to record the state of neurocentral suture closure in cervical, dorsal, and caudal vertebrae and in the scapula-coracoid suture. These terms will be used here.

The growth spurt experienced by *Tyrannosaurus rex* makes this species unique, even among other tyrannosaurids (Erickson et al. 2004). It is possible that the extremely accelerated growth rate between 10 and 20 years of age resulted in pre-caudal neurocentral sutures that never fused. Even in FMNH PR 2081, ontogenetically the oldest recorded *T. rex* (Erickson et al. 2004), sutures between the centra and neural spines are clearly visible (closed) through the entire pre-sacral and sacral series, as well as the first 15 (anterior) caudals (Brochu 2003; Takehito Ikejiri, pers. comm., July 2003, and pers. obs.). A second adult specimen, BHI 3033 (ontogenetic age not determined), shows closed (but visible) neurocentral sutures through caudal 10, and a sub-adult, TCM 2001.90.1 (nearly adult size – ontogenetic age not determined), has closed sutures through at least caudal 15 (series interrupted). Across species lines, a sub-adult (but nearly full-sized) *Gorgosaurus* (TCM 2001.89.1) shows open sutures through caudal 12, closed sutures through caudal 17, with sutures also visible on caudal 19 and 20, and fused sutures on the remaining caudals. An adult *Albertosaurus* (TMP 81.10.1) shows closed sutures through caudal 11.

By contrast, BMR P2002.4.1 has visible (closed) neurocentral sutures on only the first 11 caudal vertebrae. Number 12 caudal and greater, and one of the three preserved dorsal vertebrae, show no sutures (fused condition). In terms of suture closure, this indicates a more advanced ontological stage for BMR P2002.4.1 than FMNH PR 2081, if they are the same species. It seems reasonable for the neurocentral suture to remain visible (closed) on *Tyrannosaurus rex* throughout its growth spurt to allow for the tremendous expansion the vertebrae have to undergo as it quickly puts on weight. It seems just as logical that in a more flattened growth curve (i.e., *Gorgosaurus*, *Albertosaurus*, etc.), these sutures could fuse early (at 12 years when half, or more, of the mass has been attained) and still allow for a limited amount of growth.

Although no similar study has been published for theropods, in Ikejiri et al. (2005), the authors use closure of the scapula-coracoid suture to determine ontogenetic stage in *Camarasaurus lentus*. In stage 1 (the youngest), the suture is open, and the two bones separate. In stage 2, the suture is closed, but there is a visible suture. The suture is completely fused in stages 3 and 4 (sub-adult and adult), with no visible suture. BMR P2002.4.1's well-preserved right scapula-coracoid is nearly completely fused, with only a faint line visible over only a portion of the lateral and medial aspects of the junction of the two bones.

By contrast, although the 28-year-old (see Erickson et al. 2004) FMNH PR 2081's pathological left scapula-coracoid suture is fused, the suture is clearly visible on the right scapula-coracoid throughout its entire length (Brochu 2003). This suture is not visible on MOR 555 (22 years old, age estimated from femur length), although preservation, imperfect cleaning, and restoration may have obscured it somewhat. The suture is visible on both scapula-coracoids of BHI 4100, a specimen somewhat smaller than MOR 555 and considerably more massive than BMR P2002.4.1 (dentary length: FMNH PR 2081 = 1050 mm; MOR 555 = 990 mm; BHI 4100 = 770 mm; and BMR P2002.4.1 = 505 mm). The scapula is completely disarticulated in the type *Tyrannosaurus rex* (CM 9380), which, from femur length, was 18 years old at time of death, and in TCM 2001.90.1, whose age estimated from growth rings was 16 (Greg Erickson, pers. comm., October 2005).

Even more troubling, if BMR P2002.4.1 is really a young *Tyrannosaurus rex*, is the fusion of the pelvis. On both sides, the ilium, ischium, and pubis are fused together, with no suture visible (Mike Henderson, pers. comm., October 2004, and pers. obs.). No other specimen of *Tyrannosaurus* is preserved in this condition. In fact, the sutures between the ilium and pubis and between the ilium and ischium are open on all known specimens. Even when they are preserved in an articulated condition, they have all separated on the sutures during preparation (FMNH PR 2081, BHI 3033 BHI 3033, MOR 555, AMNH 5027, etc.; note that, although these sutures do not fuse, the articular surfaces are very scalloped, presumably strengthening the pelvis against stresses gathering at the acetabulum). The sutures joining the pubes and joining the pubis to the ischium are closed and may actually fuse in larger individuals (e.g., FMNH PR 2081, BHI 3033, MOR 555, and AMNH 5027). The suture between the ischia remains open in "Jane" and in all tyrannosaurs.

The fact that BMR P2002.4.1's pelvis is fused is problematic for an animal that would supposedly increase its mass ten fold within the next 10 years. Skeletal sutures remain open during growth to allow for skeletal changes. That these sutures remain open is particularly important if the individual needs to increase the size of an opening (e.g., a fenestra, neural arch, acetabulum, etc.). Although it is possible to increase the surface area of an opening through remodeling, as in the case of most foramina, it is much simpler to increase the surface area of a skeletal "hole" by simply adding bone at the sutures. It is for this reason that mammal pelves do not fuse until late in ontogeny (Walker and Liem 1994). In reptiles (including crocodilians) that continue to grow throughout their lifetimes, pelves never completely fuse (Romer 1956).

Pelvic bones fuse in many extant bird species as they reach adult, or near-adult, size (pers. obs.). Partial (e.g., *Carnotaurus*, Tykoski and Rowe 2004; *Coelophysis*, Colbert 1989; and *Allosaurus*, Mark Lowen, pers. comm., October 2005) to complete fusion (e.g., *Avimimus*, Osmólska et al. 2004; *Ceratosaurus*, Gilmore 1920; and *Syntarsus*, Raath 1990) has been noted for some other adult non-avian theropods. Before BMR P2002.4.1, complete fusion of the pelvis had not been recorded for any tyrannosaur. Ornithomimid pelves fuse in a manner similar to BMR P2002.4.1's, but only as adults (Makovicky et al. 2004). Given the complete fusion of the pelvis, it would be prudent to consider the possibility that BMR P2002.4.1 (and therefore *Nanotyrannus*) may have followed a

Table 2.1. Characters possibly explained by ontogeny

Character	*Tyrannosaurus rex*	*Nanotyrannus lancensis*
Nasal-maxilla suture	strongly scalloped	smooth
Serrated premaxilla teeth	yes	no
Curved fibula	no	yes
Ilium profile	high	low
Cervical vertebrae	short	long
First dorsal rib	capitulum longer than tuberculum	tuberculum longer than capitulum
Rear limb proportion	femur longer than tibia	tibia longer than femur
Humerus lenth: femur length	29%	39%

different logistic growth curve than *Tyrannosaurus* (and even *Daspleto-saurus*, *Gorgosaurus*, and *Albertosaurus*), one that leveled out sooner, perhaps terminating growth as early as 12 years of age.

There are a number of characters that seem to separate the group containing CMNH 7541, LACM 28471, BHI 6235 and BMR P2002.4.1, *Nanotyrannus lancensis*, from *Tyrannosaurus rex*. These include such differences as might arguably be related to changes brought about by ontogenetic development. It is important to attempt to exclude those ontogenetic differences from any list of characters that define a taxon. These differences may include ratios of the length of the femur compared to that of the tibia, skull, humerus, and so on; absence or presence of certain muscle scars; relative robustness of skeletal elements; and so forth (see Table 2.1).

Given the current state of our knowledge of tyrannosaurs, some ontogenetic differences are easy to isolate; others may not be so readily identifiable. The following argument attempts to exclude differences due to ontogenetic stage.

Characters That Separate *Nanotyrannus* from *Tyrannosaurus*

Skeletal Characters

BMR P2002.4.1 includes the only confirmed post-cranial material attributed to *Nanotyrannus lancensis*. Therefore, the skeletal characters described below are derived from this single specimen. As research continues, additional characters will undoubtedly be added to this list.

ANTERIOR ILIAC HOOK One of the first things that struck me as different about BMR P2002.4.1 was noticed during excavation. The anterior ends of the ilia expand ventrally, as they do in all tyrannosaurs. Unlike *Tyrannosaurus*, however, this ventral expansion includes a posteriorly facing "hook," which will be referred to as the "anterior iliac hook." This anterior iliac hook is also a character for the ornithomimosaur *Gallimimus* (ZPAL MgD-I/94) and the tyrannosaur *Gorgosaurus* (TCM 2001.89.1 and AMNH 5458; see Fig. 2.4). It does not occur in *Albertosaurus* (ROM 807 and FMNH PR 87469), *Daspletosaurus* (NMC 8505 and FMNH PR 308), or, of course,

2.4. Ilia of A) *Nanotyrannus*, BMR P2002.4.1; B) *Gorgosaurus*, TCM 2001.89.1; and C) *Tyrannosaurus*, BHI 3033. Note the anterior iliac hook present on *Nanotyrannus* and *Gorgosaurus* but absent on *Tyrannosaurus*.

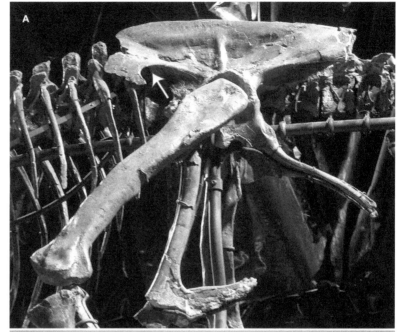

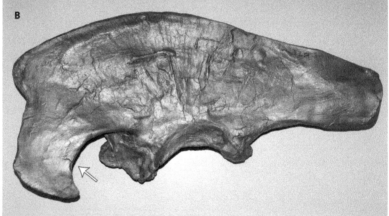

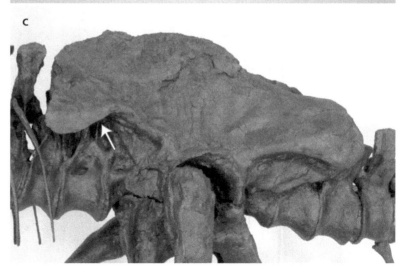

Tyrannosaurus (CM 9380, AMNH 5027, FMNH PR 2081, BHI 3033, BHI 6230, MOR 555, MOR 007, TCM 2001.90.1, etc.).

FUSED PELVIS As mentioned above, BMR P2002.4.1's pelvis is fused. Although in adult *Tyrannosaurus* the joints between the pubes and between the pubis and ischium may sometimes fuse, the sutures between the ilium and pubis, and ilium and ischium, never fuse. These joints, however, *are* fused in *Nanotyrannus* (BMR P2002.4.1). Although fusion of the pelvis has not been previously recorded for any tyrannosaur, BMR P2002.4.1's pelvis is fused in manner similar to adult ornithomimosaurs, where it is considered a character of the clade (Makovicky et al. 2004; see also discussion above, titled Ontogenetic Stage).

GLENOID BMR P2002.4.1's well-preserved and fused scapula-coracoid (see Fig. 2.5) carries a rather unusual glenoid that is not repeated in any other specimen referred to the genus *Tyrannosaurus* (CM 9380, AMNH 5027, FMNH PR 2081, BHI 4100, BHI 6230, MOR 980, MOR 555, LACM 23844, LACM 23845, etc.). In *Tyrannosaurus* the glenoid is a concavity oriented caudoventrally. In BMR P2002.4.1, in addition to the caudoventral articular surface, the glenoid also has a lateral component. This is similar to the condition found in ornithomimosaurs (Makovicky et al. 2004) and is nearly identical to that of *Struthiomimus sedens* (BHI 1266), but has not been seen in adult or sub-adult tyrannosaurs.

Interestingly, a similar glenoid has been seen in a single, disassociated scapula (TMP 86.144.1) attributed to *Albertosaurus* (Michael Parrish, pers. comm., September 2005; and Philip Currie, pers. comm., September 2005) that is approximately 75 percent the size of BMR P2002.4.1's. It is assumed to have come from an *Albertosaurus* because it is from the Dry Island bone bed, but, because it is from a bone bed, the possibility remains that it is from another species (e.g., an ornithomimosaur). The scapula-coracoid articulation of LACM 23845 (the smallest individual *Tyrannosaurus rex* available for this study) is not fused (open) and is nearly identical to BMR P2002.4.1's in size. The glenoid of LACM 23845 shows no indication that it ever possessed a lateral component (see also Carr and Williamson 2004).

Cranial Characters

The most convincing evidence of the validity of *Nanotyrannus lancensis* is preserved in the skulls of the type *Nanotyrannus* (CMNH 7541) and BMR P2002.4.1. These two specimens share characters not found in any of the larger specimens that are unequivocally attributable to *Tyrannosaurus rex*. For the purposes of this analysis, I am not attempting to list all the cranial characters that separate *Nanotyrannus* from *Tyrannosaurus*, only some of the most obvious.

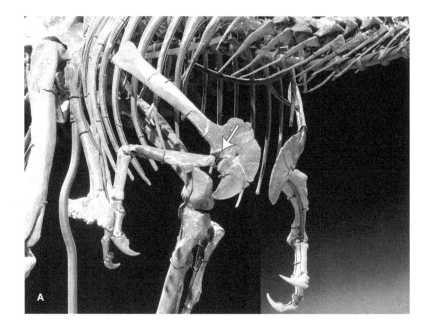

2.5. BMR P2002.4.1's scapula has a very unusual glenoid. A) BMR P2002.4.1's shoulder girdle (arrow). B) Comparison of *Tyrannosaurus* and *Nanotyrannus* glenoids.

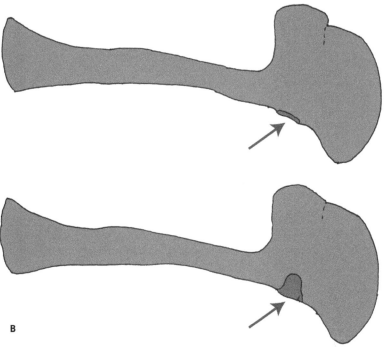

ANTORBITAL FOSSA The antorbital fossa, on both the type *Nanotyrannus* (CMNH 7541) and BMR P2002.4.1, forms a shallow depression slightly below the lateral surface of the maxilla. A thin ridge of bone rises along the dorsal margin of the posteroventral extension of the maxilla above the posterior alveoli. This thin ridge (the ventral antorbital maxillary ridge) is approximately one half of the dorsoventral dimension of the posteroventral maxillary extension. The ventral antorbital maxillary ridge extends

2.6. Comparison of right maxillae for A) *Nanotyrannus*, BMR P2002.4.1; B) *Gorgosaurus*, TCM 2001.89.1; and C) *Tyrannosaurus*, BHI 3033 (mirrored).

well past the last alveolus and passes under the jugal as it articulates to the maxilla. The constriction at the base of the ventral antorbital maxillary ridge forms the ventral border of the antorbital fossa (see Fig. 2.6). The top of the ridge is the ventral border of the antorbital fenestra. This condition closely resembles that of *Gorgosaurus* (TCM 2001.89.1).

In *Tyrannosaurus* the antorbital fossa is a very deep depression in even the smallest individuals (e.g., BHI 4100 and LACM 23845). In addition, the ventral antorbital maxillary ridge disappears well before the maxilla-jugal suture. The ventral borders of the antorbital fossa and the antorbital fenestra are only partially bounded by the antorbital maxillary ridge. This condition is identical to that of adult *Tarbosaurus* (ZPAL MgD-1/4) and is replicated in juvenile *Tarbosaurus* MPC 107/5 (Philip Currie, pers. comm., October 2005) (see Fig. 2.7).

MAXILLARY FENESTRA The maxillary fenestra (the second antorbital fenestra of Osborn 1912) in adult *Tyrannosaurus* and *Tarbosaurus* contacts the anterior and posterior borders of the antorbital fossa. This is also true for juvenile *Tarbosaurus* MPC107/5 (Philip Currie, pers. comm., October 2005) (see Fig. 2.7). In *Nanotyrannus* (CMNH 7541 and BMR P2002.4.1), however, the maxillary fenestra touches neither border and is centered within the shelf bounded by the edge of the antorbital fossa dorsally, anteriorly, and ventrally and posteriorly by the antorbital fenestra (see Figs. 2.6 and 2.7). The position of the maxillary fenestra within the antorbital fossa in *Nanotyrannus* most closely resembles *Gorgosaurus* (TCM 2001.89.1) and *Appalachiosaurus* (RMM 6670) and is a defining character for those genera (Carr et al. 2005).

VOMER The vomer of *Nanotyrannus* (BMR P2002.4.1) more closely resembles that of *Gorgosaurus* (TCM 2001.89.1) than it does that of *Tyrannosaurus* (BHI 3033; see Fig. 2.8). Although it is possible that the large spatulate anterior portion of the vomer could broaden ontogenetically, it is hard to explain why the central portion of the shaft of BMR P2002.4.1's vomer, if it were a juvenile *T. rex*, is greater in the vertical dimension than the adult BHI 3033's.

QUADRATOJUGAL Certainly one of the most interesting elements found with both the type *Nanotyrannus* and BMR P2002.4.1 is the quadratojugal. In BMP R2002.4.1 the quadratojugal is disarticulated and pristinely preserved (see Fig. 2.9). This bone alone shows four characters that separate *Nanotyrannus lancensis* from *Tyrannosaurus rex*. In lateral view, the dorsal edge of the ascending process preserves a central and a posterior notch (also present in the type, CMNH 7541). A similar central dorsal notch occurs in *Gorgosaurus* (TCM 2001.89.1) but not in *Tyrannosaurus* (BHI 3033, etc.). Although TCM 2001.89.1's quadratojugals do not preserve the posterior notch, Currie's 2003 illustration of TMP 91.36.500 clearly shows that *Gorgosaurus* has a posterior dorsal notch like *Nanotyrannus*. This posterior dorsal notch is absent in *Tyrannosaurus*.

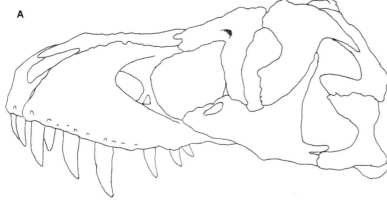

2.7. Tyrannosaur skull comparisons. A) *Tyrannosaurus rex,* BHI 3033. B) *Nanotyrannus lancensis,* BMR P2002.4.1. C) *Tarbosaurus bataar* (adult), PAL MgD-I/4. D) *Tarbosaurus bataar* (juvenile). See also Figure 2.6.

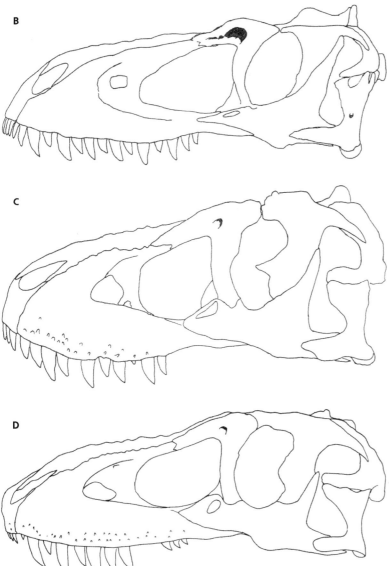

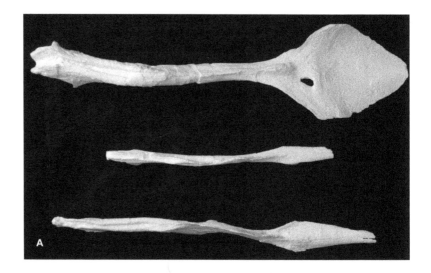

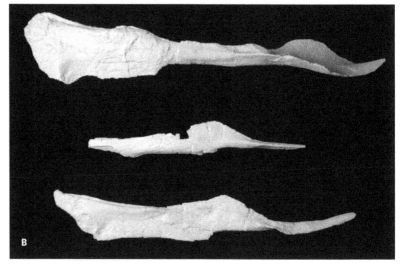

Interestingly enough, Carr and Williamson (2004) utilized the presence of a notch on the dorsal border of the ascending process of the quadratojugal of *Tyrannosaurus* to separate this genus from *Tarbosaurus, Albertosaurus,* and *Gorgosaurus* (Carr and Williamson's *Albertosaurus libratus*). This notch is absent in the quadratojugal of *Nanotyrannus* (BMR P2002.4.1) A fourth character of the quadratojugal, related to the respiratory system, is discussed later.

LACHRYMAL One of the characters uniting *Tyrannosaurus* with *Tarbosaurus,* but separating it from other tyrannosaurs, like *Daspletosaurus, Gorgosaurus* and *Albertosaurus,* is the absence of a cornual process on the lachrymal (Carr 1999; Carr and Williamson 2004). Although this portion of the lachrymal is missing (and restored) in the type *Nanotyrannus,* BMR P2002.4.1 preserves a pronounced cornual process, or lachrymal horn, separating it from *Tyrannosaurus.* In addition, the lachrymal is 7-shaped

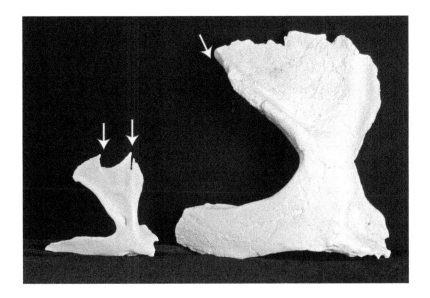

2.9. Lateral view of the left quadratojugal for (*left*) *Nanotyrannus*, BMR P2002.4.1 and (*right*) *Tyrannosaurus*, BHI 3033.

in *Tyrannosaurus*, as it is in *Tarbosaurus*, and T-shaped in *Nanotyrannus*, as it is in *Albertosaurus*, *Gorgosaurus*, and *Daspletosaurus* (Carr et al. 2005; see also Fig. 2.10).

QUADRATE BMR P2002.4.1's quadrate preserves the dorsal articular surface (this is one half of the ball-and-socket joint allowing streptostyly). The dorsal articular surface (which inserts into a socket on the ventroposterior surface of the squamosal) is ball shaped. This contrasts with the quadrate for *Tyrannosaurus rex* (BHI 3033), which has a double articular surface – two ball-shaped articulations joined by a saddle corresponding to a double socket on the squamosal. *Daspletosaurus* (Currie 2003) and *Albertosaurus* (BHI 6234) quadrates resemble that of *Nanotyrannus*, while *Gorgosaurus* (TCM 2001.89.1) has a double dorsal articular surface like *Tyrannosaurus rex* (see Fig. 2.11).

V-2 CRANIAL NERVE Osborn (1912:8) noted an opening at the junction of the maxilla and premaxilla that he called "the third antorbital fenestra." This opening, called the subnarialis foramen by Carr (1999) and Brochu (2003), is here considered to be the maxillary exit for cranial nerve V-2. The V-2 cranial nerve (the second branch of the fifth cranial or trigeminal nerve) is the superior maxillary nerve and is the sensory nerve for the upper lip and cheeks in humans (Gray 1901). In *Tyrannosaurus rex* this nerve exits at the junction of, and is equally bounded by, the maxilla and premaxilla (see Fig. 2.12A). This is also the case in *Allosaurus* (Madsen 1976). In *Alligator mississippiensis* (BHI 6240), the maxillary exit foramen for cranial nerve V-2 is quite small, and, although it is near the maxilla-premaxillary suture, it is completely bounded by the maxilla.

In *Nanotyrannus* the V-2 cranial nerve exit is completely bounded by the maxilla and is associated with a deep fossa projecting anterioventrally from the foramen (Fig. 2.12B). This difference between *Tyrannosaurus*

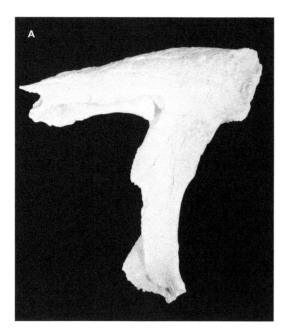

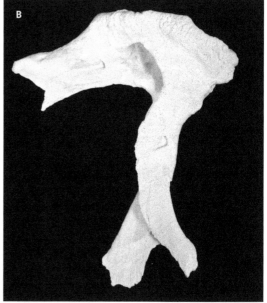

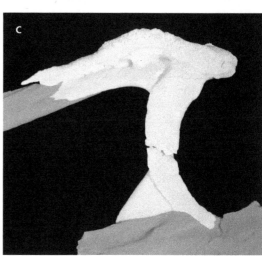

2.10. Lateral view of left lachrymals for A) *Tyrannosaurus*, BHI 3033; B) *Gorgosaurus*, TCM 2001.89.1; and C) *Nanotyrannus*, BMR P2002.4.1.

rex and the type *Nanotyrannus* (CMNH 7541) was first noted by Brochu (2003). This placement of the V-2 exit foramen is similar in the allosaurid *Sinraptor* (Currie and Zhao 1993). The condition also seems to occur in *Gorgosaurus*, which has a large exit foramen at approximately the same site as *Nanotyrannus*, also accompanied by a deep anterioventrally facing fossa (Fig. 2.12C). Carr's (1999) composite drawing for *Gorgosaurus* (*Albertosaurus*) *libratus* (AMNH 5664, CMN 11315, TMP 91.36.500, USNM 12815) shows a *T. rex*–like "foramen nasalis" and a second large foramen that Carr calls one of the "foramina neurovasculares." However, Currie (2003), in his drawing of *Gorgosaurus libratus* (TMP 91.36.500), shows only a single opening (with fossa) bordered entirely by the maxilla, duplicating the condition of the *Gorgosaurus* sp. (TCM 2001.89.1) found in Figure 2.12.

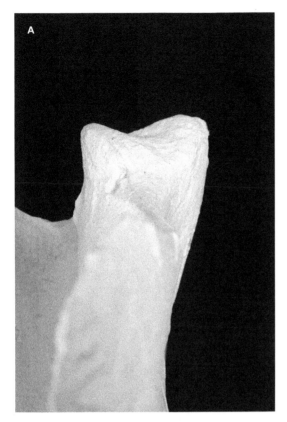

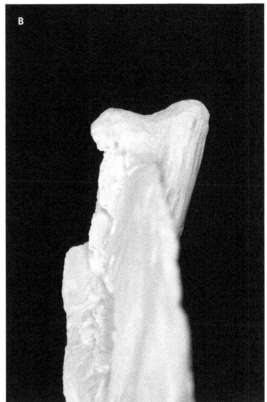

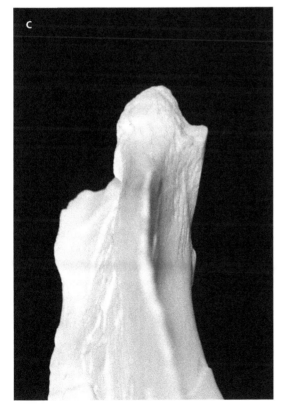

2.11. Cranial aspect of the dorsal articular surface (condyle) of the right quadrate for A) *Tyrannosaurus,* BHI 3033; B) *Gorgosaurus,* TCM 2001.89.1; and C) *Nanotyrannus,* BMR P2002.4.1.

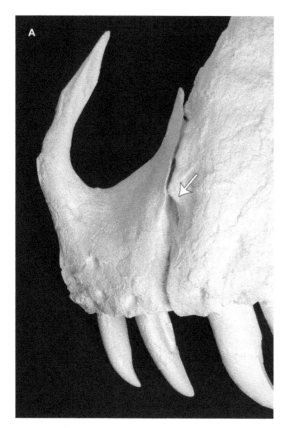

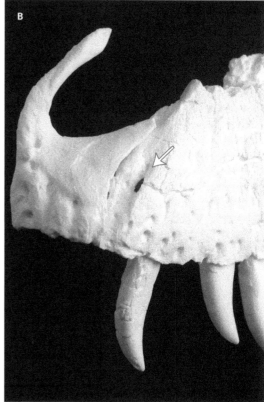

2.12. The V-2 cranial nerve facial opening for (A) *Tyrannosaurus*, BHI 3033, lies at the suture between the maxilla and premaxilla. For (B) *Nanotyrannus*, BMR P2002.4.1, and (C) *Gorgosaurus*, TCM 2001.89.1, it lies caudal to the maxilla-premaxillary suture and is completely bounded by the maxilla.

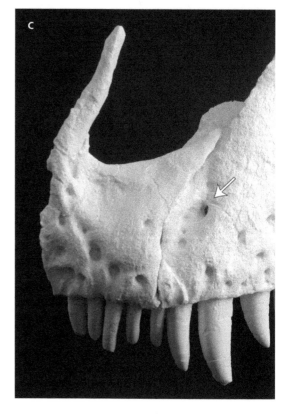

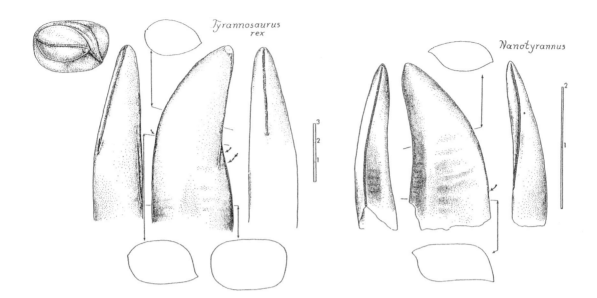

Tyrannosaurus rex

Nanotyrannus

TOOTH MORPHOLOGY Bakker et al. (1988:24) noted that "*Nanotyrannus* differs from all other tyrannosaurids, except possibly *Alioramus*, in retaining the primitive theropod condition of maxillary teeth that are compressed strongly from side to side, very wide front-to-back relative to the crown height." Illustrating isolated maxillary teeth referred to *Nanotyrannus* and *Tyrannosaurus*, Bakker et al. demonstrated a clear difference in cross section between the two genera (see Fig. 2.13). For this study I measured the length and width of the fourth maxillary tooth at the base of the crown for a number of taxa. Dividing the length by the width, *Nanotyrannus* did indeed give the lowest value (see Table 2.2). Larson (1999) also provided evidence of feeding behavior differences between *Nanotyrannus* and *Tyrannosaurus*, presumably arising from this difference in tooth morphology.

However, Carr (1999) implies that in his growth series for *Albertosaurus libratus* (*Gorgosaurus libratus*) and *Daspletosaurus torosus* the crown width/length ratio increases with age, unfortunately without corroborating data. Carr and Williamson (2004) do successfully link "the Jordan theropod" (LACM 28471) to the type of *Nanotyrannus* (CMNH 7541), indicating that similar tooth cross sections imply the same taxon. Their analysis implies that tooth cross sections have utility in assessing relationships. In contrast, Farlow et al. (1991) noted that crown length versus crown width is basically a linear relationship implying a lack of utility for species recognition.

In the initial description of the upper jaw of CMNH 7541 (the type of *Nanotyrannus*) as *Gorgosaurus lancensis*, Gilmore (1946:199) noted: "In the front of the series there are ten teeth of reduced size, rod like and with flattened sides. As far as the teeth can be examined, they appear to be in full accord with those described by Leidy (1868) under the genus name *Aublysodon*." Bakker et al. (1988:24) also noted that "the first maxillary

Table 2.2. Fourth maxillary tooth (measured at base of crown in millimeters)

	Tyrannosaurus rex BHI 3033	Tyrannosaurus "x" MOR 008	Tarbosaurus ZPAL MgD-I/4	Gorgosaurus TCM 2001.89.1	Nanotyrannus BMR P2002.4.1
Width	31.5	37.6	20.7	16.3	12.2
Length	50.3	49.9	36.6	22.0	23.5
Width/length	0.63	0.75	0.57	0.74	0.52

tooth has a form like that of the premaxillary teeth." The holotype of *Nanotyrannus*, BMR P2002.4.1, and LACM 28471 have four tooth positions in each premaxillae and also have incisiform first maxillary teeth. In *Nanotyrannus*, the first maxillary teeth are equal in size and morphology to those of the premaxillae. They differ from those of the maxilla in that they lack serrations (non-denticulate), are D-shaped in cross section (instead of ellipsoid), and are greatly reduced in size (see Fig. 2.14). This is not the condition in *Tyrannosaurus*, where the first maxillary tooth is virtually the same size and shape as the second. This condition is, however, a long-recognized character for *Gorgosaurus* (Lambe 1917; Gilmore 1946; Russell 1970; Carr 1999).

The first dentary tooth of *Nanotyrannus* also presents a problem if we are to believe that CMNH 7541, LACM 28471, and BMR P2002.4.1 are juvenile *Tyrannosaurus rex*. In *T. rex* the first dentary tooth is incisiform and, if found loose, could easily be mistaken for a tooth from the premaxilla. It is D-shaped in cross section and is approximately the same size as those from the premaxilla, somewhat reduced in size from the immediately succeeding dentary teeth. This differs from *Gorgosaurus* (TCM 2001.89.1), which has a subconical first dentary tooth (not D-shaped), approximately the same size and shape as those succeeding it. *Nanotyrannus* (LACM 28471 and BMR P2002.4.1) presents a completely different condition, with a conical (not D-shaped) first dentary tooth that is greatly reduced in size from those succeeding it (see Fig. 2.15), and even from those of the premaxilla.

TOOTH COUNTS Bakker et al. (1988) noted that the type of *Nanotyrannus* (CMNH 7541) has 15 maxillary teeth, whereas *Tyrannosaurus rex* (TMP 81.6.1, AMNH 5027, LACM 23844) possesses only 12. BMRP 2002.4.1 has 15 tooth positions in the right maxilla and 16 in the left. The specimens of *Tyrannosaurus rex* examined for this study vary from 11 to 12 tooth positions in the maxilla (see Table 2.3a), and four specimens that have been tentatively assigned to a new species (Larson 2008) have 12 to 13 maxillary teeth. BMR P2002.4.1, with the only countable *Nanotyrannus* lower alveoli, has 17 tooth positions in each dentary. *T. rex* has 12 to 13 and *Tyrannosaurus* "x" has 13 to 14 dentary tooth positions. All tyrannosaurs, *Nanotyrannus* included, have four alveoli in each premaxilla (see Table 2.3).

Carr (1999:514) attempts to resolve this tooth count problem through ontogenetic change: "The maxilla loses three or four [up to six] teeth from the rostral end of the tooth row" as the animal grows up. Carr cited

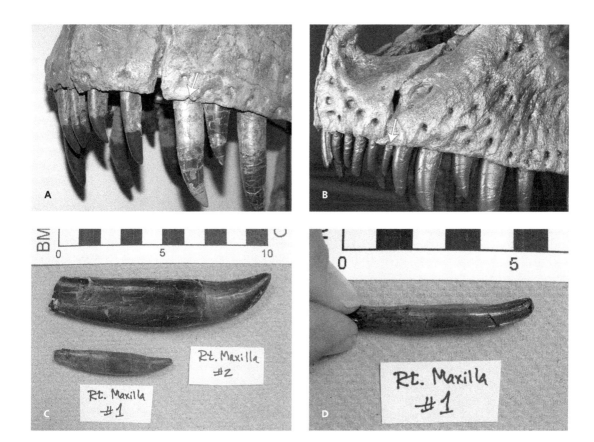

2.14. The first maxillary tooth of A) *Tyrannosaurus,* BHI 3033, is approximately the same size and morphology as the second; B) *Gorgosaurus,* TCM 2001.89.1, is reduced in size and is incisiform; C) *Nanotyrannus,* BMR P2002.4.1, is greatly reduced in size; and D) is incisiform.

Albertosaurus (*Gorgosaurus*) *libratus* as an example. "A similar pattern of loss of tooth positions is present in the maxilla of *A. libratus*, dropping from 16 to 13 alveoli" (Carr 1999:514), although his table actually shows a variation of from 13 to 15 alveoli and the 10 specimens listed do not show a pattern of loss from smallest to largest (see Table 2.4)

Few studies of single species theropod collections include samples of alveolar counts. One notable exception is Madsen's 1976 monograph on *Allosaurus*. Most of his data came from the disarticulated *Allosaurus* specimens of the Cleveland-Lloyd Quarry. Madsen lists 36 measured maxillae whose alveoli counts varied from a low of 14 (one maxilla) to a high of 17 (two maxillae), with most specimens falling between 15 and 16 alveoli. Premaxilla tooth counts remained constant at 5 (for 45 premaxilla), and for 43 dentaries the alveoli counts varied from a low of 14 (one dentary) to a high of 17 (12 dentaries). For all elements, there is no pattern of alveolar loss or gain (the smallest maxilla has 16 and the largest 16, the smallest dentary has 15 and the largest 16), indicating that the differing number of alveoli are individual variation, completely unrelated to ontogeny (see also Currie 2003). These results were also confirmed by Colbert (1990:89) in his assessment of variation within *Coelophysis bauri*: "The variation in the number of maxillary teeth ranges between 22 and 26, and seems to be largely independent of size."

2.15. The first dentary tooth of *Tyrannosaurus rex*, BHI 3033, is (A) smaller than the second and third and (B) shows serrations on both lingual and occlusal aspects, giving the characteristic D-shaped cross section found also in the premaxillary teeth (incisiform). In contrast, (C) *Nanotyrannus*, BMR P2002.4.1, has a reduced (non-incisiform) second and greatly reduced first dentary tooth, which also is not incisiform (D).

Carr's (1999:514) assertion that *Tyrannosaurus rex* lost tooth positions as it matured was also based upon published work on crocodilians: "Among other archosaurs, ontogenetic tooth loss has been reported by Mook (1921), Wermuth (1953), and Iordansky (1973)." However, those references do not precisely bolster Carr's assertion.

Mook (1921) actually makes no reference to ontogenetic tooth loss. Instead, he refers to a decrease in the number of teeth beneath the orbit as crocodiles grow and skull proportions change. This is not tooth loss but, rather, a reduction in the relative size and a shift in position of the orbit as the animals mature. Maxillary tooth counts remain unchanged through ontogeny.

Iordansky (1973:487) makes a brief reference to tooth loss in his section on ontogeny: "In some species (*Crocodylus cataphractus, C. porosus, C. siamensis, Tomistoma schlegelii*) the 2nd premaxillary tooth is lost by juvenile animals." This single sentence reference to the "second" premaxillary tooth is clarified by Wermuth (1953), who refers to the loss of the second premaxillary tooth in *alter* (older) crocodiles. But Wermuth attributes this to a durchbrochen (breaking through or piercing) of the premaxilla, at the second alveolus, by the first tooth of the dentary: "Da sich die endgultigen Verhaltnisse in der Praemaxillarybezanung erst im spatern Alter herausbilden, findet man bei jungeren Exemplaren solcher

Table 2.3A. *Tyrannosaurus rex* tooth counts

Specimen Number	Dentary Tooth Row Length (mm)	Maxillary Tooth Counts		Premaxillary Tooth Counts		Dentary Tooth Counts	
		R	L	R	L	R	L
BHI 4100	480	-	-	4	-	-	13
TMP 81.6.1	490	12	12	4	4	13	13
MOR 1125	490	12	12	-	-	14	-
AMNH 973 = CM 9380	510	12	12	-	-	13	13
BHI 4182	530	-	-	-	-	14	14
LACM 23844	540	-	-	-	4	-	13
MOR 980	540	11	11	4	4	13	13
AMNH 5866 = NHM R8001	est. 550	-	-	-	-	13	13
NMMP 1013.1	est. 560	-	-	-	-	-	13
CM 1400	560	12	-	-	-	-	-
BHI 3033	570	11	11	4	4	13	13
RSM P2523.8	590	-	11	4	4	14	14
MOR 555	600	-	12	-	-	13	-
UCMP 118742	est. 600	12	-	-	-	-	-
BHI 2033 = FMNH PR 2081	620	12	12	4	4	13	13

Table 2.3B. *Tyrannosaurus "x"* tooth counts

Specimen Number	Dentary Tooth Row Length (mm)	Maxillary Tooth Counts		Premaxillary Tooth Counts		Dentary Tooth Counts	
		R	L	R	L	R	L
AMNH 5027	520	12	12	4	4	14	14
SDSM 12047	est. 520	12	12	-	-	15	15
SAMSON	540	13	13	4	4	-	15
MOR 008	est. 550	12	12	-	-	-	-

Table 2.3C. *Nanotyrannus lancenis* tooth counts

Specimen Number	Dentary Tooth Row Length (mm)	Maxillary Tooth Counts		Premaxillary Tooth Counts		Dentary Tooth Counts	
		R	L	R	L	R	L
CMNH 7541	est. 260	15	15	4	4	-	-
BMR P2002.4.1	330	15	16	4	4	17	17

Arten, deren praemaxillare Zahnzahl im Alter reduziert ist, den spater obliterierenden" (The final premaxillary tooth arrangement only materializes late in life. Therefore, in younger individuals, one still finds the second–later in life obliterated–tooth; Wermuth 1953:398, translation by Klaus Wesphal).

Using extant phylogenetic bracketing (Witmer 1995), crocodilian (ancestral) ontogenetic tooth position loss would provide a way to test Carr's theory. On the descendant side, living birds do not suffer ontogenetic tooth loss; the numbers remain constant, at zero. For this test, 42 *Alligator mississippiensis* skulls (some still living) were measured, and elemental tooth (or alveoli) counts were made (see Table 2.5). In this living archosaur (all specimens indisputably from a single species), the

Specimen	Maxillary Tooth Count	Skull Length (mm)
AMNH 5664	15	678
ROM 1247	14	750
USNM 12814	13	795
TMP 83.36.100	15	-
CMN 2270	15	-
CMN 12063	14	-
UA 10	13	900
AMNH 5336	13	962
AMNH 5458	14	990
CMN 2120	13	1000

Table 2.4. *Gorgosaurus (Albertosaurus) libratus* tooth counts (after Carr 1999)

maxillary tooth count varies from 14 to 16, premaxillary from 4 to 5, and the dentary from 19 to 20. The absence of any pattern clearly demonstrates that, for this extant species, variation in tooth counts is simply a function of individual variation and this modern archosaur does *not* lose tooth positions as it matures.

Many authors (e.g., Osborn 1905, 1906, 1912; Gilmore 1920, 1946; Russell 1970; Bakker et al. 1988; Paul 1988; Molnar 1991; Larson and Donnan 2002; Currie 2003; Currie et al. 2003; Hurum and Sabath 2003; Larson 2008) use tooth counts as tyrannosaurid taxa-defining characters. Researchers also use tooth counts (and ranges) as defining characters for fossil and living crocodiles (e.g., Gilmore 1911; Mook 1923a, 1923b, 1925; Wermuth 1953, Langston 1965; Erickson 1976; Tchernov 1986; Norell et al. 1994). The fact that they are used as characters shows at least perceived consistency. Although Carr (1999:514) advises caution "because tooth counts appear to be sensitive to ontogenetic and individual variation," the case for ontogenetic variation remains undemonstrated. As long as a larger sample is used to determine range, tooth counts seem to maintain their taxonomic utility.

If we apply the concept of range of tooth (alveoli) counts and deviation from the mean (standard deviation), utilizing the tooth count data from other archosaurs, to the question of whether or not *Nanotyrannus* is a juvenile *Tyrannosaurus rex*, we have an additional test of Carr's (1999) hypothesis. Table 2.6 provides values for a wide range of archosaurs, including *Alligator mississippiensis*, *Allosaurus fragilis*, and a number of tyrannosaurid species. Note that when *Nanotyrannus* is combined with *Tyrannosaurus* (Carr's 1999 hypothesis) the deviation from the mean is more than for the primitive *Coelophysis*, nearly twice that for *Allosaurus*, and two and one half (or more) times greater than that for any tyrannosaurid species (and the modern *Alligator mississippiensis*).

Table 2.7 gives ranges of tooth counts for 20 species of extant crocodilians. Notice that for living species, standard deviation of elemental tooth counts never exceeds a value of 1. Remarkably, even if we combine the tooth counts for the 11 described species of *Crocodylus*, the standard deviation is only 1.5. This strongly suggests that an anomalously high standard

Table 2.5. *Alligator mississippiensis* tooth counts

Specimen	Skull Length (mm)	Right Maxilla	Left Maxilla	Right Premaxilla	Left Premaxilla	Right Dentary	Left Dentary
BHI 6237	40	15	15	5	5	20	20
RG Fluffy 4	52	16	15	5	5	19	19
RG Fluffy 5	72	15	15	5	5	20	19
RG Fluffy 2	85	15	15	4	5	20	20
RG Fluffy 1	110	15	15	5	5	20	19
RG Fluffy 3	110	15	15	5	5	20	19
BHI 6260	130	15	15	5	5	19	20
BHI 6263	130	15	15	5	5	20	20
BHI 6270	130	15	15	5	4	20	20
BHI 6255	132	15	15	5	5	20	19
BHI 6267	132	15	15	4	4	20	20
BHI 6269	132	15	15	5	5	20	19
BHI 6251	135	15	15	5	5	19	19
BHI 6256	135	15	15	5	5	20	20
BHI 6265	135	15	15	5	4	20	20
BHI 6266	135	14	15	5	5	19	20
BHI 6268	135	15	15	4	5	20	19
BHI 6264	137	15	15	5	5	20	20
BHI 6254	138	14	15	5	5	19	20
BHI 6259	138	15	15	5	5	19	20
BHI 6245	140	15	15	5	5	19	19
BHI 6253	145	15	15	5	5	20	20
BHI 6252	150	15	15	5	5	19	19
BHI 6258	150	15	15	4	5	20	20
BHI 6261	150	15	15	5	5	20	20
BHI 6262	168	15	15	5	5	20	20
BHI 6244	169	14	14	5	5	19	19
BHI 6246	170	14	15	4	5	20	20
BHI 6257	170	15	15	5	5	20	20
BHI 6247	175	14	14	4	5	19	20
BHI 6250	175	15	15	5	5	19	20
BHI 6243	238	15	15	5	5	19	20
BHI 6239	240	15	15	5	5	20	20
KU 195568	245	15	15	5	5	19	-
BHI 6238	245	15	15	4	4	20	20
GRP 0501 0045	322	15	15	5	5	20	20
KU 322	345	15	15	5	5	-	-
KU 19538	355	16	15	5	5	19	20
TCM 2000.37.1.2	405	15	15	5	5	20	19
RG Sherman	450	16	est. 16	4	4	19	20
TCM 2005.25.2.1	460	15	15	4	5	19	20
RG-Brutus	500	15	15	5	5	20	20
BHI-6240	510	16	16	5	5	20	20

Table 2.6. Standard deviation of tooth counts

Species	Maxilla			Premaxilla			Dentary		
	# Teeth	Range	SD	# Teeth	Range	SD	# Teeth	Range	SD
Alligator mississippiensis	14–16	2	1.0	4–5	1	0.5	19–20	1	0.5
Coelophysis bauri	22–26	4	2.0	4	0	0	25–27	2	1.0
Allosaurus fragilis	14–17	3	1.5	5	0	0	14–17	3	1.5
Albertosaurus sarcophagus	13–15	2	1.0	4	0	0	13–15	2	1.0
Daspletosaurus torosus	14–16	2	1.0	4	0	0	16–17	1	0.5
Gorgosaurus libratus	13–15	2	1.0	4	0	0	15–17	2	1.0
Tarbosaurus baatar	12–13	1	0.5	4	0	0	14–15	1	0.5
Tyrannosaurus rex	11–12	1	0.5	4	0	0	12–13	1	0.5
Tyrannosaurus "x"	12–13	1	0.5	4	0	0	13–14	1	05
Tyrannosaurus rex + "x"	11–13	2	1.0	4	0	0	12–14	2	1.0
Nanotyrannus lancensis	15–16	1	0.5	4	0	0	17	0	0
Tyrannosaurus + Nanotyrannus	11–16	5	2.5	4	0	0	12–17	5	2.5

Table 2.7. Variation in tooth count for living crocodilians (in part after Wermuth 1953)

Species	Tooth Counts, Maxilla	Maxilla SD	Tooth Counts, Premaxilla	Premaxilla SD	Tooth Counts, Dentary	Dentary SD
Alligator sinensis	13–14	0.5	5	0	18–19	0.5
Melanosuchus niger	13–14	0.5	5	0	17–18	0.5
Caiman latirostris	13–14	0.5	5	0	17–19	1.0
Caiman crocodilus	14–15	0.5	5	0	18–20	1.0
Paleosuchus trigonatus	15–16	0.5	4	0	21–22	0.5
Paleosuchus palpebrosus	14–15	0.5	4	0	21–22	0.5
Tomistoma schlegelii	15–16	0.5	4–5	0.5	19–20	0.5
Gavialis gangeticus	23–24	0.5	5	0	25–26	0.5
Osteolaemus tetraspis	12–13	0.5	4	0	14–15	0.5
Crocodylus niloticus	13–14	0.5	5	0	14–15	0.5
Crocodylus palustris	13–14	0.5	5	0	14–15	0.5
Crocodylus acutus	13–14	0.5	5	0	15	0
Crocodylus moreletii	13–14	0.5	5	0	15	0
Crocodylus siamensis	13–14	0.5	4–5	0.5	15	0
Crocodylus rhombifer	13–14	0.5	5	0	15	0
Crocodylus porosus	13–14	0.5	4–5	0.5	14–15	0.5
Crocodylus novae-guineae	13–14	0.5	5	0	15	0
Crocodylus intermedius	13–14	0.5	4–5	0.5	15–16	0.5
Crocodylus cataphractus	13–14	0.5	4–5	0.5	15–16	0.5
Crocodylus johnsoni	14–16	1.0	5	0	15	0
Combined *Crocodylus* (11 spp.)	13–16	1.5	4–5	0.5	14–16	1.0

deviation of 2.5 for the combined tooth counts of *Nanotyrannus* and *Tyrannosaurus* (see Table 2.6) is due to an unnatural joining of two taxa.

Cranial Characters Related to the Respiratory System

During the course of the examination of the specimens attributable to *Nanotyrannus* and *Tyrannosaurus*, I noticed a number of characters that were inconsistent between *Nanotyrannus* and *Tyrannosaurus*, related to

Table 2.8. Comparative measurements of *Struthio, Tyrannosaurus,* and *Nanotyrannus* (in mm)

Element	Adult *Struthio* BHI 6242	Juvenile *Struthio* BHI 6243	*Tyrannosaurus* BHI 3033	*Nanotyrannus* BMR P2002.4.1
Femur	295	135	1310	720
Tibiotarsus	545	212	1070	800
Metatarsal III	485	190	680	550
Skull	200	115	1400	710

the diverticula (air sacs) of the respiratory system. These characters consist of alterations of the cranial skeleton by invasion or depression of the bones by the air sacs (Britt 1993; Larson 1997b). In order to verify whether or not these inconsistencies constitute taxon-defining characters, it is important to understand how these features change as an individual grows from a juvenile into an adult. Do existing pneumatic foramina close and/or new ones open during ontogeny?

To test this hypothesis, I examined extant ostrich (*Struthio camelus*) adult (BHI 6241 and BHI 6242) and juvenile (BHI 6243) skulls and skeletons. The developmental stage of the two ostrich skeletons are roughly comparable to that of an adult *Tyrannosaurus rex* (BHI 3033) and BMR P2002.4.1, if BMR P2002.4.1 were a juvenile *T. rex*. Measurements for the specimens used in this study are given in Table 2.8.

In all cases there is a one-to-one correlation between the juvenile and adult *Struthio* pneumatic foramina (pneumatopores) and pneumatic fossa. The only changes that could be detected from juvenile to adult pneumatopores were occasional multiple perforations (same site) in the adult. In no instance was there a closure of a juvenile pneumatopore in the adult or even the opening of a new site from the juvenile to the adult. In all cases there was a one-to-one correlation of pneumatopores and pneumatic fossas between the adult and juvenile specimens (see Fig. 2.16).

Clearly pneumatic features like pneumatopores and pneumatic fossae can develop early in ontogeny in avian theropods. But how early did these features manifest themselves in birds' theropod ancestors? Even as unhatched embryos, theropods show the marks of the diverticula of the respiratory system. Pneumatopores in the vertebrae can be present even at this early stage of development (pers. obs.; see Fig. 2.17).

POST-ORBITAL The medial aspect of BMR P2002.4.1's post-orbital sports a deep central depression, or fossa. This fossa has a clear boundary and may represent the impression of diverticula of the respiratory system. A similar depression is seen in *Gorgosaurus* (TCM 2001.89.1) and *Daspletosaurus* (Currie 2003). *Albertosaurus* (BHI 6234) has a depression that is less developed, and this fossa is absent in *Tyrannosaurus* (BHI 3033) and *Tarbosaurus* (ZPAL MgD-I/4; see Fig. 2.18).

QUADRATOJUGAL The type of *Nanotyrannus* (CMNH 7541) and BMR P2002.4.1 have large pneumatic foramina on the central portion of the

2.16. There is a one-to-one correspondence in pneumatopores between juvenile and adult *Struthio camelus* skeletons. A) Femora. B) Coracoids. C) Ribs. D) Vertebrae. E) Skulls. F) Lower jaws.

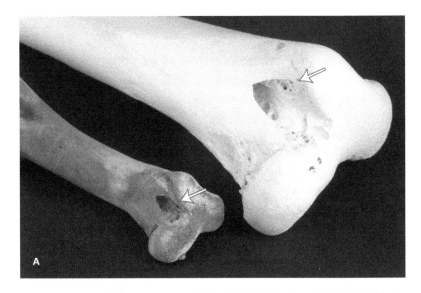

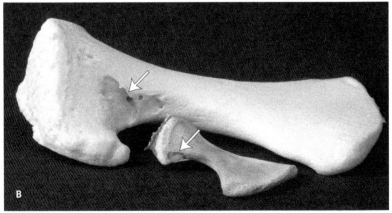

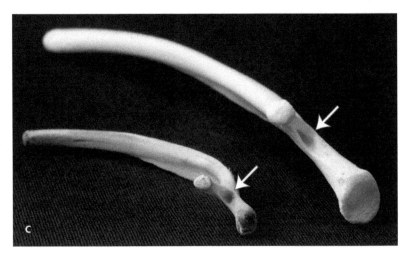

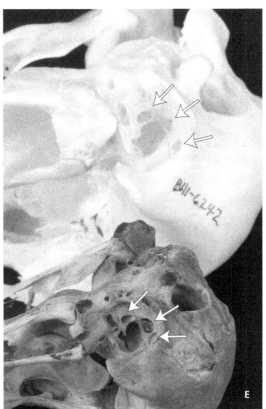

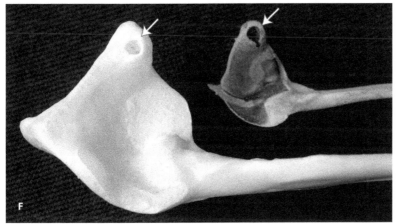

lateral aspect of the quadratojugal. These are the only known pneuma-
tizations of the quadratojugal for *any* theropod (see Figs. 2.19 and 2.9).
This character seems unique to *Nanotyrannus lancensis*.

SQUAMOSAL The central portion of the anterior aspect of the squamosal
of *Tyrannosaurus rex* is perforated by a very large pneumatopore. This
pneumatopore is completely absent in *Nanotyrannus* (see Fig. 2.20).
Instead, the central portion of the anterior aspect of the squamosal of

2.17. Pneumatopores develop early in ontogeny. Even this (A) Late Cretaceous non-avian theropod (oviraptorid) embryo (BHI 6402) shows (B) well-developed pneumatopores in the dorsal vertebrae (arrows).

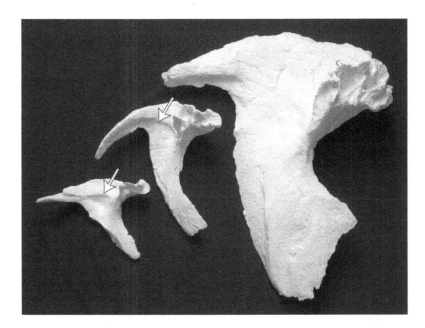

2.18. Left postorbitals (*left to right): Nanotyrannus*, BMR P2002.4.1; *Gorgosaurus*, TCM 2001.89.1; and *Tyrannosaurus*, BHI 3033. Note the central medial fossa in *Nanotyrannus* and *Gorgosaurus* (arrows), absent in *Tyrannosaurus*.

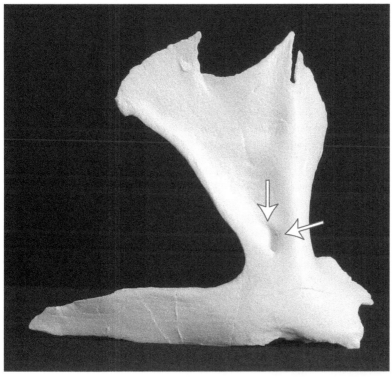

2.19. *Nanotyrannus* quadratojugal. Note the central pneumatopore (see also Fig. 2.9 for comparison to *Tyrannosaurus*).

Nanotyrannus lancensis (BMR P2002.4.1) has a deep fossa (presumably a pneumatic fossa). This condition is similar in *Daspletosaurus* (Currie 2003) and *Albertosaurus* (BHI 6234). In contrast, *Tarbosaurus* (ZPAL MgD-I/4) more closely resembles *Tyrannosaurus* (BHI 3033, BHI 4100, etc.). *Gorgosaurus* also resembles *Tyrannosaurus rex*, although in *Gorgosaurus* (TCM 2002.89.1) the pneumatopore is, relatively, much smaller.

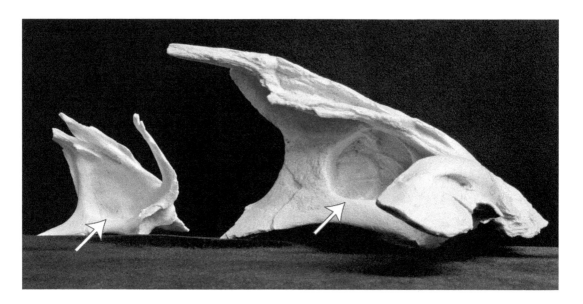

2.20. Anterior view of (*left*) *Nanotyrannus*, BMR P2002.4.1, and (*right*) *Tyrannosaurus*, BHI 4100, left squamosals. Note the large pneumatopore on *Tyrannosaurus* that is absent on *Nanotyrannus*.

LACHRYMAL In all tyrannosaurs, a distinctive pneumatic foramen (the lateral lachrymal pneumatopore) is found at the junction of the dorsal and vertical ramus of the 7-shaped (or in some cases T-shaped) lachrymal. This pneumatic foramen is at the posterior dorsal corner of the antorbital fossa. In *Tyrannosaurus rex* the lateral lachrymal foramen is relatively small, approximately one fourth the width of the vertical ramus or one fourth the height of the dorsal ramus. In *Nanotyrannus lancensis* the lateral lachrymal foramen is relatively much larger than that for *T. rex*, encompassing more than one half the width of the vertical ramus and more than one half the height of the dorsal ramus. In the case of BMR P2002.4.1 (a supposed juvenile *T. rex*), the dimensions of the lateral lachrymal foramen literally exceed those of an adult *T. rex* (BHI 3033). In *T. rex* the lateral lachrymal foramen is a single structure, but in *Nanotyrannus* there is a multiple perforation (see Fig. 2.21). The number of perforations seem to increase with age, as the smallest specimen (BHI 6235) has only two perforations and the largest, BMR P2002.4.1, has three or more. An increase in the number of perforations (associated with a pneumatopore) with age is consistent with what happens ontogenetically with some pneumatopores in *Struthio* (see above).

In medial view *Tyrannosaurus rex* (BHI 3033) has a large pneumatic foramen (the medial lachrymal pneumatopore) anterior to and at the dorsal margin of a thin ridge that descends the vertical ramus diagonally (this ridge terminates at the ventral anterior border of the vertical ramus). Although a similar ridge is present in *Nanotyrannus*, there is no evidence of a medial lachrymal pneumatopore on either BMR P2002.4.1 or the isolated lachrymal (BHI 6235; see Fig. 2.22).

ECTOPTERYGOID In their 2005 description of a new tyrannosaurid from Alabama, Carr et al. separated *Appalachiosaurus* and *Albertosaurus* from *Daspletosaurus*, *Tarbosaurus*, and *Tyrannosaurus* based, in part, upon

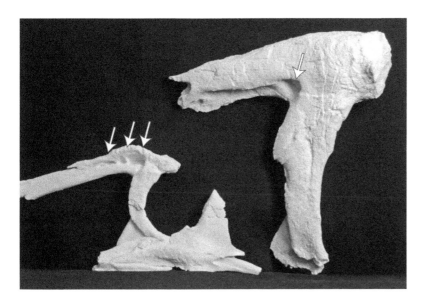

2.21. Left lachrymals, lateral view: (*left*) *Nanotyrannus*, BMR P2002.4.1, and (*right*) *Tyrannosaurus*, BHI 3033. (The *Nanotyrannus* lachrymal was cast with portions of the left maxilla and jugal still attached.) Note the large multiple pneumatopore perforating the lachrymal at the junction of the dorsal and vertical rami on *Nanotyrannus,* which is actually smaller and solitary on the adult *Tyrannosaurus.* The lachrymal of *Nanotyrannus* also has a cornual process (lachrymal horn), not found in *Tyrannosaurus.*

a character found in the ectopterygoid. In *Tyrannosaurus* (BHI 3033), *Tarbosaurus* (ZPAL MgD-I/4), and *Daspletosaurus* (NMC 8506), the edge of the large pneumatopore, on the posterior ventral surface of this hook-shaped bone, is bounded by a thick lip. That thick lip is not present on *Appalachiosaurus* (RMM 6670) or *Albertosaurus* (BHI 6234), and it is not found on *Nanotyrannus* (BMR P2002.4.1), further separating it from *Tyrannosaurus rex* (see Fig. 2.23).

JUGAL In lateral view, the jugals of tyrannosaurs all have a large pneumatic foramen (the anterolateral jugal pneumatopore) located at the posterior ventral corner of the antorbital fossa (Currie 2003). The orientation and shape of the anterolateral jugal pneumatopore have been used by other authors as defining characters (Carr 1999; Currie 2003; and Currie et al. 2003). In *Tyrannosaurus rex* (BHI 3033) the anteriolateral jugal pneumatopore opens anteriolaterally. In *Nanotyrannus lancensis* (CMNH 7541 and BMR P2002.4.1), the anteriolateral jugal pneumatopore opens dorsally (see Fig. 2.24).

In 1988 Bakker et al. designated an isolated theropod skull, the holotype of *Gorgosaurus lancensis* Gilmore (1946), as the type specimen for a new genus, *Nanotyrannus*, while retaining the species *lancensis*. The designation became contentious over the years, culminating with Carr's 1999 study, reassigning *Nanotyrannus* as a junior synonym of *Tyrannosaurus rex*. Based upon this conclusion, Carr erected a growth series for *Tyrannosaurus rex* that, in addition to *Nanotyrannus*, included a smaller specimen (LACM 28471), first described by Molnar (1978) as "the Jordon theropod."

However, this study demonstrates that these two specimens, plus an isolated lachrymal (BHI 6235), and a recently discovered, 50 percent complete skull and skeleton known as BMR P2002.4.1 constitute a valid

Conclusion

2.22. Left lachrymals in medial view: (*left*) *Tyrannosaurus,* BHI 3033, and (*right*) *Nanotyrannus,* BMR P2002.4.1. Note the large pneumatopore on *Tyrannosaurus* near the junction of the dorsal and vertical rami, absent in *Nanotyrannus.*

2.23. Left ectopterygoids: (*left*) *Nanotyrannus,* BMR P2002.4.1, and (*right*) *Tyrannosaurus,* BHI 3033. Note the lip bordering the anteromedial border of the large pneumatopore on *Tyrannosaurus,* absent in *Nanotyrannus.*

clade. *Nanotyrannus lancensis* is separated from *Tyrannosaurus rex* by more than a score of cranial and post-cranial characters. These characters include features related to the dentition, pneumatics, and pectoral and pelvic girdles (see Table 2.9). BMR P2002.4.1 also shows an ontogenetic stage of development equal to or surpassing the largest-known individuals of *Tyrannosaurus rex.* Although *Nanotyrannus* may be a sister taxon to *Tyrannosaurus,* it clearly stands alone, and small juvenile skulls and skeletons of *Tyrannosaurus rex* have yet to be discovered.

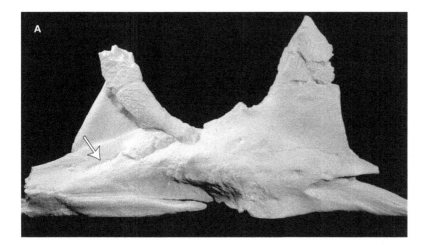

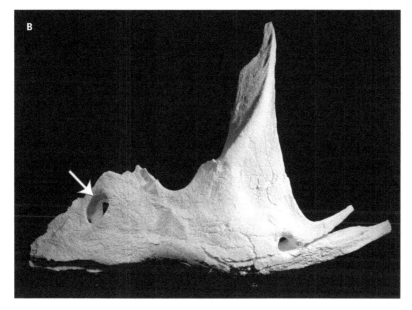

2.24. Left jugal, lateral aspect for (A) *Nanotyrannus*, BMR P2002.4.1; and (B) *Tyrannosaurus*, BHI 3033. Note the orientation of the large anterior lateral pneumatopore.

Acknowledgments

I greatly appreciate the unparalleled access to BMR P2002.4.1 granted by Lew Crampton, Mike Henderson, and Scott Williams of the Burpee Museum. However, this paper would not have come about without the enticing hypothesis proposed by Tom Carr (1999). Tom's revolutionary idea jump-started a new round of investigations, focusing well-deserved attention upon a little-known but extremely fascinating dinosaur. Thanks, Tom!

Literature Cited

Anderson, J. F., A. Hall-Martin, and D. A. Russell. 1985. Long-bone circumference and weight in mammals, birds, and dinosaurs. *Journal of Zoology* 207:53–61.

Bakker, R. T., M. Williams, and P. Currie. 1988. *Nanotyrannus,* a new genus of pygmy tyrannosaur, from the Latest Cretaceous of Montana. *Hunteria* 1(5):1–28.

Britt, B. B. 1993. Pneumatic postcranial bones in dinosaurs and other archosaurs. Ph.D. dissertation, University of Alberta, Calgary.

Brochu, C. A. 2003. Osteology of *Tyrannosaurus rex:* insights from a nearly

Table 2.9A. Skull Characters separating *Nanotyrannus* from *Tyrannosaurus*

Skull Characters	*Tyrannosaurus rex*	*Tyrannosaurus "x"*
Antorbital fossa	very deep	deep
Ventral antorbital maxillary ridge meets jugal	no	no
Maxillary fenestra reaches rostral margin of antorbital fossa (Carr et al. 2005)	yes	yes
Maxillary fenestra reaches ventral margin of antorbital fossa (Carr et al. 2005)	yes	approaches
Vomer expansion	lateral	lateral
Posterior dorsal quadratojugal notch	no	no
Central dorsal quadratojugal notch	no	no
Anterior dorsal medial notch in quadratojugal (Carr and Williamson 2004)	yes	?
Lachrymal horn (cornual process)	absent	absent
Lachrymal shape	7	7
Quadrate-squamosal articulation	double	double
Cranial nerve V-2 bounded by	maxilla and premaxilla	maxilla and premaxilla
Anterior maxilla fossa at cranial nerve V-2	maxilla and premaxilla	maxilla and premaxilla
Tooth cross section at base of crown	ovate	ovate
Fourth maxillary tooth length/width (at base of crown)	1.76	1.23
Fourth dentary tooth length/width (at base of crown)	1.38	1.34
First maxillary tooth small and incisiform	no	no
D-shaped first dentary tooth	yes	?
First dentary tooth reduced	yes	slightly
Maxillary tooth count	11–12	12–13
Dentary tooth count	13–14	14–15
Medial post-orbital fossa	no	no
Foramina on lateral aspect (center) of quadratojugal	absent	absent
Anterior squamosal pneumatic foramina	very large	very large
Lateral lachrymal pneumatic foramina	small	very small
Medial lachrymal pneumatic foramina	present, large	present, large
Ectopterygoid pneumatic foramina bounded by thick lip (Carr et al. 2005)	yes	yes
Jugal pneumatic foramina	anterolateral facing	anterolateral facing

Table 2.9B. Skeletal characters separating *Nanotyrannus* from *Tyrannosaurus*

Skeletal Characters	*Tyrannosaurus rex*	*Tyrannosaurus "x"*
Anterior iliac hook	absent	absent
Fused pelvis	no	no
Lateral component of glenoid	absent	absent

Tarbosaurus bataar	Tarbosaurus bataar juvenile	Nanotyrannus lancensis	Gorgosaurus sp.	Albertosaurus sp.
deep	deep	very shallow	shallow	shallow
no	no	yes	yes	yes
yes	yes	no	no	no
yes	yes	no	no	approaches
lateral	?	dorsoventral	dorsoventral	?
no	no	yes	yes	no
no	no	yes	yes	no
no	?	no	no	no
absent	absent	present	present	present
7	7	T	T	T
?	?	single	double	single
maxilla and premaxilla	maxilla and premaxilla	maxilla only	maxilla only	?
maxilla and premaxilla	maxilla and premaxilla	maxilla only	maxilla only	?
compressed	compressed	compressed	ovate	compressed
1.68	?	2.12	1.36	?
1.39	?	1.66	1.23	?
no	no	yes	yes	?
?	?	no	no	?
slightly	?	greatly	no	?
12–13	13	15–16	13–15	13–15
15	?	17	15–17	13–15
no	?	yes	yes	yes
absent	absent	large pneumatic	small	absent
present	?	absent	small	absent
small	small	multiple, large	large	large
present	?	absent	small	absent
yes	?	no	no	no
anterolateral facing	anterolateral facing	dorsolateral facing	anterolateral facing	?

Daspletosaurus	Tarbosaurus	Nanotyrannus	Gorgosaurus	Albertosaurus
absent	absent	present	present	absent
no	no	yes	no	no
absent	absent	present	absent	present in juvenile?

complete skeleton and high-resolution computed tomographic analysis of the skull. *Society of Vertebrate Paleontology Memoir* 7:1–138.

Carpenter, K. 1992. Tyrannosaurids (Dinosauria) of Asia and North America; pp. 250–268 in N. Mateer and P. J. Chen (eds.), *Aspects of Nonmarine Cretaceous Geology.* China Ocean Press, Beijing.

Carr, T. D. 1999. Craniofacial ontogeny in Tyrannosauridae (Dinosauria, Coelurosauria). *Journal of Vertebrate Paleontology* 19(3):497–520.

Carr, T. D., and T. E. Williamson. 2004. Diversity of Late Maastrichtian Tyrannosauridae (Dinosauria: Theropoda) from western North America. *Zoological Journal of the Linnaean Society* 142:479–523.

Carr, T. D., T. E. Williamson, and D. R. Schwimmer. 2005. A new genus of tyrannosaurid from the Late Cretaceous (Middle Campanian) Demopolis Formation of Alabama. *Journal of Vertebrate Paleontology* 25(1):119–143.

Colbert, E. H. 1989. The Triassic dinosaur *Coelophysis. Museum of Northern Arizona Bulletin,* no. 57. Museum of Northern Arizona, Flagstaff.

Colbert, E. H. 1990. Variation in *Coelophysis bauri;* pp. 81–90 in K. Carpenter and P. J. Currie (eds.), *Dinosaur Systematics: Approaches and Perspectives.* Cambridge University Press, Cambridge.

Currie, P. J. 2003. Cranial anatomy of tyrannosaurid dinosaurs from the Late Cretaceous of Alberta, Canada. *Acta Palaeontologia Polonica* 48(2):191–226.

Currie, P. J., and X. J. Zhao. 1993. A new carnosaur (Dinosauria, Theropoda) from the Jurassic of Xinjian, People's Republic of China. *Canadian Journal of Earth Sciences* 30:2037–2081.

Currie, P. J., J. H. Hurum, and K. Sabath. 2003. Skull structure and evolution in tyrannosaurid dinosaurs. *Acta Palaeontologia Polonica* 48(2):227–234.

Erickson, B. R. 1976. Osteology of the early Eusuchian crocodile *Leidyosuchus formidabilis,* sp. nov. *Monograph of the Science Museum of Minnesota,* vol. 2: Paleontology.

Erickson, G. M., P. J. Makovicky, P. J. Currie, M. A. Norell, S. A. Yerby, and C. A. Brochu. 2004. Gigantism and comparative life-history parameters of tyrannosaurid dinosaurs. *Nature* 430:772–775.

Farlow, J. O., D. L. Brinkman, W. L. Abler, and P. J. Currie. 1991. Size, shape, and serration density of theropod dinosaur lateral teeth. *Modern Geology* 16:161–198.

Gilmore, C. W. 1911. A new fossil alligator from the Hell Creek Beds of Montana. *Proceedings U.S. National Museum,* vol. 41, no. 1860.

Gilmore, C. W. 1920. Osteology of the carnivorous dinosauria in the United States National Museum, with special reference to the genera *Antrodemus (Allosaurus)* and *Ceratosaurus. Bulletin of the U. S. National Museum* 110:1–159.

Gilmore, C. W. 1946. A new carnivorous dinosaur from the Lance Formation of Montana. *Smithsonian Miscellaneous Collection* 106:1–19.

Gray, H. 1901. *Anatomy, Descriptive and Surgical.* Running Press, Philadelphia.

Holtz, T. R., Jr. 2004. Tyrannosauroidea; pp. 165–183 in D. B. Weishampel, P. Dodson, and H. Osmólska (eds.), *The Dinosauria,* 2nd ed. University of California Press, Berkeley.

Hurum, J. H., and K. Sabath. 2003. Giant theropod dinosaurs from Asia and North America: skulls of *Tarbosaurus bataar* and *Tyrannosaurus rex* compared. *Acta Palæontologia Polonica* 48:2, 161–190.

Ikejiri, T., V. Tidwell, and D. L. Trexler. 2005. New adult specimens of *Camarasaurus lentus* highlight ontogenetic variation within the species; pp. 154–179 in V. Tidwell and K. Carpenter (eds.), *Thunder Lizards: The Sauropodomorph Dinosaurs.* Indiana University Press, Bloomington.

Iordansky, N. N. 1973. The skull of the Crocodilia; pp. 201–262 in C. Gans and T. S. Parsons (eds.), *Biology of Reptilia,* vol. 4. Academic Press, New York.

Lambe, L. M. 1917. The Cretaceous theropodous dinosaur *Gorgosaurus. Memoir of the Geological Survey of Canada* 100:1–84.

Langston, W., Jr. 1965. Fossil crocodilians from Colombia and the Cenozoic history of the crocodilia in South America. *University of California Publications in Geological Sciences,* vol. 52.

Larson, P. L. 1997a. The king's new clothes: a fresh look at *Tyrannosaurus rex;* pp. 65–71 in D. L. Wolberg, E. Stump, and G. D. Rosenberg (eds.), *Dinofest International: Proceedings of a Symposium Sponsored by Arizona State University.* Academy of Natural Sciences, Philadelphia.

Larson, P. L. 1997b. Do dinosaurs have class? Implications of the avian respiratory system; pp. 105–111 in D. L. Wolberg, E. Stump, and G. D. Rosenberg (eds.), *Dinofest International: Proceedings of a Symposium Sponsored by Arizona State University.* Academy of Natural Sciences, Philadelphia.

Larson, P. L. 1999. Guess who's coming to dinner; *Tyrannosaurus* vs. *Nanotyrannus:* variance in feeding habits. Abstract.

Journal of Vertebrate Paleontology. 19(3, suppl.):58A.

Larson, P. L. 2008. Variation and sexual dimorphism in *Tyrannosaurus rex;* pp. 102–128 in P. Larson and K. Carpenter (eds.), Tyrannosaurus rex, *the Tyrant King.* Indiana University Press, Bloomington.

Larson, P. L., and K. Donnan. 2002. *Rex Appeal: The Amazing Story of Sue, the Dinosaur That Changed Science, the Law and My Life.* Invisible Cities Press, Montpelier, Vermont.

Leidy, J. 1868. Remarks on a jaw fragment of *Megalosaurus. Proceedings of the Academy of Natural Sciences of Philadelphia.* 20:197–200.

Madsen, J. H., Jr. 1976. *Allosaurus fragilis:* a revised osteology. *Bulletin of the Utah Geological Survey* 109:1–163.

Makovicky, P. J., Y. Kobayashi, and P. J. Currie. 2004. Ornithomimosauria; pp. 137–150 in D. B. Weishampel, P. Dodson, and H. Osmólska (eds.), *The Dinosauria,* 2nd ed. University of California Press, Berkeley.

Molnar, R. E. 1978. A new theropod dinosaur from the Upper Cretaceous of central Montana. *Journal of Paleontology* 52:73–82.

Molnar, R. E. 1991. The cranial morphology of *Tyrannosaurus rex. Palaeontographica, Section A* 217:137–176.

Molnar, R. E., and K. Carpenter. 1989. The Jordan theropod (Maastrichtian, Montana, U.S.A.) referred to the genus *Aublysodon. Geobios* 22:445–454.

Mook, C. C. 1921. Individual and age variation in the skulls of recent Crocodilia. *Bulletin of the American Museum of Natural History* 44:51–66.

Mook, C. C. 1923a. Skull characters of *Alligator sinense* found. *Bulletin of the American Museum of Natural History* 48(16):553–562.

Mook, C. C. 1923b. A new species of alligator from the Snake Creek Beds. *American Museum Novitates* no. 73, 13 pp.

Mook, C. C. 1925. A revision of the Mesozoic crocodilia of North America. *Bulletin of the American Museum of Natural History* 51:319–432.

Norell, M. A., J. Clark, and J. H. Hutchison. 1994. The late Cretaceous *Brachychampsa Montana* (Crocodylia): new material and putative relationships. *American Museum Novitates* no. 3116, 26 pp.

Osborn, H. F. 1905. *Tyrannosaurus,* and other Cretaceous carnivorous dinosaurs. *Bulletin of the American Museum of Natural History* 21:259–266.

Osborn, H. F. 1906. *Tyrannosaurus,* Upper Cretaceous carnivorous dinosaur. *Bulletin of the American Museum of Natural History* 22:281–296.

Osborn, H. F. 1912. Crania of *Tyrannosaurus* and *Allosaurus. Memoirs of the American Museum of Natural History, n.s.,* 1:1–30.

Osmólska, H., P. J. Currie, and R. Barsbold. 2004. Oviraptorosauria; pp. 165–183 in D. B. Weishampel, P. Dodson, and H. Osmólska (eds.), *The Dinosauria,* 2nd ed. University of California Press, Berkeley.

Paul, G. S. 1988. *Predatory Dinosaurs of the World: A Complete Illustrated Guide.* Simon & Schuster, New York.

Raath, M. A. 1990. Morphological variation in small theropods and its meaning in systematics: evidence from *Syntarsus rhodesiensis;* pp. 91–105 in K. Carpenter and P. J. Currie (eds.), *Dinosaur Systematics: Approaches and Perspectives.* Cambridge University Press, Cambridge.

Rozhdestvenskii, A. K. 1965. Growth changes in Asian dinosaurs and some problems of their taxonomy. *Palæontological Journal* 3:95–103. (In Russian)

Romer, A. S. 1956. *Osteology of the Reptiles.* University of Chicago Press, Chicago.

Russell, D. A. 1970. Tyrannosaurs from the Late Cretaceous of western Canada. *Publications in Paleontology* 1:1–34.

Tchernov, E. 1986. Évolution des crocodiles en Afrique du Nord et de l'Est. *Travaux de Paleontologie Est-Africaine sous la direction de Yve Coppens.* Editions du Centre National de la Recherche Scientifique, Paris.

Tykoski, R. S., and T. Rowe. 2004. Ceratosauria; pp. 47–70 in D. B. Weishampel, P. Dodson, and H. Osmólska (eds.), *The Dinosauria,* 2nd ed. University of California Press, Berkeley.

Walker, W. F., and K. F. Liem. 1994. *Functional anatomy of the Vertebrates.* Saunders College Publishing, Fort Worth, Texas.

Wermuth, H. 1953. Systematik der rezenten krokodile. *Mitteilungen der Zoologische Museum Berlin* 29:375–514.

Witmer, L. M. 1995. The extant phylogenetic bracket and the importance of reconstructing soft tissue in fossils; pp. 19–33 in J. J. Thomason (ed.), *Functional Morphology in Vertebrate Paleontology.* Cambridge University Press, Cambridge.

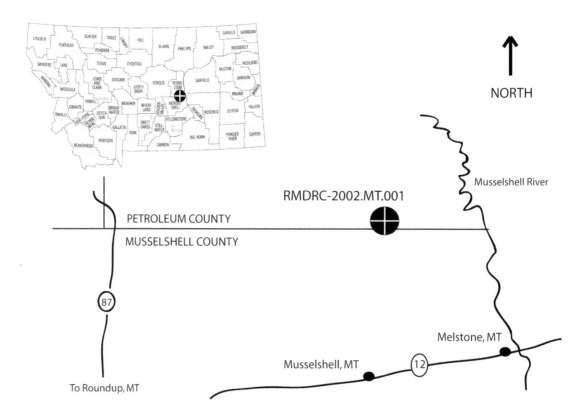

NORTH

RMDRC-2002.MT.001

Musselshell River

PETROLEUM COUNTY

MUSSELSHELL COUNTY

87

Melstone, MT

Musselshell, MT 12

To Roundup, MT

3.1. Site location map for RMDRC 2002.MT-001. The specimen was discovered just north of the Petroleum and Musselshell county line, approximately 15 miles northwest of the town of Melstone, Montana.

Preliminary Analysis of a Sub-adult Tyrannosaurid Skeleton from the Judith River Formation of Petroleum County, Montana

3

Walter W. Stein and Michael Triebold

In 2002 we discovered an enigmatic theropod skeleton approximately 15 miles northwest of the town of Melstone, Montana, along the Petroleum/Musselshell county line. Based upon the size, robustness, and interpreted stratigraphic position (the lower third of the Hell Creek Formation) of the exposed elements, the skeleton, at the time, was thought to be a sub-adult *Tyrannosaurus rex*. Recent, detailed geologic mapping in the area, however, places this site within the lower third of the Judith River Formation, and analysis of the recovered skeletal elements leave more questions than answers. Study continues to present.

To date, approximately 20–25 percent of the skeleton has been recovered, including approximately 20–30 percent of the skull. Major elements of the skeleton consist of the left femur, both ischia, several cervical and dorsal vertebrae, thoracic and cervical ribs, and many gastralia. Important skull elements recovered include both dentaries, the right squamosal, left lachrymal, left postorbital, left ectopterygoid, left pterygoid, left quadratojugal, and a portion of the right jugal. The elements were completely disarticulated, poorly preserved, and encased in sideritic ironstone concretions, and they show strong evidence for both pre-depositional weathering and possible dispersal from scavenging or predation (tooth marks and shed tyrannosaur teeth).

Recent analysis of the recovered elements shows many similarities to both *Tyrannosaurus* and *Daspletosaurus*, but neither is a perfect match. As a result, assignment to a specific generic taxon is not conclusive at this time. Three possibilities, however, exist:

1) The specimen represents a large-bodied new genus and species of Late Cretaceous tyrannosaur more closely aligned with *Tyrannosaurus* and *Daspletosaurus* than to the albertosaurines (Carr 2005) *Albertosaurus*, *Gorgosaurus* and *Nanotyrannus*.

2) The specimen represents a new, larger, more advanced species of *Daspletosaurus*

3) The specimen represents a very large individual, or robust sexual morphotype of *Daspletosaurus torosus*.

Abstract

Institutional abbreviations AMNH, American Museum of Natural History, New York; BHI, Black Hills Institute of Geological Research, Hill City, South Dakota; BMR, Burpee Museum of Natural History, Rockford, Illinois; FMNH, Field Museum of Natural History, Chicago; MOR, Museum of the Rockies, Bozeman, Montana; RMDRC, Rocky Mountain Dinosaur Resource Center, Woodland Park, Colorado; RTMP, Royal Tyrrell Museum of Palaeontology, Drumheller, Alberta; TCM, The Children's Museum, Indianapolis, Indiana.

Further study of this specimen by skilled anatomists is necessary for final taxonomic classification.

Introduction

Enigmatic large theropod skeletons from the Late Cretaceous of North America are nothing new (Currie 2003; Carr and Williamson 2004). Paleontologists have been wrestling with the questions of taxonomic classification, sexual dimorphism, ontogenetic variation, and individual variation of tyrannosaurs ever since the first *Tyrannosaurus rex* tooth was discovered near Denver, Colorado, in 1874 (Carpenter and Young 2002). Since that lone tooth discovery, many skeletons, skulls, and isolated elements of these amazing creatures have been collected, described, studied, and fought over. The majority of these remains are often fragmentary and open to much debate. Whether it is the validity of genera such as *Nanotyrannus* (Bakker et al. 1988; Carr 1999; Larson 2013), *Dinotyrannus* (Olshevsky and Ford 1995), *Aublysodon* (Leidy 1868; Molnar and Carpenter 1989), or *Tarbosaurus* (Maleev 1955; Paul 1988; Currie et al. 2003) or whether to split *Tyrannosaurus rex* into two separate species or sexual morphotypes (Bakker, pers. comm., 2005; Larson 2008), the theoretical battles will continue to rage for years to come. That is the nature of the beast and the nature of dinosaur paleontology. Each year, however, additional material is discovered, and with this new material we get a better understanding of the taxonomy, evolutionary relationships, and variations of one of the most amazing groups of carnivores to ever walk the planet. Since we cannot travel back in time to answer the questions of variation, it is only through additional discoveries (such as this one) and large datasets that we get closer to the truth.

In 2002, we discovered a new enigmatic tyrannosaurid skeleton in Petroleum County, Montana, and entered it into the pantheon of tyrannosaurid skeletons. This specimen, informally known as "Sir William" (RMDRC 2002.MT-001; also known as AMNH 30564, a single bone histology thin section), displays many interesting characters that make it quite unusual. Like many tyrannosaur specimens it inspires more questions than answers. This paper is an attempt to document the discovery, collection, geology, taphonomy, and preliminary taxonomic classification of the Sir William discovery so that others may become aware of its existence. This paper is not designed to be a complete and formal description of this specimen or new genus (if this proves to be the case).

The Discovery

During the summer of 2002, Triebold Paleontology, Inc., acquired a fossil lease on private land in West Central Montana approximately 15 miles northwest of the town of Melstone. This ranch consists of 43,000 acres +/- in Petroleum and Musselshell Counties, northwest of the Ivanho Dome (Porter and Wilde 1999; Vuke and Wilde 2004). The majority of the ranch consists of sparsely vegetated rolling hills and flat-lying grasslands with limited outcrops. In areas where the topography is more

3.2. View of the outcrop facing northeast. Bone fragments were found weathering out over a 10 m section of steeply dipping, heavily cemented sandstone.

rugged, outcrops of dark brown to gray shales with calcareous nodules containing ammonites (*Baculites* sp., *Placenticeras* sp.) typical of the Bearpaw Shale were frequently observed.

On June 27, 2002, we were conducting a field reconnaissance of the ranch, searching for marine vertebrates in the Bearpaw Shale. As we traveled south across the ranch we noticed a very narrow band of terrestrial rocks outcropping on a low ridgeline trending from the northeast to the southwest. This ridgeline ran near the southern margin of the ranch along the Petroleum/Mussellshell county line, approximately three-fourths of a mile south of Salt Sage Coulee (Fig. 3.1). Most of the land surrounding this ridgeline was covered in grasses and topsoil, but the Bearpaw Shale was clearly exposed along Salt Sage Coulee and in some of the lower, less-vegetated slopes. At a few places along the ridgeline small patches of badlands were exposed, and these appeared, in every aspect, to be typical Hell Creek Formation sandstones, mudstones, ironstones, shales, and bentonitic ash layers. Knowing the abundance of dinosaur remains in the Hell Creek Formation, we decided to investigate the area on foot. Dinosaur bone fragments and micro-site fossil material were quickly found at several places. This, too, appeared to be consistent with typical Hell Creek Formation fauna and flora. After a short exploration of this area, several bones were discovered eroding out of the base of a tall, southwest facing hill, on the northeast side of a narrow drainage cutting through the ridgeline. The vertebrate material was eroding out of a sandstone/ironstone horizon that was steeply dipping to the northwest (Fig. 3.2). Bone was emerging from this rock layer at various places over a distance of approximately 10 m until the rock layer dipped below the surface of the drainage. A large debris field trailed downhill from this. Bone in the debris field was highly weathered, and much of it appeared to have been

3.3. Right dentary fragment (BCT-004) of RMDRC 2002. MT-001 after preparation.

completely or partially encased in large sideritic ironstone concretions that also trailed downhill in large chunks. Some of the weathered bone appeared to be hadrosaurian, but others appeared to be theropod. One severely weathered bone that was in situ had tyrannosaurid teeth and tooth fragments trailing down the hill below it. After careful examination, it was determined to be the tip of a theropod right dentary (BCT-004; see Fig. 3.3). As a result, we labeled this site "Bethel College Theropod" (BCT).

We plotted the BCT area on a U.S. Geological Survey (USGS) 1:24,000 topographic map and spent the remainder of the day excavating around some of the severely damaged bones on the weathered edge in an attempt to find something conclusive. After a short time another bone was uncovered, and this turned out to be a theropod quadratojugal. Both of these elements appeared to be tyrannosaurid but much smaller than your typical full-grown adult *Tyrannosaurus rex*. At the time we believed we had discovered a juvenile *T. rex* or a *Nanotyrannus lancensis* skeleton, but prior commitments prevented us from excavating the site immediately.

The Excavation, 2002

Excavation of the specimen at the BCT quarry began in earnest on August 17, 2002. The weathered debris were collected, a site journal and quarry map begun, and geological observations and descriptions were initiated. Vertebrate, invertebrate, and plant fossils found on the weathered surface were collected. Excavation began with simple hand tools, hammer and chisel, dental picks, brushes, and X-Acto knives, but these often proved futile against the hard, strongly cemented sandstone matrix. Soon, pneumatic tools were employed, including jackhammers for removing

overburden and air scribes for detailed work. Since the material was coming out at the base of a large hill and dipped steeply under ground, 1–8 m of overburden had to be removed above the site. Initially, this was done with picks, shovels, jackhammers, and a small skid steer.

Major elements recovered in the first two weeks of work included several vertebrae fragments, the remains of the tibia and astragulus (in the float), several theropod teeth, many rib and gastralia fragments, some nicely preserved cervical vertebrae (BCT-005, 006, 007, and 028), some dorsal centra, a badly weathered ischium fragment (BCT-014), a badly weathered scapula coracoid (BCT-029), a well-preserved pterygoid (BCT-026), and the left dentary (BCT-009). The dentary was missing most of its teeth and showed both pre-depositional (abrasion and breakage due to fluvial transport) and post-depositional weathering (crushing, root damage, iron pyrite damage, selenite gypsum damage, exposure, etc.). Many of the fossils recovered in the first week consisted of pre-depositionally broken chunks of theropod bone found isolated and completely surrounded by matrix.

Geology, Taphonomy, and Paleoenvironment

Soon after beginning our excavation it became apparent that we had more than one horizon producing vertebrate material. Bone was emerging from the quarry wall at several spots spanning 10 m of lateral distance and about 1 m of vertical distance. A total of three bone-producing horizons were identified in the quarry, and these were labeled the "A," "B," and "C" horizons, respectively (Fig. 3.4).

The "A" horizon consisted of a medium orange-brown, highly cemented, medium-grained, lithic-wacke sandstone with abundant, large, irregular, black, orange-brown and red-brown sideritic ironstone concretions. This horizon was the highest bone-producing rock layer in the quarry but did not contain any material referable to RMDRC 2002.MT-001. Isolated hadrosaur (ulna, fibula, and vertebrae), hypsilophodontid (femur), dromaeosaur (claws, vertebrae, teeth), crocodile (teeth, including one large, 3-inch-long tooth from *Deinosuchus*), turtle (one large trionychid turtle braincase, shell fragments), fish (vertebrae, *Myledaphus* teeth), and other fossils were also found in this horizon. The "A" horizon appeared to pinch out to the southeast but thickened to the northwest until it became 90 percent ironstone and only 10 percent unaltered sandstone. On average, the layer was about 30–40 cm thick.

The "B" horizon consisted of white to buff, tan to light orange-brown, very fine grained to fine grained, strongly cemented, cross-laminated lithic-wacke sandstone, lithic-arenite sandstone, and at least one prominent greenish-gray to tan claystone. The claystone did not exceed 3 cm in thickness. The prominent claystone layer was relatively flat and pinched out quickly to both the southeast and the northwest after spanning a lateral distance of approximately 6 m along the outcrop. Many of the best-preserved bones were found contained in the sandstone approximately

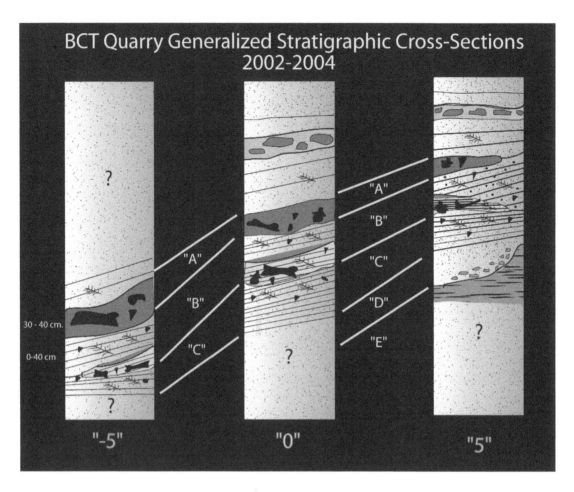

BCT Quarry Generalized Stratigraphic Cross-Sections
2002-2004

"A"
"B"
"C"
"D"
"E"

"A"
"B"
"C"

30 - 40 cm.

0-40 cm

?

?

?

?

?

"-5"

"0"

"5"

3.4. Stratigraphy of the BCT site at various points along the quarry wall.

12 cm below this thin clay layer. This area we informally called the "hot zone." The lower "B" horizon contained the majority of the elements attributable to RMDRC 2002.MT-001. Most of the skeletal elements were found along the contact of the "B" and "C" horizons (B/C contact). Many of these bones were fully encased in irregular orange-brown to red sideritic ironstone concretions. The concretions usually occurred when a large cluster of bones were piled on top of one another. Other elements were only partially encased. This typically occurred when only two or three bones were tangled together. Some elements were not encased at all, usually when the bone was not broken badly and/or was resting by itself. These were often the best-preserved elements.

Many of the elements clearly had pre-depositional breaks with jagged edges. Sometimes the broken sections of the bones were separated only by a centimeter or two of sandstone or ironstone. Other times they were separated by over 40 cm but were clearly once part of the same element. The pre-depositional breaks were often the focal point for the development of the ironstone concretions. Tangled around and between the elements were often large logs, sticks, and other organic debris that had been coalified. Non-carbonized leaf impressions were also found

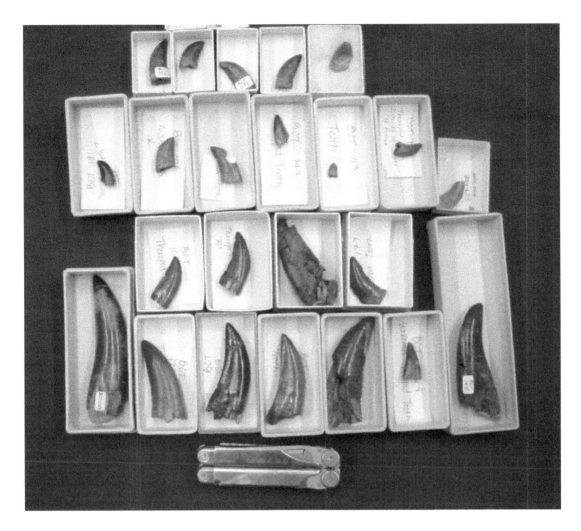

within this zone, but these have not been studied in any detail to date. Occasionally, small chunks of low-grade amber (usually no larger than a pea) were also found in this layer.

As previously noted, the majority of the elements from this zone can be attributed to the RMDRC 2002.MT-001 specimen; however, other vertebrate material was recovered, including hadrosaur (quadrate, maxilla, small vertebrae), dromaeosaur (teeth), troodontid (teeth), ceratopsian (worn teeth), crocodile (teeth), fish (vertebrae and teeth), and mammal (teeth) fossils. Of interest was the high number of small to mid-sized tyrannosaur teeth found at the site. Over 50 tyrannosaur teeth and tooth fragments have been recovered to date from the hot zone. These are finely serrated, mediolaterally compressed teeth that appear in all aspects to be *Albertosaurus* teeth (Fig. 3.5). These outnumbered ornithischian teeth by a ratio of 3:1. Some of these may have come from the Sir William specimen (the larger more robust teeth) as the skeleton was disarticulated and buried, but most did not. The average height of these teeth is approximately 2–4 cm, but some of the isolated ones are quite large, exceeding

3.5. Some of the more than 50 shed tyrannosaur teeth discovered at the BCT quarry. Some of these may have broken from Sir William's jaws postmortem, but most were from another carnivore or carnivores that had been feeding on the carcass.

10 cm. The majority of the broken theropod bone fragments and other unidentifiable fragments were also found within the "B" horizon but not necessarily along the B/C contact.

The "C" horizon was a white to buff colored, medium to fine grained, strongly cemented, laminar feldspathic arenite sandstone. Beds within this horizon were thinly laminated and separated at their tops by a thin layer of black organic debris (carbonized plant matter – inertinite). This horizon appeared to thicken both to the southeast and to the northwest, though not much of the section was exposed to the northwest. A variety of small microfossils were discovered within the "C" horizon. These consisted mostly of non-dinosaurian taxa, such as turtle, small amphibian, reptile, fish, and the occasional unidentifiable dinosaur bone fragment. On rare occasions a worn ceratopsian tooth, hadrosaur tooth, or theropod tooth was found in the "C" horizon. The upper beds of the "C" horizon also seemed to contain a few bones attributable to the Sir William specimen, again indicating several stages of deposition over the course of a few seasons, though these may have been compacted into the sand at a later date.

Below the "C" horizon on the southeast end of the outcrop were two layers we called the "D" and "E" horizons. These sections were devoid of all macrofossils and microfossils but did contain flat, shale pebble rip-ups. The "D" horizon consisted of a buff to tan colored, medium to fine grained, highly cemented, feldspathic arenite with well-rounded, poorly sorted (2–70 mm), greenish-gray to tan, clay shale pebble clasts that were matrix supported. This was interpreted to be the scour channel sand resting upon the greenish-gray "E" horizon mud and clay shale. The "E" horizon is considered to be an overbank floodplain mud. Both the "D" and the "E" horizons were not exposed on the northwest end of the outcrop, but it is assumed that they are present below the surface.

The fossil-bearing horizons of the BCT quarry dipped strongly (approximately 10–20°) to the northwest, where they disappeared under a thick layer of topsoil. Approximately a half mile to the north along the banks of Salt Sage Coulee, we encountered typical Bearpaw shale, containing ammonites. A fault was predicted to lie somewhere between the outcrops of Salt Sage Coulee and BCT, given that the Fox Hills Formation was nowhere to be found. Further exploration to the south revealed a narrow band of sandy soil that was interpreted (at the time) to be the top of the Fox Hills Formation. No more than a half mile to the south just over the county line additional outcrops of medium and dark brown shale were discovered. These also appeared to be shales of the Bearpaw. In fact, these shales extended throughout the 3½-mile-wide valley to the south of BCT, along Howard Coulee. On the opposite side of the valley (further south), more terrestrial rocks of what appeared to be the Hell Creek Formation were exposed along another northeast/southwest-trending ridgeline. This led us to conclude that the BCT site was located on the northern flank of a breached anticline. At the time we had been working only with an inferior and out-of-date Montana geologic highway map that did not show any real detail but did place us in the Upper Cretaceous. Based on our field observations, however, we didn't have any reason to think that the BCT site wasn't in the lower ⅓ of the Hell Creek Formation.

It wasn't until the spring of 2004 that we began to doubt our initial interpretation. After further research it was discovered that the Montana Geological Survey had recently completed a more detailed geologic map of the area and had recently published this online (Porter and Wilde 1999; Vuke and Wilde 2004). These maps showed that the original structural interpretation of the site was correct; that is, the faults we predicted were in the right places, and we were on the northern flank of a breached anticline. The trouble was that the shale discovered in Howard Coulee was not the Bearpaw Shale as we had assumed, but the Claggett Shale, a rock unit neither of us was familiar with. This placed the overlying terrestrial rocks and the BCT quarry in the lower to middle ⅓ of the Judith River Formation and not the lower ⅓ of the Hell Creek Formation.

In order to confirm the specimen's stratigraphic position and give Sir William an approximate time of death, we supplied Rocktell Services, Inc., of Calgary, Alberta, with rock samples taken from deep within the quarry from the "B" and "C" horizons. These were processed and analyzed for fossil pollen grains by Ed Davies, a palynologist for Rocktell. The samples provided were mostly barren of pollen fossils, and those that were present (*Anulisporites*, *Schizosporis grandis*) were reported to be long-ranging forms that were not age diagnostic. Davies suggested that the *Anulisporites* was probably reworked from older sediments at the time of BCT deposition. *Schizosporis* is a type of green algae common to freshwater lakes, possibly indicating the presence of stagnant water. Davies did, however, note the presence of a significant amount of inertinite, which is the remains of fossilized charcoal and indicative of frequent forest fires in the area.

Analysis of the Geology, Taphonomy, and Paleoenvironment

Based upon the data collected and the stratigraphic observations, it appears that a small- to medium-sized, moderate velocity, fluvial system (most likely a crevasse splay) scoured into and on top of the surrounding floodplain ("E") horizon. As current velocities and carrying capacity waned, the "C" horizon was deposited. This may have had a minor seasonal component, as evidenced by the thin layers of carbonized plant matter (inertinite) covering the tops of the thin beds. After several seasons of deposition, our tyrannosaur died very near to the fluvial system, just upstream of this low spot on the floodplain. The quantity of the bone fragments, the large number of theropod teeth over a short area, and the pre-depositional angular breaks on many of Sir Williams's bones indicate that RMDRC 2002.MT-001 was fed upon by other theropods. Some of the bones from RMDRC 2002.MT-001 were then washed into the depression in the floodplain, concentrating them in a small area (the hot zone). Over the course of a season or two, flow rates continued to increase and carried additional elements of the remaining carcass downstream, where they were deposited in the depression along with plants, logs, and microfossils. Some of this debris may have been surficially exposed, and additional pre-depositional weathering and breakage occurred. Soon this deposition began to wane, and a stagnant pool of water covered the site, as evidenced by the thin layer of mudshale overlying the hot zone. Deposition began again several seasons later, depositing the bones of the upper "B" horizon and the "A" horizon.

The majority of RMDRC 2002.MT-001 elements do not show a strong preferred orientation since most are piled on top of one another. Some of the more scattered and isolated elements away from the hot zone, however, do tend to point in a southeasterly direction, possibly indicating that current direction was from northwest to southeast.

The Excavation Continues, 2003–2004

In late May 2003 excavation of the skeleton commenced again. A larger team with better equipment and more powerful air tools were brought to the site. Despite the more powerful equipment, progress was slow due to the nature of the highly cemented sandstone. By the end of June, however, the team had recovered many additional elements and three large jackets containing an unknown quantity of bone. These included jackets containing cervical and dorsal vertebrae, fragments of sacral vertebrae, gastralia, ribs, a left ischium (BCT-076), and several skull elements, including a nice lachrymal (BCT-082), postorbital, jugal fragment, squamosal (BCT-088), and possibly parts of the braincase. Due to the nature of the sandstone it was impossible to identify much of what was collected while in the field (this would be revealed during preparation of the specimen from 2004 to 2010). The majority of the identifiable bones appeared to be gastralia and rib fragments. Many of RMDRC 2002.MT-001's dorsal ribs were broken pre-depositionally, and the shafts are missing or found in pieces. The rib heads are often the only thing that is preserved intact. One rib head in particular (BCT-110; see Fig. 3.6) was clearly broken before final burial, probably as a result of predatory or scavenging activity since the type of twisting/torque that would cause such a break would seem an unlikely result of tectonic stresses or stream transport.

In July 2003 our quarry had reached its limits without the use of heavier earthmovers. In order to reach further into the bone bed we employed a local excavating company to remove over 8 m of overburden. After this was completed and the last 1 m of rock overlying the site was gone through using hand and pneumatic tools, we encountered additional bones. These included a nicely preserved femur (BCT-115), gastralia, dorsal centra, and more theropod teeth. Spacing between bones, however, began to increase to the point where over 3 m of rock had to be removed between major elements. Excavation continued through to September, but bone density declined even further. Some additional work was done later in the summer of 2004, but no additional major elements of RMDRC 2002.MT-001 were recovered, and other priorities caused us to abandon the site.

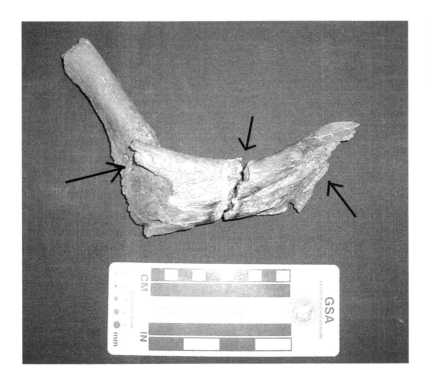

3.6. BCT-110, one of RMDRC 2002.MT-001's dorsal rib heads that had been snapped, twisted, and broken prior to burial.

Preparation

Preparation of the recovered elements began in earnest during the winter of 2003 and continued sporadically through the spring of 2010. The majority of the preparation was completed using mechanical techniques (air scribes, hand tools, and micro-abrasion), though chemical techniques were attempted on some of the elements (fragments) with poor results. Preservation of the elements has made it a difficult, frustrating, and challenging project. Many of the bones were fully encased in sideritic ironstone concretions, making them difficult to see and extract. Often these concretions contained several bones lying on top of one another. Fragments of gastralia were found overlying skull elements, and vertebrae were found overlying gastralia. Untangling the bones from one another required thousands of man-hours of work, by multiple technicians, over the last seven years.

As of spring 2010, all of the larger, more complete elements have been prepared and accessioned into the RMDRC collections. There still remain, however, several boxes containing isolated bone fragments and microfossils from the quarry. Some of these will no doubt yield additional information on the life and death of Sir William.

Analysis

To date, approximately 20–25 percent of the skeleton and 20–30 percent of the skull has been recovered (Fig. 3.7). The majority of the skull elements recovered is from the left side of the posterior portion of the skull. Maxilla, pre-maxilla and nasals were not recovered, and the braincase or elements thereof are so poorly preserved that identification is inconclusive.

None of the elements were in articulation, but some, like the dentaries, were closely associated.

The centra of most of the recovered dorsal vertebrae were disarticulated from their neural arches, potentially showing a weakly fused vertebra, a characteristic often seen in juvenile and occasionally in sub-adult animals. Greg Erickson (Erickson et al. 2004) thin sectioned one of Sir William's gastralia, revealing the animal's annual growth rings. (This thin section is currently accessioned as AMNH 30564.) From this specimen, he concluded that Sir William was approximately 15 years old when it died. This age places RMDRC 2002.MT-001 as a sub-adult, perhaps just reaching sexual maturity, in all of the genera (*Tyrannosaurus*, *Daspletosaurus*, *Albertosaurus*, and *Gorgosaurus*) that Erickson studied.

As stated previously, many of the elements show jagged breaks, scours, and other pre-depositional damage. Some of this damage appears to represent bite marks from a large- to mid-sized carnivore with serrated teeth, most likely, *Albertosaurus*. One such gouge in a fragment from the posterior portion of the right (?) dentary or jugal (BCT-084-C) appears to be a tooth mark from another tyrannosaur. The bone tissue around this gouge does not show any evidence of healing or scar tissue (Fig. 3.8). Another gastralia fragment (Fig. 3.9) shows a striated scrape across its surface, as if a serrated tooth scratched across it. Both ends of these bones had been broken prior to deposition and were missing. These probable bite marks, the presence of large quantities of broken bone fragments, the presence of large quantities of elements missing shafts (ribs and gastralia) or sections, the presence of an abundance of worn tyrannosaur teeth and tooth fragments, and so forth, indicate to us that the specimen was either killed by or scavenged by other tyrannosaurs. It also indicates that tyrannosaurs had no problem feeding on other tyrannosaurs.

Taxonomic Analysis

Up until the spring of 2004, we had been operating on the assumption that we had recovered a juvenile or sub-adult *Tyrannosaurus rex*. As preparation of the skeletal elements progressed, certain bones did not

3.8. Tooth marks. This photo shows a broken section of jugal from RMDRC 2002.MT-001 that appears to have another tyrannosaur bite mark across its surface. This and other evidence suggests that the Sir William specimen was either killed or scavenged by another large tyrannosaur or a group of tyrannosaurs.

3.9. Tooth marks on a small fragment of gastralia. This was most likely caused by a serrated tyrannosaur tooth scraping across its surface.

seem to fully correspond to the described, accepted anatomy of this species, particularly as published in Brochu (2002). These differences led us to question our initial hypothesis.

The elements with the most interesting and diagnostic features that we used to tentatively classify the RMDRC 2002.MT-001 specimen were the dentaries (BCT-004 and BCT-009) and an exceptionally well preserved left lachrymal (BCT-082). Additional elements of taxonomic significance

3.10. Photograph comparing the teeth of Sir William (right dentary) with one of the many shed *Albertosaurus* teeth recovered at the BCT site.

include the left femur, the ischia, cervical vertebrae, the quadratojugal, pterygoid, squamosal, postorbital, and jugal.

Dentaries

Both dentaries of RMDRC 2002.MT-001 were recovered. The right dentary (BCT-004; (Fig. 3.10), was highly weathered and damaged due to post-depositional weathering and exposure. The remains consisted of only the anterior one fourth, approximately 16 cm in length. Two teeth were found preserved in the jaw. These were small (2 cm in height) and not fully erupted. The first tooth was in alveolus #3 and was surprisingly well preserved. The second tooth was found in alveolus #6 and was badly weathered. A third tooth (BCT-007) was found only a short distance downhill from the weathered remains, along with several tooth fragments, and is assumed to have originated from the dentary. Surprisingly (given the size of the dentaries themselves), all of the recovered teeth directly attributable to RMDRC 2002.MT-001 were less than 2 cm in height. The teeth are all laterally compressed, like *Nanotyrannus*, *Albertosaurus*, and *Gorgosaurus*, but not as compressed as in these genera. Their bases were thicker and their tips tended to be more rounded off, like *Tyrannosaurus* and *Daspletosaurus*. Both the mesial and distal carina are relatively straight and serrated. These are unlike *Nanotyrannus* or *Albertosaurus*, which tend to have a strongly pronounced sigmoidal curvature to their distal serrated edge. Like other tyrannosaurs (Carr and Williamson 2004), the denticles of the distal carina are larger and slightly more robust than those of the mesial carina. Unlike *Daspletosaurus*, but similar to most other tyrannosaurids (Carr and Williamson 2004), the mesial carina does not reach all the way down to the crown base. The denticles of the distal carina do, however, reach the base of the crown in tooth #3.

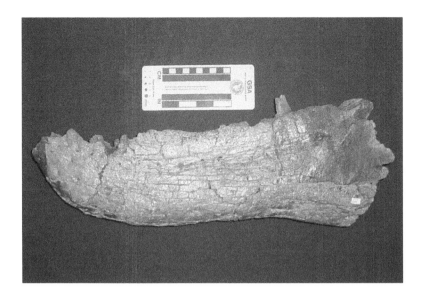

3.11. The complete left dentary of RMDRC 2002.MT-001. This dentary shows several features similar to *Tyrannosaurus rex,* including 13 aveoli.

The left dentary (BCT-009; see Fig. 3.11) is far more complete than the right. The left dentary distinctly shows 13 alveoli, a condition common to *Tyrannosaurus rex* (11–14; Larson 2008; Tsuihiji et al. 2011) and occasionally seen in *Albertosaurus* (Currie, Hurum, and Sabath 2003; Larson 2008). Most of the other tyrannosaurid genera have much larger tooth counts in the dentary including, *Gorgosaurus* (15–17), *Daspletosaurus* (16–17), *Alioramus* (17–18; Currie, Hurum, and Sabath 2003), and *Nanotyrannus* (17; Bakker et. al 1988). Carr (1999) has suggested that the tooth count changes throughout a tyrannosaur's ontogeny and that it cannot be accurately used to determine genus. Other authors, however, have disagreed with several parts of his assessment (Currie 2003; Hurum and Sabath 2003; Larson 2008), and a recent description of an extremely well preserved juvenile *Tarbosaurus* skull (Tsuihiji et al. 2011) confirms that dentary tooth counts do not vary with growth stage. If this condition is universal within the Tyrannosauridae, then it indicates that Sir William cannot be assigned to *Daspletosaurus, Gorgosaurus,* or *Nanotyrannus* as described and is most likely closely affiliated with *Tyrannosaurus.*

In the dentary, all but five of the alveoli are nearly circular in shape and of nearly uniform diameter. The first alveolus is smaller and points forward slightly, similar to most specimens referred to as *T. rex* (Larson 2008). The second alveolus is oval in shape and about 20 percent smaller in diameter than the third through tenth alveoli. This condition is unlike most *T. rex* specimens and shows similarities to what Larson (2008) calls *Tyrannosaurus* "x," a hypothetical new species or morphotype of *Tyrannosaurus* whose second alveolus is significantly smaller than that of *T. rex.* (In *T. rex* the third alveolus tend to be quite large.) In RMDRC 2002.MT-001, the 11th and 12th alveoli are about the same diameter as the second and begin to mediolaterally flatten into an oval. The 13th alveolus is approximately the same size as tooth socket #1. The left dentary also has

Table 3.1. Comparison of dentary characters used in this study (all measurements in cm)

Character: Dentary	"Sir William," RMDRC 2002. MT-001	Tyrannosaurus BHI 3033	Tyrannosaurus MOR 980	Gorgosaurus TCM 2001.89.1	Nanotyrannus BMR P2002.4.1	Daspletosaurus RTMP 94.143–1
Tooth count	13	13	13	15	17	18
Tooth row length	41.5	55.0	54.5	37.0	31.5	28.0*
Total length	62 est.	86.0	88.0 est.	55.0	50.3	36.0*
Width at tooth #6	5.0	7.3	7.6	3.5	2.5	not measured
Ratio of thickness to length	.081	.085	.086	.064	.050	not measured
Height at tooth at #6	12.5	16.7	16.3	10.0	7.3	6.0*
Ratio of width to height at tooth socket #6	0.40	0.44	0.47	0.35	0.34	not measured
Ratio of height at #6 to tooth row length	0.30	0.30	0.30	0.27	0.23	0.21*
Ventral surface point of convex curvature	posterior to midline of tooth row	approximately at midline of tooth row	approximately at midline of tooth row	posterior to midline of tooth row	posterior to midline of tooth row	posterior to midline of tooth row
Curvature of labial margin	relatively straight	strongly curved	strongly curved	strongly curved	slight curvature	relatively straight

* Estimated from illustration of RTMP 94.143–1 in Currie (2003).

two teeth within alveoli; however, neither of these is complete enough or well preserved enough for analysis. One tooth can be found within alveolus #11, but it is badly weathered and shows no distinguishing characters (no serrations, missing tip, etc.). The only thing of note is that it appears to have very little curvature. Another tooth can be found just starting to erupt at the 13th aveolus, though what has erupted is poorly preserved and too small for much description.

The length of the dentary to the broken posterior end is 51.0 cm, giving an estimated total dentary length of approximately 62–64 cm. Unlike those of the primitive, gracile tyrannosaurs, the dentary itself is quite thick and massive. The thickness of the dentary at tooth socket #6 is 5.0 cm. The ratio of maximum thickness at tooth socket #6 to maximum length of the dentary is.081. This compares well with an adult *Tyrannosaurus rex* (BHI 3033, MOR 980) at.085/.086, varies significantly from *Gorgosaurus* (TCM 2001.89.1) at.064, *Daspletosaurus* (RTMP 94.143–1) at 0.21, and *Nanotyrannus* (BMR P2002.4.1) at.050, all gracile specimens. The intermandibular symphysis of both dentaries is not well defined like other tyrannosaurines (Currie 2003), providing flexibility of movement in RMDRC 2002.MT-001's jaw. Several rows of foramen can be found on the

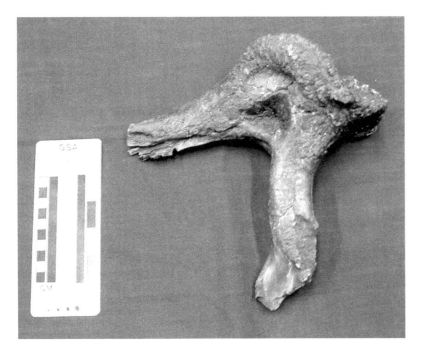

3.12. Left lachrymal of RMDRC 2002.MT-001. This specimen is extremely rugose with a prominent horn and is unlike any other tyrannosaur lachrymal we could compare it to. This strongly suggests we are dealing with a new genus of tyrannosaur from the Late Cretaceous.

left dentary; the largest are located near the anterior portion of the dentary near the tooth margin below alveolus #1. (See Table 3.1.)

Lachrymals

The lachrymal bone is one of the most diagnostic elements in a tyrannosaur skull. The shape, orientation, structures, position of the lachrymal horn, angle of the rami, and number and position of pneumatic foramen are important characters in determining or defining each genus and are particularly interesting in RMDRC 2002.MT-001, whose lachrymal is unlike any of the established tyrannosaur genera (see Fig. 3.12).

The lachrymal horns in *Albertosaurus*, *Gorgosaurus*, and *Nanotyrannus* are positioned far forward and anterior to the anterior margin of the descending ramus, whereas the lachrymal horn of *Daspletosaurus* is positioned directly above the anterior margin of the descending ramus. *Tyrannosaurus*, in contrast, does not have a well defined "horn" but, rather, a thickening or swelling of the posterior portion of the dorsal ramus (Carr and Williamson 2004). Based upon this, it would seem that the more "advanced" the genera is, the more posteriorly shifted the horn, ridge, or crest. If correct (as shown below), this would place RMDRC 2002.MT-001 in a more advanced position that of *Albertosaurus*, *Gorgosaurus*, or *Nanotyrannus* but less advanced than that of *Tyrannosaurus*.

The lachrymal horn of RMDRC 2002.MT-001 is positioned like that of a *Daspletosaurus*; that is, the midline (apex) of the horn is located directly above the anterior margin of the descending ramus (Russell 1970), but its shape and size are unlike any of the genera we studied (the shape for *Daspletosaurus* is described as triangular by Russell [1970]). RMDRC 2002.

Character: Lachrymal	"Sir William," RMDRC 2002. MT-001	*Tyrannosaurus* BHI 3033
Height of descending ramus	24.1	37.5
Length of dorsal ramus	25.0	35.8
Ratio of height to length rami	0.96	1.05
Midline lachrymal horn	above	posterior
Height of lachrymal horn	5.2	not pronounced
Length of lachrymal horn	10.7	not pronounced
Ratio of length to height of horn	2.06	not pronounced
Symmetry of horn	symmetric	asymmetric
Shape of horn	rounded	sloped
Ratio of horn height to total height ramus	0.22	not pronounced
Relative rugosity of horn	extreme	low-moderate
Maximum diameter of lateral lachrymal foramen	4.5	3.8
Ratio of diameter L.L.C. to length of dorsal ramus	0.18	0.11
Thickness of lateral surface of descending ramus	thick	mediolaterally compressed
# of pneumatic foramen on anterior margin	2	1
Anterior foramen face	anteriorly	medially
# of pneumatic foramen on posterior margin	2	0
Posterior foramen face	posteriorly	N/A

Table 3.2. Comparison of lachrymal characters used in this study

Note: N/A = not applicable.

MT-001's lachrymal horn is very prominent, severely rugose, and nearly symmetrical and rounded in lateral view. This is different from *Gorgosaurus*, whose lachrymal horn is prominent but shorter, asymmetric and more flattened (almost box-like) on top; *Albertosaurus*, whose lachrymal horn is reported to be rectangular in lateral aspect (Russell 1970) or triangular; and *Nanotyrannus*, whose lachrymal horn displays a lower, less prominent, but symmetrical mound.

Not only is the lachrymal horn of RMDRC 2002.MT-001 dramatically different from any of the other genera, but the shape and position of the pneumatic foramen on both the descending and dorsal rami are unusual as well. To begin with, the lateral lachrymal foramen (L.L.F.) on RMDRC 2002.MT-001 tends to be triangular in shape, large by ratio, and simple in structure. The lateral lachrymal foramen of *Tyrannosaurus* (BHI 3033, MOR 555, MOR 980) and *Daspletosaurus* are more rounded and smaller by ratio than that of Sir William. The L.L.F. of the other genera tended to be much larger by ratio, more complex, and irregular in shape. (See Table 3.2.)

Both the anterior and posterior margins of RMDRC 2002.MT-001's lachrymal have pneumatic foramen. Two prominent, well-defined, medium-sized, anteriorly facing pneumatopores are located on the anterior surface. The most dorsal of these was located proximally to the L.L.F. The two pneumatopores were separated by 1.8 cm. of bone. A possible third pneumatic foramen may also be found a short distance ventral to the second, but minor crushing in the area may be mimicking this. Two additional pneumatopores can be found on the posterior margin of the

Tyrannosaurus MOR 555	Tyrannosaurus MOR 980	Gorgosaurus TCM 2001.89.1	Albertosaurus BHI 6234	Nannotyrannus BMR P2002.4.1
36.8	30 +	24.2	17.1	10.5 +
39.0	32.2+	20.6	22.4	18.3
0.94	incomplete	1.17	0.76	incomplete
posterior	posterior	anterior	anterior	anterior
not pronounced	not pronounced	4.5	3.4	1.5
not pronounced	not pronounced	13.0	8.2	7.4
not pronounced	not pronounced	2.89	2.41	4.93
asymmetric	asymmetric	asymmetric	asymmetric	symmetric
sloped	sloped	box-like	rounded	long low rounded
not pronounced	not pronounced	0.19	0.20	incomplete
low-moderate	low-moderate	moderate	low	moderate
3.6	2.8	5.5	2.6	4.1
0.09	incomplete	0.23	0.15	0.22
mediolaterally compressed	mediolaterally compressed	mediolaterally compressed	mediolaterally compressed	mediolaterally compressed
1	1	2	3	2?
medially	medially	medially	medially	45°
0	0	2?	2	2?
N/A	N/A	posteriorly	posteriorly	posteriorly

descending ramus. These are also well-developed, small-medium-sized, oval openings that face posteriorly.

On *Tyrannosaurus rex*, the anterior pneumatopores seem to have migrated medially and dorsally (as if twisted) and then merged into one large pneumatic foramen. This foramen faces medially. On *Gorgosaurus*, there are also two pneumatopores on the anterior surface, but these, as in *Tyrannosaurus*, are also shifted to the back side of the lachrymal and are facing medially. Interestingly, on *Nanotyrannus* there are two possible pneumatopores, but these are smaller and not well defined. If they are indeed pneumatopores (we were working off of a cast of BMR P2002.4.1), they face neither medially nor anteriorly but 45° to both. Interestingly, some of the lachrymal characters like the symmetry of the lachrymal horn and the number of the pneumatic foramen on the anterior and posterior sides of the descending ramus match well with *Nanotyrannus*. These features alone, however, are probably not enough evidence to support a close association with *Nanotyrannus* since the other characters are so different. Time constraints prevented us from directly observing this feature on *Albertosaurus* or *Daspletosaurus*; however, one illustration in Currie (2003), for RTMP 94.143–1, does show two anterior lachrymal "ducts" in a similar position and size to that of Sir William. *Tyrannosaurus* did not have any pneumatopores on the posterior surface. Both *Gorgosaurus* and *Nannotyrannus* did appear to have two small pneumatopores on the posterior margin of the descending ramus, but crushing may have mimicked this as well.

According to Larson (2013; and pers. comm., 2005), pneumatopores on extant birds such as ostriches do not change position or number with ontogeny. He argues that the position and number of these pneumatopores can be used to help identify genera as well. If he is correct, RMDRC 2002.MT-001 can not be considered a *Tyrannosaurus rex*, not only because of the morphological differences but also because the shape, size, number, and position of the pneumatopores are so different.

Another striking difference in RMDRC 2002.MT-001's lachrymal is the thickness of the lateral surface of the descending ramus. In all of the genera we studied, the descending ramus appeared almost twisted, with the anterior margin facing medially. The lateral surface of the descending ramus was mediolaterally flattened in *Tyrannosaurus*, *Gorgosaurus*, *Albertosaurus*, and *Nanotyrannus* as a result. The lachrymal of Sir William, however, is not "twisted" in this fashion, and the anterior surface faces anteriorly, rather than medially. This caused the lateral surface to be thick and shaft-like, as opposed to flattened.

Several authors point out that the advanced tyrannosaurs, like *Tyrannosaurus rex*, *Tarbosaurus bataar*, and *Daspletosaurus torosus*, all have lachrymals that show a strongly acute angle between the dorsal and descending rami (Holtz 2000). RMDRC 2002.MT-001 is different in that the rami are nearly perpendicular to each other (approximately 80°). This may indicate that RMDRC 2002.MT-001's skull was not as tall as those genera.

Age/Size

The size-to-age ratio of RMDRC 2002.MT-001 does not match that of any other tyrannosaur genera based upon Erickson's growth curves (2004; see Fig. 2.3, in this volume). Specimens of similar age, such as *Gorgosaurus* RTMP 73.30.1, estimated to be 14 years old at its time of death; *Albertosaurus* RTMP 86.64.01, estimated to be 15 years old at its time of death; and *Daspletosaurus* AMNH 5438, estimated to be 17 years old at its time of death, were all significantly smaller in size and mass estimates than that of RMDRC 2002.MT-001. Even the older specimens of those genera that Erickson studied were much smaller, with the exception of FMNH PR 308, a 21-year-old *Daspletosaurus* estimated by Erickson to be a mere 30 kg heavier. This indicates that unless individual or sexual variation varied greatly within those genera, RMDRC 2002.MT-001 can not be classified as one of those genera. It also indirectly shows that if RMDRC 2002.MT-001 is a fully grown adult, as some have quietly suggested, that it must have grown and reached sexual maturity at a faster rate than any other tyrannosaur known to date. (See Table 3.3.)

Currie (2003) has shown that in *Gorgosaurus libratus*, the lachrymal horn actually changes its morphology through ontogeny. He concludes that for *Gorgosaurus* the lachrymal horn is poorly developed in juveniles, sharp and pronounced in young adults, and more massive and less pronounced in older, more mature adults. This again supports the hypothesis

Specimen	Age (Years)	Body Mass (kg)
"Sir William," RMDRC 2002.MT-001 (AMNH 30564)	15	1761
Gorgosaurus RMTP 94.12.602	18	1105
Gorgosaurus RTMP 73.30.1	14	747
Gorgosaurus RTMP 99.33.1	14	607
Albertosaurus RTMP 81.10.1	24	1142
Albertosaurus RTMP 86.64.01	15	762
Daspletosaurus FMNH PR 308	21	1791
Daspletosaurus AMNH 5438	17	1518

Table 3.3. Comparison of measured ages and estimated body masses of various tyrannosaur specimens used in this study (redrawn from Erickson et al. 2004)

that the robustness and rugosity of the lachrymal horn may be a character related to sexual maturity in sub-adults and not necessarily a character of old age. If correct, the extreme rugosity and robustness of RMDRC 2002. MT-001's lachrymal horn is evidence for its sub-adult "teenager" status.

Our brief analysis of the RMDRC 2002.MT-001 specimen indicates that its eventual taxonomic classification will be difficult at best and contentious at worst owing to the specimen's state of preservation and its unusual cranial features. Some features—such as tooth count, dentary proportions, lachrymal proportions, body proportions, and robustness—seem to match well with *Tyrannosaurus rex*. Lachrymal shape, position of the lachrymal horn, and position and orientation of pneumatic foramen and other characters do not. Position of the lachrymal horn matches with reported specimens of *Daspletosaurus*, but time prevented direct observation of any *Daspletosaurus* material to personally evaluate the shape, size, number, and position of the pneumatic foramen. RMDRC 2002.MT-001 seems to align with the albertosaurines *Albertosaurus* and *Gorgosaurus* and the enigmatic *Nanotyrannus* in the size, shape, and number of the pneumatopores on both the anterior and posterior margins of the descending ramus of the lachrymal. These lachrymals, however, showed a significant "twist" or shifting mediodorsally of the anterior pneumatopores, which is vastly different from those of RMDRC 2002.MT-001. This condition is seen in its extreme with the advanced *Tyrannosaurus rex*, where only a single, deep well-defined pneumatopore exists on the medial aspect of the lachrymal. Each of these also show a mediolaterally flattened condition as a result, a condition dramatically different from that of RMDRC 2002.MT-001. The age-to-body-size ratio of RMDRC 2002.MT-001 makes it larger than any known albertosaurine from the Judith River Formation. Its age-to-body-size ratio is also greater than that of the *Daspletosaurus* specimens surveyed in the Erickson (2004) growth-study paper, with the exception of a much older (21-years-old) *Daspletosaurus* (FMNH-PR 308), which was estimated to be only 30 kg heavier.

Based upon all of these observations, we conclude that three probable taxonomic classifications exist for RMDRC 2002.MT-001:

Conclusions

1) RMDRC 2002.MT-001 represents a new genus and species of tyrannosaur, closer to *Daspletosaurus* and *Tyrannosaurus* than to the albertosaurines *Albertosaurus* and *Gorgosaurus* or to the enigmatic *Nanotyrannus*;

2) RMDRC 2002.MT-001 represents a larger undescribed species of *Daspletosaurus*; or

3) RMDRC 2002.MT-001 represents a much larger sexual morphotype of *Daspletosaurus torosus*.

To summarize, the RMDRC 2002.MT-001 specimen represents a large-bodied, robust tyrannosaurid genus whose dentary features, proportions, and size match that of *Tyrannosaurus rex* but whose other cranial features appear to have more in common with *Daspletosaurus*. Additional work by others is needed to determine where this enigmatic specimen should be placed.

Acknowledgments

We would like to thank all of the people who have helped in the preparation of this paper, including the staff of the Rocky Mountain Dinosaur Resource Center and Triebold Paleontology, Inc., for their hard work and dedication in the preservation of this specimen, especially Anthony Maltese and Jacob Jett, who took over the remaining preparation after WWS left RMDRC; the staff of the Black Hills Institute for their assistance with research; Pete and Neal Larson for their advice and support; Ken Carpenter, Phil Currie, Robert Bakker, Chris Ott, and others for their most valuable opinions and advice; Bill Simpson at the Field Museum and Tetsuto Miyashita of the Royal Tyrell with their assistance with research; Scott Williams and Michael Henderson, who helped put the tyrannosaur conference together and who allowed us to present even though our abstract was submitted several months after the deadline; and, most important, our wives and families, who have put up with us throughout this time-intensive project.

Literature Cited

Bakker, R. T.,M. Williams, and P. J. Currie. 1988. *Nanotyrannus,* a new genus of pygmy tyrannosaur, from the latest Cretaceous of Montana. *Hunteria* 1(5):1–30.

Brochu, C. 2002. Osteology of *Tyrannosaurus rex:* Insights from a nearly complete skeleton and high-resolution computed tomographic analysis of the skull. *Journal of Vertebrate Paleontology,* memoir 7.

Capenter, K., and D. B. Young. 2002. Late Cretaceous dinosaurs from the Denver Basin, Colorado. *Rocky Mountain Geology* 37(2):237–254.

Carr, T. D. 1999. Craniofacial ontogeny in Tyrannosauridae (Dinosauria, Coelurosauria). *Journal of Vertebrate Paleontology* 19(3):497–520.

Carr, T. D. 2005. Cranialfacial ontogengy in Tyrannosauridae. *Journal of Vertebrate Paleontology* 19:497–520.

Carr, T. D., and T. E. Williamson. 2004. Diversity of Late Maastrichtian Tyrannosauridae (Dinosauria: Theropoda) from western North America. *Zoological Journal of the Linnean Society* 142:479–523.

Currie, P. J. 2003. Cranial anatomy of tyrannosaurid dinosaurs from the Late Cretaceous of Alberta, Canada. *Acta Paleontologica Polonica* 48(2):191–226.

Currie, P. J., J. H. Hurum, and K. Sabath. 2003. Skull structure and evolution in Tyrannosaurid dinosaurs.

Acta Paleontologica Polonica 48(2):227–234.

Erickson, G. M., P. J. Makovicky, P. J. Currie, M. A. Norell, S. A. Yarby, and C. A. Brochu. 2004. Gigantism and comparative life-history parameters of tyrannosaurid dinosaurs. *Nature* 430:772–775.

Holtz, T. R., Jr. 2000. Tyrannosauridae: Tyrant Dinosaurs. Tree of Life Web Project (website). Available at http://tolweb.org/Tyrannosauridae/15896. Accessed September 24, 2012.

Hurum, J. H., and K. Sabath. 2003. Giant theropod dinosaurs from Asia and North America: skulls of *Tarbosaurus bataar* and *Tyrannosaurus rex* compared. *Acta Paleontologica Polonica* 48(2):161–190.

Larson, P. 2008. Variation and sexual dimorphism in *T. rex;* pp. 103–128 in P. Larson and K. Carpenter (eds.), Tyrannosaurus rex, the Tyrant King. Indiana University Press, Bloomington.

Larson, P. 2013. The case for *Nanotyrannus*; pp. 14–53 in J. Michael Parrish, Ralph E. Molnar, Philip J. Currie, and Eva B. Koppelhus (eds.) *Tyrannosaurid Paleobiology.* Indiana University Press, Bloomington.

Leidy, J. 1868. Remarks on a jaw fragment of *Megalosaurus. Proceedings of the Academy of Natural Sciences of Philadelphia.* 20:197–200.

Maleev, E. A., 1955. New carnivorous dinosaurs from the Upper Cretaceous of Mongolia. *Doklady, Akademii Nauk, SSSR* 104(5):779–783.

Molnar, R. E., and K. Carpenter. 1989. The Jordan theropod (Mastrichtian, Montana, U.S.A.) referred to the genus *Aublysodon. Geobios* 22(4):445–454.

Olshevsky, G., and T. L. Ford. 1995. The origin and evolution of the tyrannosaurids, Part 1. *Dino Frontline* 9:92–119.

Paul, G. S., 1988. *Predatory Dinosaurs of the World.* Simon & Schuster, New York.

Porter, K. W., and E. W. Wilde. 1999. Geologic map of the Musselshell 30 × 60 quadrangle. Montana Bureau of Mines and Geology open file #386.

Russell, D. A., 1970. Tyrannosaurs from the Late Cretaceous of western Canada. *National Museum of Natural Sciences, Publications in Paleontology* 1:1–34.

Tsuihiji, T., M. Watabe, K. Tsogtbaatar, T. Tsubamoto, R. Barsbold, S. Suzuki, A. Lee, R. Ridgely, Y. Kawahara, and L. Witmer. 2011. Cranial osteology of a juvenile specimen of *Tarbosaurus bataar* (Theropoda, Tyrannosauridae) from the Nemegt Formation (Upper Cretaceous) of Bugin Tsav, Mongolia. *Journal of Vertebrate Paleontology* 31(3):1–21.

Vuke, S. M., and E. W. Wilde. 2004. Geologic map of the Melstone 30 × 60 quadrangle. Montana Bureau of Mines and Geology open file #513.

Functional Morphology and Reconstruction

2

4.1. a) Earliest known mention of a relationship between sharpness of a blade and the transverse curvature of the cutting edge. As seen in cross section, the sharp edge comes to a point, while the dull edge ends in a rounded curve.
b) Earliest known mention of the individual cutting action of serrations

a) From Abler (1973:9);
b) From Abler (1973:23).

a

When a knife gets dull—
this is what happens:

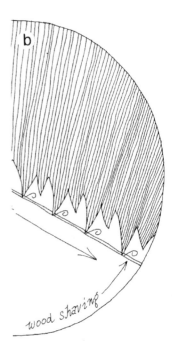

b

wood shaving

Internal Structure of Tooth Serrations

William L. Abler

4

Serrations on the teeth of vertebrates are functional, but they have only
a few characteristic external shapes that are only a partial aid to identification. Internal structure of serrations can be both functional and characteristic. Thus, serrations easily differentiated on the basis of internal
structure include those of a phytosaur (with internal peak), *Dimetrodon*
(with rounded interior), *Troodon* (with radiating interior tubules), and
Albertosaurus (with inter-serrational loop). A physical model is described
here as an aid to understanding internal structure of serrations that include an inter-serrational loop (ampulla) as protection against breaking
under pressure. Serrations of *Tyrannosaurus rex* will be considered.

Abstract

Interest in the precise mechanism of cutting is both sparse and recent.
As a result, the function of serrations and serrated teeth remains only
partially understood. The oldest descriptions I know of, showing a relationship between the sharpness of a blade and some physical quantity, or
of serrations in action, are barely 30 years old (Abler 1973:9, 23; see Fig.
4.1). But when it comes to serrated teeth, the phrases "razor-sharp" and
"cuts meat like a serrated steak knife" have proven so irresistible that more
refined detail (Frazzetta 1988; Farlow et al. 1991) and controlled experi-
ments (Abler 1992, 2001) have failed to dislodge them. Figure 4.2 shows
a more emotional appeal, a thumb being pressed harmlessly against the
putative cutting edge of a serrated *Albertosaurus* tooth.

Introduction

More questions than answers remain. To begin, the serrations on the
teeth of a shark (*Hemipristis*) and a carnivorous dinosur (*Troodon*) show
distinctive chisel and ratchet shapes, respectively. But serrations on the
teeth of a mammal-like reptile (*Dimetrodon*) and of two tyrannosaurids
(*Albertosaurus* and *Tyrannosaurus rex*) show a nearly identical cubic
shape. Out of five species of *Dimetrodon* (Romer and Price 1940:table
2) examined, four (*D. angelensis*, *D. gigashomogenes*, *D. grandis*, and
D. loomisi) had serrations on their teeth, but one (*D. limbatus*) had only a
smooth keel. What clue this difference holds to the lives of the various *Dimetrodon* species, and the function of the serrations, remains to be seen.

**Structure of
Serrations: Exterior**

The general relationship between serration density, or size (x), and
tooth size (y) may be roughly described by a single equation (Farlow et
al. 1991), the hyperbola

4.2. Fingers being pressed against the serrated edges of an *Albertosaurus* tooth. Note lack of injury to fingers. Note serrations visible between fingers, below. Don't try this with a razor or serrated steak knife.

$$y = ax^b,$$

where $a = 75.97$, and $b = -0.62$ (Selby 1975:391). Serration density and serration size are mutually reciprocal and represent equivalent measures of the same thing. For example, "2 serrations to the millimeter" (density) is equivalent to "each serration is ½ millimeter in width" (size). Density is the easier, more accurate measure for the width of structures whose boundaries may be obscured by matrix, uneven illumination, translucent surfaces, and small size. Even if it emerged at first by serendipity, the hyperbola relationship (Farlow et al. 1991:fig. 7) between tooth fore-and-aft basal length and serration density reflects a naturally occurring relationship between serration size and tooth size.

A single equation suggests the action of a single genetic program that operates across species. And the ease with which serrations may appear also suggests a single genetic program. But Farlow et al. refer to possible numerical differences between large and small teeth (1991:170) and suggest that these may have biological correlates. What is more, the rightward extension of the hyperbola (the horizontal asymptote, at approximately $y = 1.1$) limits the size of the serrations, while the upward extension (the vertical asymptote, at approximately $x = 0.1$) limits the size of the tooth. Since S-shaped curves exist (Selby 1975:393) that would regulate just the size or density of the serrations in relation to the size of the tooth, the two ends of the hyperbola may be under separate genetic and evolutionary control, confirming the idea of separate programs for large and small teeth. For example, in large teeth, there may be a maximum useful serration size, while in small teeth there may be a minimum useful number of serrations.

Structure of Serrations: Interior

The serration interiors present further puzzles. While the tooth serrations of most fossil reptiles that I have examined are constructed on the model

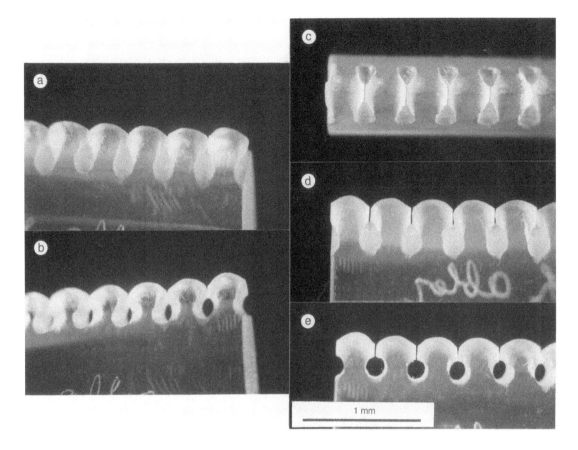

of an enamel cap covering a dentine base (e.g., phytosaur, *Albertosaurus*, and *Dimetrodon*), even that formula has at least one exception (*Troodon*). The exception may be related to the exceptional size of the serrations in relation to the size of the tooth. Thus a pure mineral cap may be too brittle for the high, exposed serration structures, or it may be difficult to form. In any case, while the exterior shapes of most serrations are not dramatically different (Farlow et al. 1991:fig. 14), many interiors are easily distinguished. Phytosaur serrations show an interior dentine peak. *Dimetrodon* serrations show an interior dentine dome. *Albertosaurus* serrations show a unique ghost of their hexagonal crystal structure, as well as the ampulla that joins neighboring serrations, making them the most sophisticated serrations known so far. The rough uniformity of serration exteriors contrasts so sharply with the characteristic interiors that exterior and interior structure may be under separate genetic control.

Since the ampulla appears to be a device for relieving stress by distributing it over an enlarged area (Abler 2001), and since the *Albertosaurus* tooth cuts meat like a dull knife, not a sharp one, the main function of *Albertosaurus* serrations may be to protect the tooth from breaking under pressure. In addition, the separate and discrete nature of *Albertosaurus* serrations suggests that they also serve to limit damage caused to the tooth by chipping and spalling–that is, a single serration may be broken away

4.3. Plastic model of the serrations on the teeth of *Albertosaurus*. The external enamel has been removed, revealing the structure of the underlying interior dentine. Note the cylindrical tunnels (ampullae), which provide stress-relief, and the narrow slots separating each serration from its neighbors: a) Three-quarter oblique view, b) Horizontal oblique view, c) Bird's-eye (perpendicular) view, d) Three-quarter view, e) Horizontal (profile) view.

4.4. Light micrograph of the external surface of serrations on the teeth of *Tyrannosaurus rex*. Serrations are of an approximately cubic type indistinguishable, to a first approximation, from serrations of *Dimetrodon* or *Albertosaurus*.
a) Labial view.
b) Posterior view.

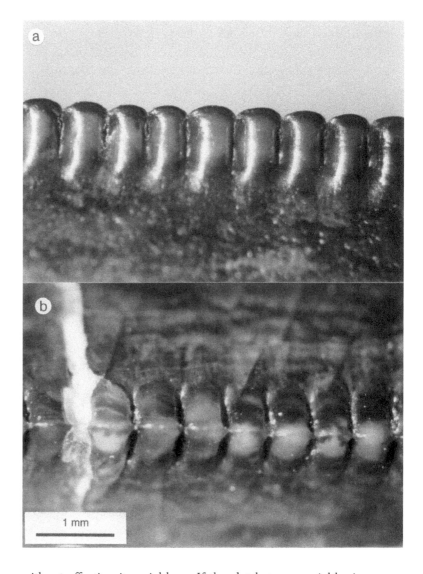

1 mm

without affecting its neighbors. If the slot between neighboring serrations served as storage for infectious bacteria, *Albertosaurus*, and possibly *Tyrannosaurus rex*, may have been obligate scavengers precisely because they were active predators. The dinosaurs may have had to eat carrion to replenish the supply of infectious bacteria that they used as weapons of attack. A more exhaustive and systematic survey of serration interior anatomy would almost certainly yield valuable results.

A Three-Dimensional Model

The three-dimensional construction of *Albertosaurus* serrations is so interesting that it demands to be understood fully but is so complicated that it resists analysis. In order to build a three-dimensional model of an object for which no three-dimensional image is available, I began with two known features. These were the ampullae, which form a row of transverse parallel tunnels beneath the tooth surface, and narrow slots

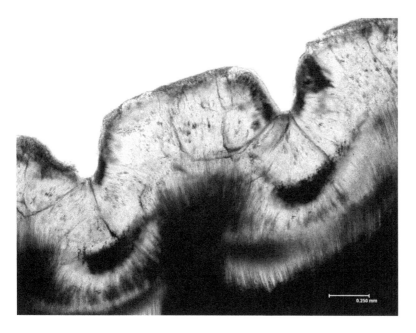

4.5. Vertical thin section through serrations of *Tyrannosaurus rex*. Three serrations are visible in the micrograph. Here, junctions between neighboring serrations exhibit a lenticular ampulla. A channel connects each ampulla to the surface, in the valley between the serrations. Note that the channel at right intersects one end, while the channel at left intersects the center of the ampulla. The hexagonal shape of the serration core is very apparent, with geometrically straight faces.

that begin at each ampulla and proceed upward to the surface, defining the individual serrations. The model (Fig. 4.3) was begun by first drilling the ampullae, then sawing the slots. The model was completed by filing the outer surface of each serration into the shape of a domed keel. In an unusual reversal of roles that is comparable to the stacking of serial sections, the plastic model provides the mental image that was not available for making the model in the first place. A replica of the enamel surface can be added to the model by stretching a sheet of plastic-wrap over the outside. Further refinements to this model may also prove interesting.

Tyrannosaurus rex

Serrations on the teeth of *Tyrannosaurus rex* can best be understood by comparison to those of *Albertosaurus*. The unique stress-relieving ampullae, first noticed in the serrations of *Albertosaurus* (Abler 2001), are also present in the serrations of *T. rex*, consistent with a close taxonomic relationship between the two animals (see Figs. 4.4 and 4.5). At a first impression, the serrations of *T. rex* appear less refined than those of *Albertosaurus*. Where some *Albertosaurus* serrations show two convexities, or even a "tail" that extends onto the tooth surface, *T. rex* serrations have a simple, unsculptured surface.

The same general impression applies to the serration interior. Serrations of *T. rex*, like those of *Albertosaurus*, possess an ampulla between and below the serrations and a channel connecting the ampulla to the surface. The channel terminates in the valley between serrations. But the serrations of *Albertosaurus* are strikingly uniform, with very round ampullae, and a channel centered both on the ampulla and on the valley between serrations. *Tyrannosaurus rex* serrations, by contrast, have ampullae that may be round or lenticular, and the channel that connects them to the surface may intersect the ampulla at the center or at one end. There may be two channels or none at all.

Such poor formation, combined with lack of uniformity, suggests that the ampullae may have become a vestigial structure in *Tyrannosaurus rex* serrations and remain present for taxonomic reasons only. The more clearly defined hexagon of the *T. rex* serration core suggests that *T. rex* may have exercised less complete control over the inorganic mineral properties of the dentine than *Albertosaurus*. Here, a finely controlled geometry may have been too metabolically expensive.

Many questions – and opportunities – remain, in addition to serration taxonomy. Why, for example, are the serrations of *T. rex* different from those of *Albertosaurus*? Also, it is possible that the ampullae of the serrations may have been invaded from the outside by bacteria or other living organisms. If so, evidence of decay products, or even of the organisms themselves, may remain, trapped inside the ampullae. Such information might offer clues to the environment in which the tyrannosaurs lived and to the interactions of these animals with their surroundings.

Acknowledgments

I thank M. Parrish, J. Farlow, W. Turnbull, and L. Schmidt for helpful comments; J. Bolt and P. Makovicky for use of specimens; W. Simpson and L. Bergwall-Herzog for technical assistance; N. Hancoca for special photographic processing; I. Glasspool for special photography; and S. Teacher for publishing information.

Literature Cited

Abler, W. L. 1973. *The Sensuous Gadgeteer: Bringing Tools and Materials to Life.* Running Press, Philadelphia. (Revised as *Shop Tactics: The Common-Sense Way of Using Tools and Working with Woods, Metals, Plastic, and Glass,* Running Press, Philadelphia, 1976.)

Abler, W. L. 1992. The serrated teeth of carnivorous dinosaurs, and biting structures in other animals. *Paleobiology* 18:161–183.

Abler, W. L. 2001. A kerf-and-drill model of tyrannosaur tooth serrations; pp. 84–89 in D. H. Tanke and K. Carpenter (eds.), *Mesozoic Vertebrate Life.* Indiana University Press, Bloomington.

Farlow, J. O., D. L. Brinkman, W. L. Abler, and P. J. Currie 1991. Size, shape, and serration density of theropod dinosaur lateral teeth. *Modern Geology* 16:161–198.

Frazzetta, T. H. 1988. The mechanics of cutting and the form of shark teeth (Chondrichthyes, Elasmobranchii). *Zoomorphology* 108:93–107.

Romer, A. S., and L. W. Price. 1940, December 6. Review of the Pelycosauria. Geological Society of America Special Papers, no. 28. 538 pp.

Selby, S. M. 1975. *Standard Mathematical Tables.* CRC Press, Cleveland, Ohio.

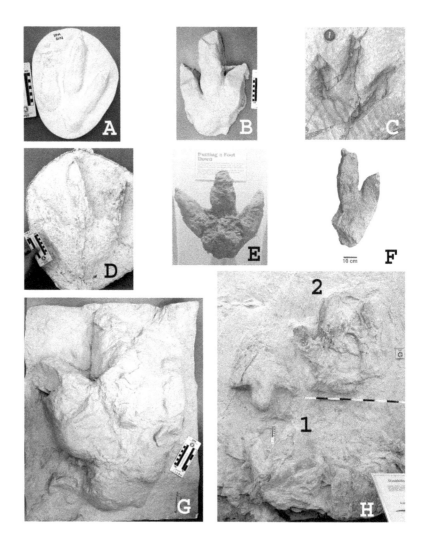

5.1. Tridactyl footprints attributed to large theropod dinosaurs; all except 5.1C are natural or artificial casts. A) YPM 2098, ichnogenus *Eubrontes;* digit II impression on left. B) UCM (formerly CU-MWC) 188.25 (ichnogenus *Megalosauripus*), Summerville Formation, Carrizo Mountains, Arizona; digit II impression on right. C) Right footprint (ichnogenus *Buckeburg-ichnus* or *Megalosauropus*), Enciso Group, Los Cayos, Spain. D) Cast of footprint from the Glen Rose Formation, bed of the Paluxy River, Somervell County, Texas; digit II impression on right. E) TMP 81.34.1, probable tyrannosaurid footprint, Dinosaur Provincial Park, Canada; digit II impression probably on right; McCrea et al. (2005), however, identified this as a large ornithopod print. F) MPC-D 100F/12, probable tyrannosaurid footprint, likely made by *Tarbosaurus.* One of the digit impressions is missing, making interpretation of the print problematic. Currie et al. (2003) identified it as a left footprint (and so the digit II impression would be on the right), inferring a greater interdigital angle between digits III and IV than between II and III. However, in many large theropod footprints the tip of digit III is directed medially; if this was true here, the print would be a right footprint, and it would be digit II, rather than IV, that is missing. G) CU-MWC 225.1, ichnogenus *Tyrannosauripus,* likely made by *Tyrannosaurus,* Raton Formation, New Mexico, digit II on left; also note likely impression of digit I behind digit II. H) Natural casts of possible *Tyrannosaurus* prints, Laramie Formation, Golden, Colorado. Possible trackway sequence indicated by prints 1 and 2 (direction of motion toward the top of the page), with the identifying numbers adjacent to the digit III impression; if this interpretation is correct, print 1 would likely be a right footprint, with digit II on the right.

Feet of the Fierce (and Not So Fierce): Pedal Proportions in Large Theropods, Other Non-avian Dinosaurs, and Large Ground Birds

5

James O. Farlow, Thomas R. Holtz, Jr., Trevor H. Worthy, and Ralph E. Chapman

The extent to which the makers of tridactyl dinosaur footprints can be identified depends on the extent to which their foot skeletons can be told apart. We examined this question for non-avian theropod dinosaurs (NATs) and large ground birds, making additional comparisons with functionally tridactyl, bipedal – or potentially bipedal – ornithischians. For birds we measured distances across the trochleae of the tarsometatarsus, and for NATs, the lengths of metatarsals II–IV. For birds, NATs, and ornithischians we measured the lengths and widths of individual phalanges and the aggregate lengths of digits II–IV.

Metatarsal, digital, and phalangeal proportions distinguish some genera among dinosaurs (including birds). At higher taxonomic levels, pedal features are useful but not infallible proxies for the systematic affinities of birds and non-avian dinosaurs. Our results suggest that the parameters commonly used to describe tridactyl dinosaur footprints can often be used to provide a minimum estimate of the number of trackmaker taxa within an ichnofauna and that similarity in footprint shape is useful but not always a trustworthy indicator of phylogenetic relationships of trackmakers.

Abstract

I submit that the characters most diagnostic for the classification of footprints as such, as well as most useful for comparison with skeletal remains, are those which reflect the bony structure of the foot.

Baird (1957:469)

Introduction

In many Mesozoic stratigraphic units, footprints of dinosaurs are much more common than dinosaur body fossils and provide our best record of the kinds of dinosaurs living in certain regions during particular times (cf. Leonardi 1989; Thulborn 1990; Lockley 1991; Schult and Farlow 1992; Gierliński 1995; Lockley and Hunt 1995; Dalla Vecchia et al. 2000; Lockley and Meyer 2000; Kvale et al. 2001; Farlow and Galton 2003; Moratalla et al. 2003; Pérez-Lorente 2003; García-Ramos et al. 2004; Farlow et al. 2006; Milner and Spears 2007; Rainforth 2007; Sullivan et al. 2009). The resolution with which such trace fossils can be used for paleoecological or biostratigraphic studies depends on the degree to which particular footprint shapes are uniquely associated with given dinosaur taxa (Baird 1957; Farlow and Chapman 1997; Farlow 2001; Smith and Farlow 2003; Farlow et al. 2006).

Institutional Abbreviations AM, Auckland Museum, Auckland; AMNH, American Museum of Natural History, New York; BHI, Black Hills Institute, Hill City, South Dakota; BMNH, Natural History Museum, London; CEU, College of Eastern Utah, Price; CM, Canterbury Museum, Christchurch; FMNH or PR, Field Museum of Natural History, Chicago; CU-MWC, University of Colorado-Museum of Western Colorado, Denver; GI, Geological Institute, Mongolian Academy of Sciences, Ulaanbaatar; GSC, Geological Survey of Canada, Ottawa; LACM, Natural History Museum of Los Angeles County, Los Angeles; MNA, Museum of Northern Arizona, Flagstaff; MOR, Museum of the Rockies, Bozeman, Montana; MPC-D, Mongolian Paleontological Center (Mongolian Academy of Science), Ulaan Baatar; NMNZ, Museum of New Zealand Te Papa Tongarewa, Wellington; PVSJ, Museo de Ciencias Naturales, Universidad Nacional de San Juan, San Juan, Argentina; QM, Queensland Museum, Brisbane; ROM, Royal Ontario Museum, Toronto; TMP, Royal Tyrrell Museum of Palaeontology, Drumheller, Alberta; TP, Thanksgiving Point Institute (North American Museum of Ancient Life), Lehi, Utah; UCM, University of Colorado-Boulder Museum of Natural History; UCMP, University of California Museum of Paleontology, Berkeley; U Illinois, University of Illinois, Urbana; UMNH, Utah Museum of Natural History, Salt Lake City; USNM, National Museum of Natural History, Smithsonian Institution, Washington, D.C.; WO, Waitomo Caves Museum, Waitomo Caves, New Zealand; YPM, Yale Peabody Museum, New Haven, Connecticut.

Tridactyl (three-toed) footprints of bipedal dinosaurs (Fig. 5.1) are often the most abundant dinosaur footprints found in track assemblages. Because dinosaur taxa are named on the basis of skeletal material, identifying the bipedal dinosaurs responsible for three-toed prints involves making correlations between footprints and foot skeletons. The "best case" of our ability to identify the makers of dinosaur tracks would therefore be that in which the shape of a footprint reflected the proportions of its maker's foot skeleton (as revealed, perhaps, by relative digit lengths, the configuration of footprint digital [toe] pads, and interdigital angles of the footprint; Smith and Farlow [2003]) with perfect fidelity. This in turn leads us to inquire as to what features, and at what taxonomic level, we could use to tell foot skeletons apart.

Although this question can be explored by a study of dinosaur foot skeletons (Farlow and Lockley 1993; Farlow and Chapman 1997; Farlow 2001; Smith and Farlow 2003), this approach is limited by the number of reasonably complete, well-preserved dinosaur feet available for study. The problem can be tackled in a more indirect fashion, however, by examining within-group and across-group variability in foot shape in ground birds—animals thought at the very least to be close relatives of dinosaurs (Feduccia 1996) and considered by most paleontologists to be extant dinosaurs (Gauthier and Gall 2001; Chiappe and Witmer 2002; Currie et al. 2003; Padian 2004). Living or recently extinct bird species are represented by many more complete foot skeletons in museum collections than are most non-avian theropod or ornithopod dinosaur species. It is therefore easier to obtain a large enough sample of foot skeletons to consider within-taxon and across-taxon variability in pedal proportions in birds than in dinosaurs.

Here we examine within-taxon and across-taxon variability in foot skeletons of non-avian theropods (NATs; Fig. 5.2) and large ground birds (both extant and extinct), and consider similarities in foot shape among NATs, ground birds, and bipedal or potentially bipedal ornithischians (Appendix 5.1). Because we are interested in pedal features that might be expressed in footprints, our study deals with the distal metatarsal and digital portions of the foot. Our analyses will examine progressively more detailed features of the dinosaurian foot (e.g., from overall digit shape to individual phalangeal shapes), but we will then take a summary "gestalt" look at the foot shapes of large theropods.

Among birds emphasis is placed on extant ratites and on moa (Dinornithiformes or Dinornithidae), a clade of recently extinct flightless bird species from New Zealand (Anderson 1989; Cooper et al. 1992; Worthy and Holdaway 2002; Bunce et al. 2003; Huynen et al. 2003; Baker et al. 2005; Worthy 2005). Of all ground birds, moa are perhaps best suited to serve as proxies for dinosaurs in an investigation of the utility of foot skeletons in discriminating among zoological taxa. As birds go, moa ranged in size from very large to enormous, with adult females of the largest species reaching body masses of 200 kg or more. Thus moa show as much size overlap with bipedal dinosaurs as any comparable avian clade. Like

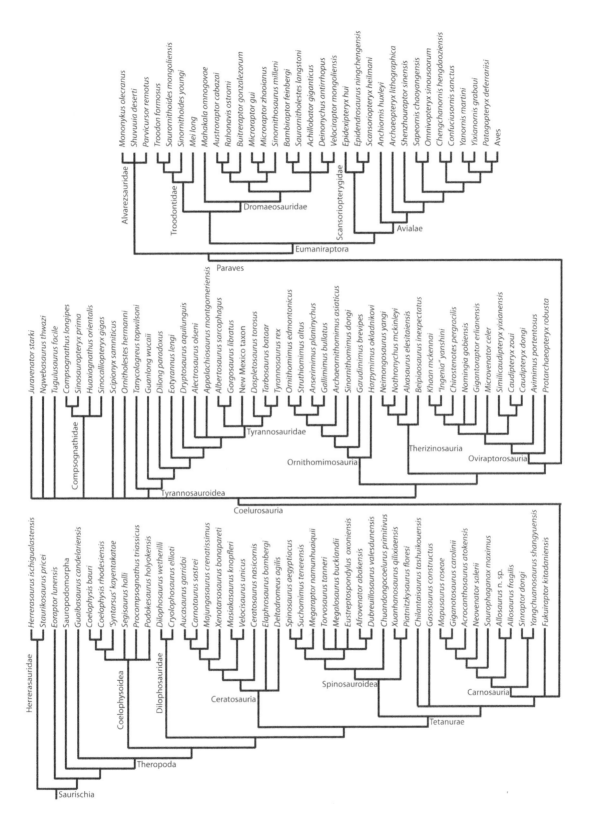

5.2. Theropod phylogeny.

many bipedal dinosaurs, moa were functionally tridactyl. Within moa, molecular data indicate that *Emeus* and *Euryapteryx* are closely related and together form a clade with *Anomalopteryx*. *Emeus* + *Euryapteryx* + *Anomalopteryx* in turn are the sister group of *Pachyornis*. All of the preceding then form the sister group of *Dinornis*, and *Megalapteryx* is the sister taxon to all the other moa (Huynen et al. 2003; Baker et al. 2005).

Considerably complicating our study is the striking lack of agreement about the relationships within and among the major neornithine bird clades (Sibley and Monroe 1990; Cooper et al. 2001, 1992; Johnsgard 1991; Cooper 1997; Houde et al. 1997; Lee et al. 1997; Cracraft 2001; Haddrath and Baker 2001; Livezey and Zusi 2001, 2007; Shapiro et al. 2002; Worthy and Holdaway 2002; Huynen et al. 2003; Mayr et al. 2003; Cracraft et al. 2004; Dyke and Van Tuinen 2004; Fain and Houde 2004; Murray and Vickers-Rich 2004; Baker et al. 2005; Ericson et al. 2006; Hackett et al. 2008; Harshman et al. 2008). Within palaeognaths, cassowaries (*Casuarius*) and emus (*Dromaius*) are generally agreed to be closely related, and the various moa are thought to represent a distinct clade. There is conflicting evidence about palaeognath relationships (depending in part on whether morphological or molecular [and which kind of molecular] data are employed). Ostriches (*Struthio*) may be closely related to rheas (*Rhea* [including *Pterocnemia*]), or they may be the sister group of all other palaeognaths. Morphological data suggest that moa are the sister group of the cassowary/emu clade plus ostriches, rheas, and elephant birds and that kiwi are the sister group of moa and all other ratites (Livezey and Zusi 2007). In contrast, molecular data group kiwi with the cassowaries and emus (Hackett et al. 2008; Harshman et al. 2008). In the face of these uncertainties, in our analysis we lump all ratites other than moa in an ad hoc group designated as "struthioniforms" but make no assumptions about the relationships among the genera in this group (apart from the closeness of *Dromaius* and *Casuarius*). Groups to which we tentatively assign neognath ground birds for the purposes of this analysis are indicated in Appendix 5.1.

Systematic assignments of non-avian dinosaur taxa generally follow Weishampel et al. (2004), with modifications as warranted (see Appendix 5.1). Our analyses emphasize large theropods, particularly comparisons of carnosaurs with tyrannosauroids (mostly tyrannosaurids).

Methods

We measured NAT, large ground bird, and selected ornithischian foot skeletons in several museums in North America, Europe, New Zealand, and Australia, taking care not to include feet that were composite assembled from the bones of more than one animal.

For a selected set of large ground birds, in which metatarsals II–IV are fused in a tarsometatarsus (TMT), we measured distances separating the centers of distal trochleae II–III, III–IV, and II–IV. The relative magnitudes of these distances might serve as proxies for such footprint features as interdigital angles and the positions of proximal ends of toe marks. The metatarsals are generally unfused in non-avian theropods, and thus we could not make the same measurements of the distal metatarsus as in birds. Instead, for a set of NATs we compared the lengths of metatarsals (MTs) II, III, and IV. Metatarsal lengths alone will not shed light on interdigital angles, but they should contribute to how far from the posterior edge of a footprint the proximal and distal ends of a toe mark will be.

We often assembled phalangeal skeletons from loose toe bones in boxes containing bones of a single individual; in many cases (particularly for digit III) we were unable to determine whether a given element was a left or a right, and so many of our data cases undoubtedly represent composites from the left and right feet of a particular animal. In some cases, our identification of which phalanx a toe bone was differed from that of whoever who had previously identified and labeled the bones.

Few foot skeletons of non-avian theropods are complete enough to provide complete data on digital dimensions. Unguals are frequently absent, and, when present, their tips are often broken. As a result, our analyses of digital and phalangeal proportions in NATs are afflicted by small sample sizes of complete foot skeletons. To mitigate this shortcoming, we analyzed digital and phalangeal proportions of NATs (and in some analyses, birds), both including and excluding unguals. As will be seen, the results are usually similar.

As in Farlow (2001), phalangeal lengths were taken from the dorsoventral midpoint of the concave proximal end of the bone to the dorsoventral midpoint of the convex distal end of the bone. Ungual lengths were measured in a straight line from the dorsoventral midpoint of the concave articular end of the bone to the tip of the ungual. Where possible, we measured lengths on both the medial and lateral sides of each phalanx and ungual and used the average of the two values. Digit lengths were calculated as the sums of individual phalanges therein. We measured the width of non-ungual phalanges as the maximum transverse dimension across the distal articular condyles.

In some moa genera (*Emeus, Euryapteryx*) there are only four phalanges in the outer toe (digit IV), rather than the usual five (including the ungual in the count) seen in most birds and bipedal dinosaurs. This raises an obvious question about the identity of the missing toebone in *Emeus* and *Euryapteryx*.

In an experimental study of the effects of blocking a bone morphogenetic protein on digit growth in domestic chickens, the distal phalanges of toes disappeared after treatment (Zou and Niswander, 1996). If this or something similar is the mechanism of phylogenetic phalangeal loss, it might be phalanx IV_5 that is missing in *Emeus* and *Euryapteryx*. However, phalanx IV_4 is usually the smallest toe bone in dinosaur and bird taxa that have five phalanges in digit IV, and it would seem an easy phylogenetic step for this bone to be lost or, perhaps, fused with phalanx IV_3. Furthermore, if phalanx IV_5 had indeed been the toe bone lost in the two moa genera, phalanx IV_4 would then have had to take the form of an ungual. Although this is certainly possible, it seems simpler just to lose

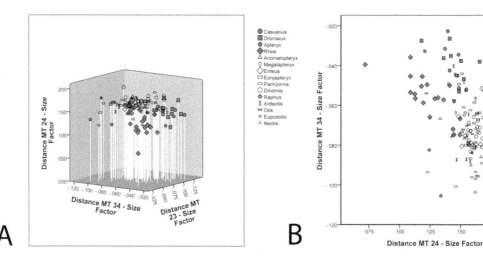

A

B

Casuarius
Dromaius
Apteryx
Rhea
Anomalopteryx
Megalapteryx
Emeus
Euryapteryx
Pachyornis
Dinornis
Raphus
Ardeotis
Otis
Eupodotis
Neotis

5.3. Comparison of the distances between the distal trochleae of the metatarsus in large ground birds: MT 23 = distance between the centers of the distal ends of the trochleae of metatarsals II and III; MT 34 = distance between the centers of the distal ends of the trochleae of metatarsals III and IV, and MT 24 = distance between the centers of the distal ends of the trochleae of metatarsals II and IV. In all cases the distances were log-transformed, and the mean of the three log-transformed distances was subtracted from each log-transformed distance as a size factor. A) Simultaneous comparison of all three inter-trochlear distances across avian genera. B) Comparison of inter-trochlear distances II–IV and III–IV across genera.

the already small fourth phalanx. We therefore assumed that IV_4 is the missing phalanx. The truth of this assumption will not affect the results of our analysis; one can simply translate the statements that follow about the loss of phalanx IV_4 to mean that there are only four phalanges in the outer toe, regardless of the exact identities of those phalanges.

All measurements were taken to the nearest millimeter. The more distal non-ungual phalanges of digit IV are the smallest bones of the foot in dinosaurs, and for small birds the lengths of phalanges IV_3 and IV_4 may be only a few millimeters long–close to our precision of measurement. For principal components analyses (PCA) we restricted our analysis of bird feet to specimens whose combined length of IV_3 and IV_4 was at least 10 mm; for discriminant analyses of extant and extinct ratites we relaxed this size restriction to increase the number of specimens in the analyses.

Data were analyzed using several techniques, all with the statistical package SPSS. In many comparisons we employed simple bivariate plots. We also did principal components analyses using a covariance matrix. Data were log-transformed prior to PCA. This required a modest data massage. For analytical purposes we assigned a value of zero for the length of IV_4 in those species with only four phalanges in digit IV and created a new length variable, defined as the sum of the lengths of phalanges IV_3 and IV_4, and used this new variable instead of lengths of those two phalanges themselves.

For cluster analyses and discriminant (including canonical variate) analyses, we removed absolute size as a confounding variable prior to analysis. This was done in two ways. One method (used for cluster analyses) was to scale all measurements (e.g., phalangeal lengths and widths) to a common value of a selected measurement (e.g., the length of phalanx III_1) and then perform the analysis on the scaled variables. For the second procedure (used for cluster and discriminant analyses), we log-transformed all the variables used in each analysis and calculated the mean of the log-transformed variables as a proxy for overall size (size

5.4. Relative metatarsal lengths in non-avian theropods, with emphasis on large predatory forms. The lengths of metatarsals (MTs) II, III, and IV were log-transformed, and the mean of the log-transformed MT lengths was

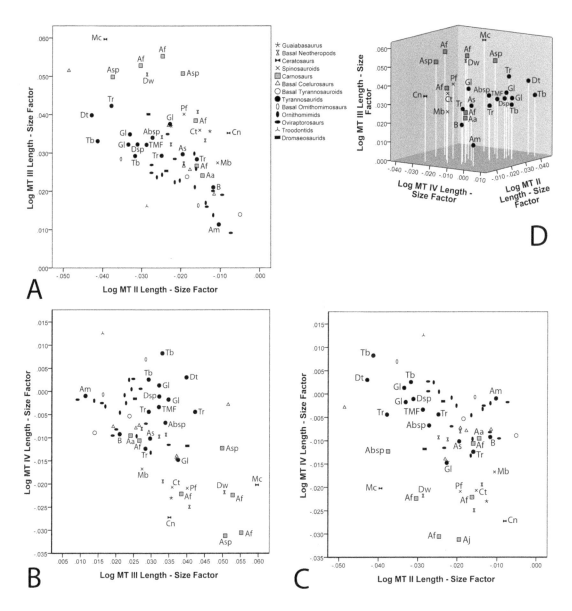

factor); the mean was then subtracted from each of the log-transformed variable measurements (McBrayer and Corbin 2007).

Canonical variate analyses (CVAs) were carried out in a forward stepwise manner, minimizing the value of Wilks's lambda, creating discriminating functions using the variables that yielded greatest separation of taxa. The success of the discriminating functions was evaluated both using all of the original data cases and also in a cross-validation (leave-one-out procedure), in which each case is classified by functions derived from all cases other than itself. Because the sample sizes of different taxa often varied considerably, we did discriminant analyses in which the prior probability that a case belonged to any particular group (taxon) was equal across groups and also in which the prior probability was adjusted according to the number of cases in each group.

computed as a size factor. This size factor was subtracted from each log-transformed MT length to create a scaled MT length. A–C) Bivariate comparisons of the scaled MT lengths. D) Simultaneous comparison of all three scaled MT lengths for large theropods only. Abbreviations for large theropod taxa as in Appendix 5.1. Note tendency for more basal theropods to have a relatively long MT III, while most coelurosaurs have a relatively long MT IV as compared with the lengths of both MT II and MT III.

The lowest systematic level at which comparisons were made was the genus, due to the lack of adequate numbers of specimens of different species within genera (particularly in non-avian dinosaurs).

Results

Metatarsal Proportions

BIRDS There is considerable scatter within genera (Fig. 5.3). However, struthioniforms other than moa (*Casuarius, Dromaius, Apteryx*, and especially *Rhea* [including *Pterocnemia*]) tend to have a relatively short MT II–IV distance. *Casuarius* and *Dromaius* have a relatively long MT III–IV distance compared with most moa. Bustards (*Ardeotis, Otis, Eupodotis*, and *Neotis*) and the dodo (*Raphus*) plot among the palaeognath points.

NON-AVIAN THEROPODS There is a slight tendency for more basal theropods (*Guiabasaurus*, basal neotheropods, ceratosaurs, spinosauroids, and carnosaurs) to have a relatively long metatarsal (MT) III (Figs. 5.4A, B, D; 5.5A, B), particularly compared with the length of MT IV. There is a marked tendency for coelurosaurs to have a relatively long MT IV compared with the lengths of MT II and III (Figs. 5.4A, C; 5.5C). As with birds, there is considerable scatter in the distribution of values of scaled MT lengths (Fig. 5.4) of individual genera (particularly noticeable for *Allosaurus, Gorgosaurus*, and *Tyrannosaurus*), resulting in significant overlap in morphospace with other large theropods.

Digit Lengths and Widths

BIRDS Principal Component (PC) 1 explains by far the greater part of data variance (Table 5.1). The overall apportioning of variance among components in this analysis (by far the greatest amount in PC 1, very small but interpretable amounts in remaining PCs) is similar to what we found in all our other digital and phalangeal PCAs and is typical of what one of us (Chapman) has observed in analyzing more than a thousand reptile and bird datasets.

Principal Component 2 contrasts specimens with long and narrow toes from those with short and broad toes (positive loadings of digit lengths vs. negative loadings of phalanx widths; see Table 5.1). Principal Component 3 contrasts forms characterized by a relatively large digit III (positive loadings of both length and width) with forms that have relatively broad digit II (negative loadings of lengths and widths of digits II and IV). The contrasts remain much the same if the PCA is run excluding the unguals (Table 5.2), except that excluding the unguals in the PCA causes a large digit III to be associated with negative rather than positive loadings on PC 3 (note reversed position of *D. novaehollandiae* with respect to PC 3 in Figs. 5.6A, B).

Some taxa separate nicely on the basis of PCs 2 and 3 (Fig. 5.6). Emus are distinguished from most other ground birds by having a relatively large digit III compared with digits II and IV, and kiwi by having

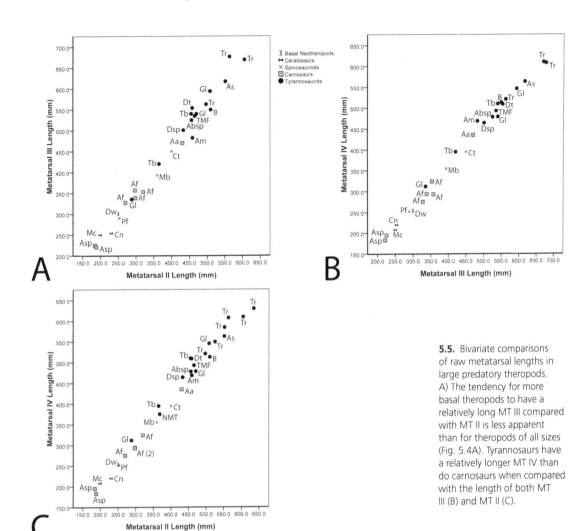

5.5. Bivariate comparisons of raw metatarsal lengths in large predatory theropods. A) The tendency for more basal theropods to have a relatively long MT III compared with MT II is less apparent than for theropods of all sizes (Fig. 5.4A). Tyrannosaurs have a relatively longer MT IV than do carnosaurs when compared with the length of both MT III (B) and MT II (C).

a relatively small digit III. Emus are most like kori bustards and unlike their close relatives, cassowaries (if unguals are included in the PCA). Cassowaries plot close to *Dinornis* and (particularly if unguals are included in the analysis) the extinct flightless goose *Cnemiornis*. Rheas are near *Pachyornis*, kiwi to *Anomalopteryx*, and the adzebill (*Aptornis*) to *Dinornis* and *Megalapteryx*.

A stepwise canonical variates (discriminant) analysis of ratite genera shows similar results (Table 5.3; Fig. 5.7). Extant "struthioniforms" have a relatively long digit III compared with moa, and kiwi and *Megalapteryx* have a relatively long digit IV and a narrow digit III. The CVA correctly assigns foot skeletons to genus at least 70 percent of the time. Most errors involve mistaking one genus of moa for another or (less frequently) one kind of extant "struthioniform" for another. A more interesting mistake is that feet of *Megalapteryx* and *Apteryx* are sometimes assigned to each other.

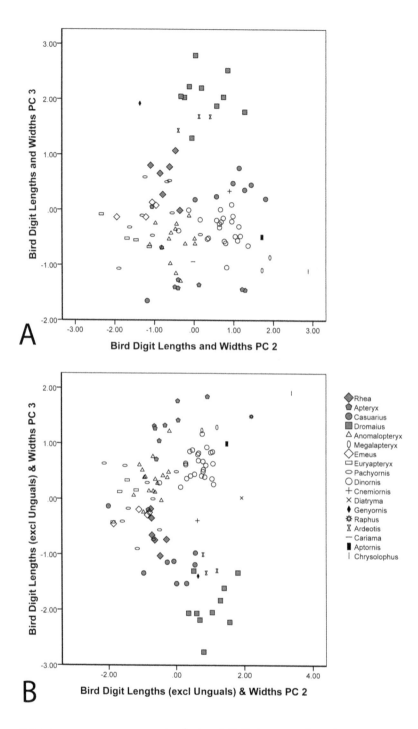

5.6. Principal components analyses (PCA) of digit lengths and widths in ground birds (Tables 5.1, 5.2). A) PCA that includes unguals in the analysis. Positive values of PC 2 are associated with relatively slender digits, and negative values with relative broad digits. Positive values of PC 3 are associated with a relatively large digit III, and negative values with relatively large digits II and IV. B) Unguals excluded. PC 2 has the same interpretation as with the PCA version that includes unguals. The same is largely true for PC 3, except that the positive or negative sign of PC 3 is reversed.

NON-AVIAN THEROPODS Analysis of digital proportions in non-avian theropods is severely hampered by the small number of specimens with complete feet, but interesting patterns nonetheless emerge (Table 5.4). As with birds, PC 1 explains nearly all of the data variance. Again as in birds, the second component involves a contrast between relatively short and broad (Fig. 5.8A, points at left of graph) and relatively long and

Parameter	PC 1 Loading (Raw [Rescaled])	PC 2 Loading (Raw [Rescaled])	PC 3 Loading (Raw [Rescaled])
Digit II length	0.193 (0.983)	0.032 (0.162)	−0.007 (−0.034)
Phalanx II2 distal width	0.273 (0.988)	−0.018 (−0.067)	−0.035 (−0.128)
Digit III length	0.182 (0.979)	0.030 (0.160)	0.019 (0.100)
Phalanx III2 distal width	0.257 (0.983)	−0.023 (−0.088)	0.041 (0.157)
Digit IV length	0.197 (0.980)	0.037 (0.182)	−0.006 (−0.031)
Phalanx IV2 distal width	0.279 (0.992)	−0.028 (−0.100)	−0.006 (−0.023)
Eigenvalues (% of variance)	0.328 (97.106)	0.005 (1.452)	0.003 (1.010)
Cumulative variance explained (%)	97.106	98.559	99.569

Table 5.1. Principal components (PC) analysis (using a covariance matrix) of log-transformed measurements of digit lengths and widths in ground birds ($N = 95$)

Note: Kaiser-Meyer-Olkin measure of sampling adequacy = 0.856; Bartlett's test of sphericity: χ^2 = 1606.424, $p < 0.001$.

Parameter	PC 1 Loading (Raw [Rescaled])	PC 2 Loading (Raw [Rescaled])	PC 3 Loading (Raw [Rescaled])
Digit II length	0.192 (0.971)	0.033 (0.166)	0.021 (0.107)
Phalanx II2 distal width	0.270 (0.986)	−0.032 (−0.118)	0.025 (0.090)
Digit III length	0.180 (0.969)	0.033 (0.179)	−0.022 (−0.118)
Phalanx III2 distal width	0.257 (0.985)	−0.011 (−0.042)	−0.041 (−0.158)
Digit IV length	0.195 (0.975)	0.039 (0.192)	0.013 (0.066)
Phalanx IV2 distal width	0.275 (0.991)	−0.030 (−0.108)	0.004 (0.016)
Eigenvalues (% of variance)	0.322 (96.449)	0.006 (1.715)	0.003 (1.026)
Cumulative variance explained (%)	96.449	98.163	99.189

Table 5.2. Principal components (PC) analysis (using a covariance matrix) of log-transformed measurements of digit lengths (excluding the ungual) and widths in ground birds ($N = 105$)

Note: Kaiser-Meyer-Olkin measure of sampling adequacy = 0.835; Bartlett's test of sphericity: χ^2 = 1609.342, $p < 0.001$.

narrow digits (Fig. 5.8A, right of graph). Although we extracted three components, there is almost no variance remaining after the second, and the third component is difficult to interpret, although the variable with the largest loading (in absolute value) on PC 3 is the length of digit IV. The pattern is similar if the PCA is done excluding the unguals (Table 5.5, Fig. 5.8B), and the importance of digit IV length for PC 3 becomes clearer (perhaps because of the greater number of data points), although of opposite sign than when the unguals are included.

The more "extreme" morphologies are all associated with small theropods (Fig. 5.8). *Bambiraptor*, *Deinonychus*, *Tanycolagreus*, and *Chirostenotes* have relatively slender toes, while *Mononykus* has rather broad toes. *Mononykus*, *Deinonychus*, and *Bambiraptor* have a relatively large digit IV compared with II and III, while *Coelophysis* has a relatively smaller IV.

Large theropods, in contrast, appear more conservative, regardless of clade (Figs. 5.9–5.11). *Allosaurus* and *Dilophosaurus* have digital proportions that differ little from those of tyrannosaurids. Digits appear to become relatively stouter with increasing size in theropods, but larger individuals of *Gorgosaurus* seem to be rather slimmer-toed than other big theropods.

5.7. Proportions of digit lengths and widths of extant "struthioniforms" and moa. A) First two discriminant functions from a stepwise canonical variate analysis (CVA; Table 5.3). Function 1 contrasts genera with a large digit III (positive values) with genera characterized by a long digit IV and a broad digit II (negative values); notice that Function 1 results in almost total separation of extant "struthioniforms" from moa. Function 2 contrasts genera characterized by a broad digit III (positive values) with genera characterized by a long digit IV (negative values). B, C) Bivariate comparisons illustrating with larger sample sizes some of the discriminating variables identified by the CVA. B) Extant "struthioniforms" tend to have a relatively longer digit III than digit IV than do moa. C) Moa tend to have a broader phalanx II2, compared with the length of digit III, than do extant "struthioniforms."

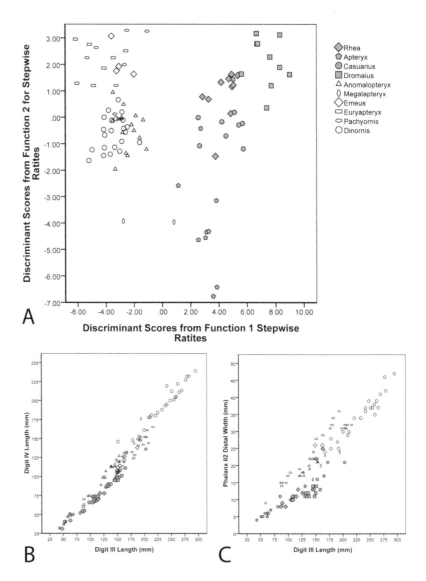

We attempted a stepwise discriminant analysis of scaled digit lengths (excluding unguals) and widths (scaled by subtracting log-transformed values of each variable from the mean of all log-transformed variables used in the analysis) of carnosaurs versus tyrannosaurids. The sample size (three carnosaur specimens, all of them *Allosaurus*, and nine tyrannosaurid specimens) was very small, and the Wilks's lambda test indicated no significant differences in mean values between carnosaurs and tyrannosaurids for any of the scaled variables.

Phalanx Lengths and Widths

BIRDS Once again, PC 1 accounts for most of the data variance, but proportionately less than in runs where phalangeal lengths are summed as digit lengths (Tables 5.1, 5.2, 5.6, 5.7). We extracted four principal

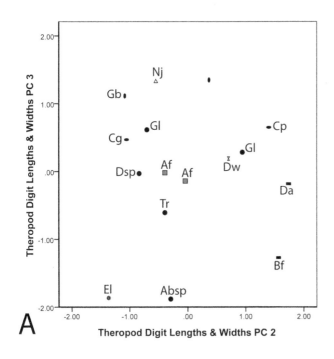

A

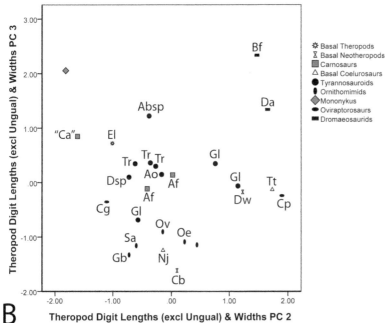

B

5.8. Principal components analyses (PCAs) of digit lengths and widths in non-avian theropods (Tables 5.4, 5.5). Acronyms labeling particular points as in Appendix 5.1. A) Unguals included. Positive values of PC 2 are associated with relatively slender digits, and negative values with relatively stout digits. Positive values of PC 3 are associated with a relative large digit IV. B) Unguals excluded. PC 2 has the same interpretation as in the with-ungual PCA. In contrast to PC 3 in the analysis that includes unguals (A), positive values of PC 3 here are associated with a large digit IV.

components in two versions of the analysis, one with (Table 5.6) and the other without (Table 5.7) unguals. In the version that includes unguals, PC 2 and PC 3 both contrast forms characterized by relatively long phalanges from more distal parts of the digits, and slender digits, with forms that have relatively long proximal phalanges and broad digits; PC 2 and PC 3 only differ in which proximal or distal phalanges have the highest loadings (Table 5.6). Principal Component 4, however, contrasts birds with

Table 5.3. Forward stepwise discriminant (canonical variate) analysis of log-transformed digit lengths (including unguals) and widths (scaled by subtracting the mean of all the log-transformed variables employed in the analysis from the value of each log-transformed variable) in genera of living and extinct ratites. In contrast to the comparable principal components analysis (Table 5.1), the requirement that all specimens have a combined length of phalanges IV3 and IV4 of at least 10 mm was relaxed here. For all variables, the p-value of the F-statistic of the Wilks's lambda test was < 0.001, indicating significant differences in the means of all variables across genera: *Rhea* (including *Pterocnemia*; N = 10); *Apteryx* (N = 9); *Casuarius* (N = 9); *Dromaius* (N = 10); *Anomalopteryx* (N = 14); *Megalapteryx* (N = 2); *Emeus* (N = 4); *Euryapteryx* (N = 7); *Pachyornis* (N = 9); *Dinornis* (N = 22). Unfortunately, the Box's M-test of the equality of group covariance matrices indicated lack of equal covariances (p < 0.001), and so the results of the analysis must be viewed with caution.

Variable[a]	Standardized canonical discriminant function coefficients (correlations between individual variables and discriminant function)			
	Function 1	Function 2	Function 3	Function 4
Digit III length	1.125 (0.818)*	0.187 (−0.513)	−0.025 (−0.074)	0.700 (0.249)
Phalanx III2 distal width	0.507 (−0.017)	0.823 (0.940)*	1.213 (0.263)	0.395 (−0.215)
Digit IV length	−0.179 (0.210)	−0.359 (−0.830)*	1.516 (0.507)	0.299 (−0.097)
Phalanx II2 distal width	−0.236 (−0.536)	0.119 (0.202)	0.518 (−0.174)	1.173 (0.801)*
Eigenvalues (% of variance)	18.988 (77.3)	3.724 (15.2)	1.393 (5.7)	0.454 (1.8)
Canonical correlation	0.975	0.888	0.763	0.559

[a]Variables were entered in the table in the order in which they were entered in the forward stepwise procedure.
*Largest absolute value of correlation between each variable and a discriminant function.

Classification results (errors in classification in bold)

Genus	Predicted Group Membership (Original Grouped Cases [Cross-Validated Grouped Cases])									
	R	Ap	C	Dr	An	M	Em	Eu	P	Di
Rhea (R)	10 (9)	0 (0)	0 **(1)**	0 (0)	0 (0)	0 (0)	0 (0)	0 (0)	0 (0)	0 (0)
Apteryx (Ap)	0 (0)	7 (7)	**1 (1)**	0 (0)	0 (0)	**1 (1)**	0 (0)	0 (0)	0 (0)	0 (0)
Casuarius (C)	**2 (2)**	0 (0)	7 (7)	0 (0)	0 (0)	0 (0)	0 (0)	0 (0)	0 (0)	0 (0)
Dromaius (Dr)	0 (0)	0 (0)	0 (0)	10 (10)	0 (0)	0 (0)	0 (0)	0 (0)	0 (0)	0 (0)
Anomalopteryx (An)	0 (0)	0 (0)	0 (0)	0 (0)	11 (11)	0 (0)	0 (0)	0 (0)	**2 (2)**	**1 (1)**
Megalapteryx (M)	0 (0)	0 **(1)**	0 (0)	0 (0)	0 (0)	2 (0)	0 (0)	0 (0)	0 (0)	0 **(1)**
Emeus (Em)	0 (0)	0 (0)	0 (0)	0 (0)	0 (0)	0 (0)	3 (1)	0 (0)	**1 (3)**	0 (0)
Euryapteryx (Eu)	0 (0)	0 (0)	0 (0)	0 (0)	0 (0)	0 (0)	**1 (1)**	5 (5)	**1 (1)**	0 (0)
Pachyornis (P)	0 (0)	0 (0)	0 (0)	0 (0)	**2 (2)**	0 (0)	**4 (3)**	**1 (3)**	1 (0)	**1 (1)**
Dinornis (Di)	0 (0)	0 (0)	0 (0)	0 (0)	**4 (4)**	0 (0)	0 (0)	0 (0)	0 (0)	18 (18)

77.1% of original grouped cases were correctly classified; 70.8% of cross-validated grouped cases were correctly classified.

Table 5.4. Principal components (PC) analysis (using a covariance matrix) of log-transformed measurements of digit lengths and widths in non-avian theropods (N = 16)

Parameter	PC 1 Loading	PC 2 Loading	PC 3 Loading
Digit II length	0.302 (0.988)	0.043 (0.140)	0.016 (0.054)
Phalanx II2 distal width	0.353 (0.988)	−0.051 (−0.142)	−0.013 (−0.038)
Digit III length	0.286 (0.991)	0.031 (0.108)	0.020 (0.069)
Phalanx III2 distal width	0.357 (0.993)	−0.040 (−0.110)	0.008 (0.022)
Digit IV length	0.287 (0.971)	0.066 (0.222)	−0.027 (−0.091)
Phalanx IV2 distal width	0.366 (0.996)	−0.024 (−0.064)	−0.003 (−0.007)
Eigenvalues (% of variance)	0.642 (97.756)	0.012 (1.798)	0.002 (0.250)
Cumulative variance explained (%)	97.756	99.554	99.804

Note: Kaiser-Meyer-Olkin measure of sampling adequacy = 0.822; Bartlett's test of sphericity: χ^2 = 267.044, p < 0.001.

Parameter	PC 1 Loading	PC 2 Loading	PC 3 Loading
Digit II length	0.343 (0.990)	0.034 (0.098)	−0.028 (−0.082)
Phalanx II2 distal width	0.377 (0.986)	−0.058 (−0.152)	0.012 (0.032)
Digit III length	0.309 (0.985)	0.044 (0.142)	−0.026 (−0.082)
Phalanx III2 distal width	0.379 (0.995)	−0.030 (−0.078)	−0.016 (−0.043)
Digit IV length	0.313 (0.973)	0.058 (0.180)	0.046 (0.143)
Phalanx IV2 distal width	0.383 (0.994)	−0.027 (−0.070)	0.013 (0.033)
Eigenvalues (% variance)	0.744 (97.634)	0.011 (1.507)	0.004 (0.546)
Cumulative variance explained (%)	97.634	99.141	99.687

Table 5.5. Principal components (PC) analysis (using a covariance matrix) of log-transformed measurements of digit lengths (excluding the unguals) and widths in non-avian theropods ($N = 27$)

Note: Kaiser-Meyer-Olkin measure of sampling adequacy = 0.862; Bartlett's test of sphericity: $\chi^2 = 451.571$, $p < 0.001$.

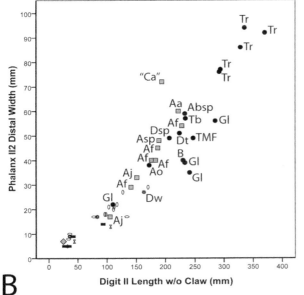

5.9. Digit II lengths and widths in non-avian theropods. Lengths are the summed lengths of individual phalanges, both including (A) and excluding (B) the unguals. Point acronym labels as in Appendix 5.1.

5.10. Digit III lengths and widths in non-avian theropods. Lengths are the summed lengths of individual phalanges, both including (A) and excluding (B) the unguals. Point acronym labels as in Appendix 5.1.

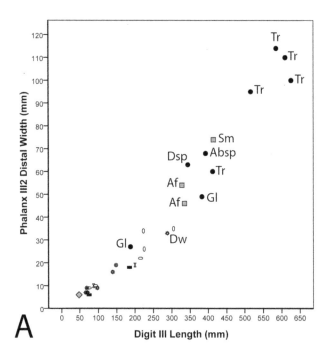

A

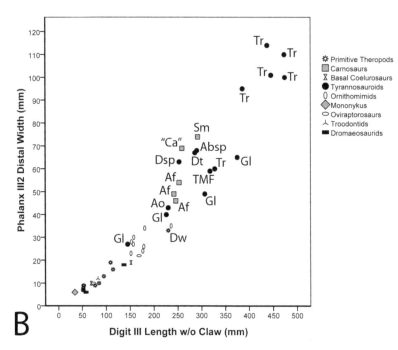

B

broad digits (particularly II and IV) and long distal phalanges with forms that have relatively long proximal phalanges. There is at least some separation of most moa from most "struthioniforms" along PC 2 (Fig. 5.12A), although kiwi plot closer to moa than to other "struthioniforms." *Apteryx, Megalapteryx, Aptornis,* and *Chrysolophus* have large positive values of PC 2 (Fig. 5.12A), whereas *Rhea, Dromaius,* and *Euryapteryx* have negative

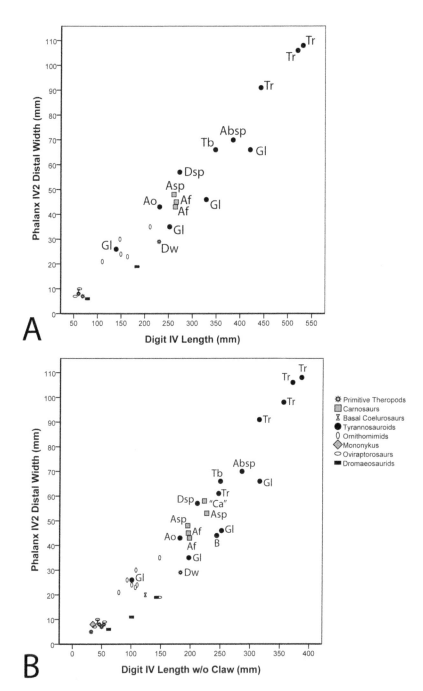

5.11. Digit IV lengths and widths in non-avian theropods. Lengths are the summed lengths of individual phalanges, both including (A) and excluding (B) the unguals. Point acronym labels as in Appendix 5.1.

values. *Casuarius* has very high positive values of PC 3, associated with the huge ungual on digit II. *Genyornis* is at the opposite extreme of PC 3, owing to its relatively tiny distal phalanges, a feature that seems to be typical of other dromornithids as well (Murray and Vickers-Rich 2004). PC 4 (Fig. 5.12C) separates emus and *Chrysolophus* (high positive values) from *Pachyornis, Euryapteryx, Apteryx,* and *Anomalopteryx.*

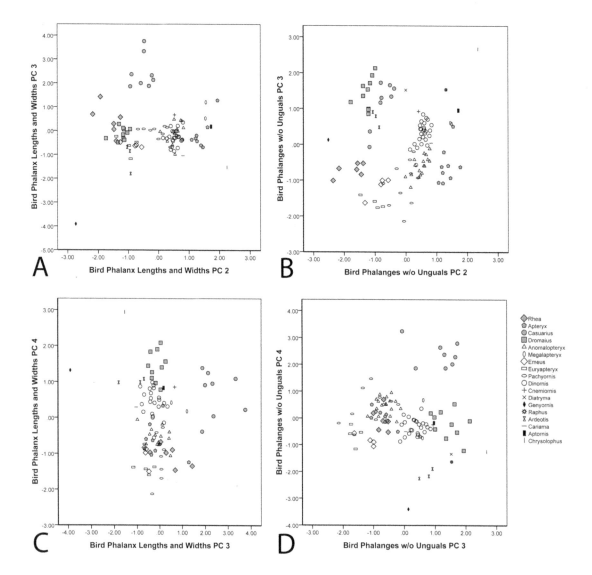

5.12. Principal components analyses (PCAs) of phalangeal lengths and widths in ground birds (Tables 5.6, 5.7), both including (A, C) and excluding (B, D) the unguals. Positive values of PC 2 are associated with relatively long distal phalanges, and negative values with relatively long proximal phalanges and broad toes. In the PCA that includes unguals, PC 3 is a similar contrast, differing only in which phalanges have the highest component loadings (A). If unguals are excluded (B), PC 3 becomes a contrast between forms with broad toes (especially digits II and IV) and a long

In the PCA that excludes unguals (Fig. 5.12B, D), PC 2, as in the PCA version that includes unguals, is a contrast between forms with long distal phalanges and slender toes and forms with long proximals and broad toes. *Chrysolophus*, *Apteryx*, *Aptornis*, and *Megalapteryx*, now joined by *Raphus*, continue to characterize the positive extreme of PC 2, while *Genyornis*, *Rhea*, *Dromaius*, *Casuarius*, *Ardeotis*, *Emeus*, and *Euryapterx* have large negative values. Thus PC 2 looks much the same whether or not unguals are included in the analysis.

Principal Component 3 has a different interpretation than in the with-unguals analysis and is now a contrast between forms with broad toes (especially digits II and IV) and a long II2 (negative values), and forms with slender toes and long phalanges other than II2 (positive values). *Chrysolophus*, *Diatryma*, *Raphus*, *Dromaius*, and *Casuarius* have high positive values of PC 3, while *Rhea*, *Emeus*, *Euryapterx*, and *Pachyornis* have negative values. Principal Component 4 (Fig. 5.12D) contrasts forms

Table 5.6. Principal components (PC) analysis (using a covariance matrix) of log-transformed measurements of phalanx lengths and the distal widths of phalanges II2, III2, and IV2 in ground birds (N = 95)

Parameter	PC 1 Loading	PC 2 Loading	PC 3 Loading	PC 4 Loading
Phalanx II1 length	0.199 (0.959)	−0.035 (−0.168)	−0.030 (−0.144)	0.012 (0.059)
Phalanx II2 length	0.192 (0.863)	0.091 (0.408)	−0.057 (−0.257)	−0.013 (−0.059)
Phalanx II2 distal width	0.271 (0.980)	−0.005 (−0.019)	−0.008 (−0.029)	−0.044 (−0.161)
Phalanx II3 length	0.203 (0.919)	0.015 (0.069)	0.081 (0.366)	<0.001 (−0.003)
Phalanx III1 length	0.204 (0.965)	−0.042 (−0.198)	−0.008 (−0.038)	0.031 (0.146)
Phalanx III2 length	0.164 (0.939)	−0.035 (−0.198)	0.016 (0.094)	0.040 (0.230)
Phalanx III2 distal width	0.253 (0.966)	−0.058 (−0.220)	−0.018 (−0.070)	−0.011 (−0.043)
Phalanx III3 length	0.157 (0.929)	0.034 (0.199)	0.028 (0.166)	0.018 (0.104)
Phalanx III4 length	0.192 (0.965)	<0.001 (−0.004)	0.032 (0.162)	−0.015 (−0.076)
Phalanx IV1 length	0.201 (0.955)	−0.052 (−0.249)	−0.014 (−0.065)	0.023 (0.108)
Phalanx IV2 length	0.210 (0.950)	0.023 (0.106)	−0.025 (−0.115)	0.043 (0.195)
Phalanx IV2 distal width	0.275 (0.978)	−0.028 (−0.100)	−0.018 (−0.065)	−0.042 (−0.150)
Combined length phalanges IV3 and IV4	0.193 (0.847)	0.114 (0.499)	−0.001 (−0.007)	0.018 (0.079)
Phalanx IV5 length	0.209 (0.964)	0.015 (0.069)	0.044 (0.201)	−0.015 (−0.068)
Eigenvalues (% variance)	0.626 (89.360)	0.034 (4.895)	0.016 (2.331)	0.010 (1.458)
Cumulative variance explained	89.360	94.255	96.586	98.044

Note: Kaiser-Meyer-Olkin measure of sampling adequacy = 0.932; Bartlett's test of sphericity: χ^2 = 3553.731, $p < 0.001$.

with relatively long distal phalanges of digits III (especially III3) and IV and broad toes (*Casuarius*), with forms characterized by relatively long phalanges of digit II and to a lesser extent the proximal phalanges of digits III and IV (*Genyornis, Ardeotis*; negative values).

Principal components analyses separate some distinctive avian pedal morphologies at the generic level, but the nearest neighbors in morphospace are not necessarily closely related genera. Thus the neognath *Ardeotis* usually plots near the paleognath *Dromaius*, while *Dromaius* is usually well away from its close relative *Casuarius*. On the other hand, *Emeus* and *Euryapteryx*, the two moa genera most closely related, tend to occur together.

Canonical variate analyses of phalangeal proportions (Tables 5.8–5.9; Fig. 5.13) do a remarkably good job of telling the various ratite genera apart. Two versions of CVA were run, one using all the phalanges of digits II–IV (Table 5.8; Fig. 5.13A, B), and a second using only the first two phalanges of digits II–IV (Table 5.9; Fig. 5.13 C, D). Dropping the more distal phalanges from the analysis increases the number of specimens at the cost of some discriminating power, but that cost is surprisingly low. Using all of the phalanges in the analysis allows 90+ percent correct classification of ratite genera (Table 5.8), while using only the first two phalanges results in correct assignment of specimens to genera 88+ percent of the time (Table 5.9). As in the discriminant analysis of overall digit lengths and widths, the first discriminant function is able to distinguish moa from most extant "struthioniforms." Errors in classification mostly involve *Emeus, Euryapteryx*, and *Pachyornis*, but *Megalapteryx* and *Apteryx* are sometimes mistaken for each other (Tables 5.8, 5.9).

II2 (negative values), and narrow-toed forms with long phalanges other than II2 (positive values). In the PCA that includes unguals (C), PC 4 is a contrast between forms with broad digits (especially II and IV) and long distal phalanges (negative), on the one hand, and forms with relatively long proximal phalanges (positive), on the other. If the unguals are excluded from the PCA (D), PC 4 contrasts forms with broad toes and long distal phalanges of digits III (especially III3) and IV (positive values) with forms that have a long digit II and long proximal phalanges of III and IV (negative values).

Table 5.7. Principal components (PC) analysis (using a covariance matrix) of log-transformed measurements of phalanx lengths (excluding the unguals) and the widths of phalanges II2, III2, and IV2 in birds ($N = 105$)

Parameter	PC 1 Loading	PC 2 Loading	PC 3 Loading	PC 4 Loading
Phalanx II1 length	0.201 (0.968)	−0.028 (−0.135)	0.010 (0.049)	−0.030 (−0.146)
Phalanx II2 length	0.189 (0.870)	0.095 (0.434)	−0.021 (−0.098)	−0.037 (−0.171)
Phalanx II2 distal width	0.268 (0.978)	−0.004 (−0.013)	−0.046 (−0.169)	0.022 (0.079)
Phalanx III1 length	0.206 (0.971)	−0.038 (−0.180)	0.028 (0.130)	−0.006 (−0.029)
Phalanx III2 length	0.164 (0.940)	−0.033 (−0.188)	0.042 (0.243)	0.016 (0.092)
Phalanx III2 distal width	0.253 (0.971)	−0.054 (−0.205)	−0.017 (−0.065)	0.002 (0.006)
Phalanx III3 length	0.154 (0.919)	0.031 (0.185)	0.026 (0.153)	0.042 (0.250)
Phalanx IV1 length	0.204 (0.963)	−0.048 (−0.227)	0.021 (0.097)	−0.013 (−0.064)
Phalanx IV2 length	0.212 (0.958)	0.026 (0.119)	0.035 (0.159)	−0.018 (−0.081)
Phalanx IV2 distal width	0.272 (0.979)	−0.024 (−0.085)	−0.046 (−0.167)	0.009 (0.032)
Combined lengths phalanges IV3 and IV4	0.190 (0.846)	0.112 (0.498)	0.017 (0.074)	0.017 (0.074)
Eigenvalues (% variance)	0.502 (89.867)	0.032 (5.759)	0.010 (1.848)	0.006 (1.020)
Cumulative variance explained	89.867	95.626	97.474	98.494

Note: Kaiser-Meyer-Olkin measure of sampling adequacy = 0.916; Bartlett's test of sphericity: $\chi^2 = 3030.069$, $p < 0.001$.

Bivariate comparisons of some of the phalanx lengths identified as having greater impact on the discriminant function allow still larger sample sizes of specimens (see Fig. 5.14; in some of these comparisons, data for *Palaeotis* and *Struthio*, which were not included in the discriminant analyses, are shown). *Rhea* and *Dromaius*, compared with moa, have a short ungual II3 as compared with the length of phalanx II1, but *Apteryx* is more like the moa in this comparison, and *Casuarius*, of course, has an enormous II3 (Fig. 5.14A). *Casuarius* (and possibly also *Palaeotis*), in contrast with other ratites, likewise has rather a short phalanx II1 compared with the length of III1 (Fig. 5.14B). "Struthioniforms" (now including *Struthio*) have a proportionately longer III2 compared with III1 than do moa other than *Megalapteryx* (Fig. 5.14C). Most "struthioniforms" have a relatively shorter combined length of phalanges IV3 and IV4, compared with the length of IV1, than do moa (Fig. 5.14D). Kiwi, in contrast, have a relatively long combined IV3 and IV4, and what is more, they fall on the same trend as *Megalapteryx*, yet another feature in which these two genera are similar. *Euryapteryx* and *Emeus* are different from other moa and more like extant "struthioniforms" in this comparison, due to the loss of one of the phalanges of digit IV (which for purposes of analysis we defined as IV4).

In a cluster analysis of phalangeal lengths and widths scaled to a common value of the length of phalanx III1 (Fig. 5.15) of large ground birds, members of the same genus usually plot in the same cluster or group of clusters, but at higher levels there are odd linkages. Although most moa genera group close together (*Emeus* and *Euryapteryx* together constitute one cluster within moa), one individual of *Megalapteryx* joins with *Apteryx* outside moa. *Ardeotis* is most like *Dromaius*, and a *Dromaius-Ardeotis-Rhea* cluster links with most moa apart from *Casuarius*.

Table 5.8. Forward stepwise discriminant (canonical variate) analysis of log-transformed phalanx lengths (including unguals) and widths (scaled by subtracting the mean of all the log-transformed variables employed in the analysis from the value of each log-transformed variable) in genera of living and extinct ratites. In contrast to the comparable principal components analysis (Table 5.6), the requirement that all specimens have a combined length of phalanges IV3 and IV4 of at least 10 mm was relaxed here. For all variables, the p-value of the F-statistic of the Wilks's lambda test was < 0.001, indicating significant differences in the means of all variables across taxa genera: *Rhea* (including *Pterocnemia*; $N = 10$); *Apteryx* ($N = 9$); *Casuarius* ($N = 9$); *Dromaius* ($N = 10$); *Anomalopteryx* ($N = 14$); *Megalapteryx* ($N = 2$); *Emeus* ($N = 4$); *Euryapteryx* ($N = 7$); *Pachyornis* ($N = 9$); *Dinornis* ($N = 22$). Unfortunately, the Box's M-test of the equality of group covariance matrices indicated lack of equal covariances ($p < 0.001$), and so the results of the analysis must be viewed with caution.

Variable[a]	Standardized Canonical Discriminant Function Coefficients (Correlations between Individual Variables and Discriminant Function)			
	Function 1	Function 2	Function 3	Function 4
III2 length	1.095 (0.592)*	0.123 (−0.198)	−0.273 (−0.015)	−0.196 (0.110)
Combined IV3 and IV4 length	0.597 (−0.124)	0.760 (0.631)*	−0.070 (−0.449)	0.178 (0.328)
II1 length	0.159 (0.083)	−0.703 (−0.528)*	−0.517 (−0.262)	0.320 (0.245)
III1 length	0.348 (0.192)	0.671 (−0.340)	0.977 (0.254)	0.897 (0.281)
IV2 length	0.286 (−0.097)	−0.043 (0.191)	0.772 (0.180)	0.735 (0.539)*
II3 length	0.882 (0.167)	0.424 (0.223)	0.743 (0.193)	−0.188 (−0.440)
III3 length	0.584 (0.224)	0.098 (0.240)	0.391 (−0.127)	−0.012 (−0.055)
II2 distal width	−0.023 (−0.357)	0.225 (−0.093)	0.379 (0.317)	−0.504 (−0.440)
III2 distal width	0.542 (−0.086)	−0.349 (−0.326)	0.591 (0.337)	0.371 (0.025)
IV1 length	0.268 (0.185)	−0.390 (−0.478)	0.261 (0.162)	−0.648 (0.052)
III4 length	0.869 (0.097)	−0.248 (−0.052)	0.096 (−0.148)	0.169 (−0.259)
Eigenvalue (% of variance)	35.767 (44.5)	24.075 (29.9)	12.684 (15.8)	4.103 (5.1)
Canonical correlation	0.986	0.980	0.963	0.897

[a]Variables were entered in the table in the order in which they were entered in the forward stepwise procedure.
*Largest absolute values of correlation between each variable and a discriminant function (several variables had their highest correlation with function 5 or higher).

Classification Results (errors in classification in bold)

Genus	Predicted Group Membership (Original Grouped Cases [Cross-Validated Grouped Cases])									
	R	Ap	C	Dr	An	M	Em	Eu	P	Di
Rhea (R)	10 (10)	0 (0)	0 (0)	0 (0)	0 (0)	0 (0)	0 (0)	0 (0)	0 (0)	0 (0)
Apteryx (Ap)	0 (0)	9 (8)	0 (0)	0 (0)	0 (0)	0 **(1)**	0 (0)	0 (0)	0 (0)	0 (0)
Casuarius (C)	0 (0)	0 (0)	9 (9)	0 (0)	0 (0)	0 (0)	0 (0)	0 (0)	0 (0)	0 (0)
Dromaius (Dr)	0 (0)	0 (0)	0 (0)	10 (10)	0 (0)	0 (0)	0 (0)	0 (0)	0 (0)	0 (0)
Anomalopteryx (An)	0 (0)	0 (0)	0 (0)	0 (0)	14 (14)	0 (0)	0 (0)	0 (0)	0 (0)	0 (0)
Megalapteryx (M)	0 (0)	0 **(1)**	0 (0)	0 (0)	0 (0)	2 (1)	0 (0)	0 (0)	0 (0)	0 (0)
Emeus (Em)	0 (0)	0 (0)	0 (0)	0 (0)	0 (0)	0 (0)	4 (3)	0 **(1)**	0 (0)	0 (0)
Euryapteryx (Eu)	0 (0)	0 (0)	0 (0)	0 (0)	0 (0)	0 (0)	0 **(2)**	7 (5)	0 (0)	0 (0)
Pachyornis (P)	0 (0)	0 (0)	0 (0)	0 (0)	0 (0)	0 (0)	0 (0)	0 (0)	8 (7)	**1 (2)**
Dinornis (Di)	0 (0)	0 (0)	0 (0)	0 (0)	0 (0)	0 (0)	0 (0)	0 (0)	0 (0)	22 (22)

99.0% of original cases correctly classified
92.7% of cross-validated cases correctly classified

In all of our analyses, *Emeus* and *Euryapteryx*, which are thought to be close relatives, plot close to each other in morphospace (Figs. 5.12–5.14). The same is true for *Dromaius* and *Casuarius*, which are also thought to be closely related, in some (Figs. 5.12B, 5.14C, D), but not all (Figs. 5.12A, C, D; 5.13; 5.14A, B) comparisons. Most "struthioniforms" can be distinguished from moa (Figs. 5.12A, B; 5.13A, B), although *Apteryx* is as much or more like *Megalapteryx* than other "struthioniforms."

Table 5.9. Forward stepwise discriminant (canonical variate) analysis of log-transformed lengths of the first two phalanges of each digit, and the distal widths of the second phalanx of each digit, all scaled by subtracting the mean of all the log-transformed variables employed in the analysis from the value of each log-transformed variable, in genera of living and extinct ratites. For all variables, the p-value of the F-statistic of the Wilks's lambda test was < 0.001, indicating significant differences in the means of all variables across genera: *Rhea* (including *Pterocnemia*; $N = 12$); *Apteryx* ($N = 10$); *Casuarius* ($N = 10$); *Dromaius* ($N = 10$); *Anomalopteryx* ($N = 18$); *Megalapteryx* ($N = 3$); *Emeus* ($N = 4$); *Euryapteryx* ($N = 8$); *Pachyornis* ($N = 12$); *Dinornis* ($N = 27$). Unfortunately, the Box's M-test of the equality of group covariance matrices indicated lack of equal covariances ($p < 0.001$), and so the results of the analysis must be viewed with caution.

Variable[a]	Standardized Canonical Discriminant Function Coefficients (Correlations between Individual Variables and Discriminant Function)			
	Function 1	Function 2	Function 3	Function 4
III2 length	0.992 (0.897)*	0.191 (0.312)	0.239 (0.139)	−0.151 (−0.255)
II2 length	0.108 (−0.273)	0.112 (0.189)	0.552 (0.785)*	0.603 (0.043)
II1 length	0.458 (0.313)	−0.845 (−0.638)*	0.545 (0.157)	0.902 (0.309)
IV2 length	0.286 (−0.135)	0.449 (0.564)	−0.100 (0.166)	1.281 (0.320)
III1 length	−0.143 (0.373)	0.873 (−0.052)	−0.309 (−0.359)	0.788 (0.209)
III2 distal width	0.546 (−0.029)	−0.210 (−0.210)	−0.398 (−0.532)	1.116 (0.075)
IV1 length	0.288 (0.363)	−0.405 (−0.317)	−0.395 (−0.359)	−0.196 (0.013)
Eigenvalue (% of variance)	28.380 (57.4)	8.795 (17.8)	8.236 (16.7)	3.151 (6.4)
Canonical correlation	0.983	0.948	0.944	0.871

[a]Variables were entered in the table in the order in which they were entered in the forward stepwise procedure.
*Largest absolute values of correlation between each variable and a discriminant function (some variables had their highest correlation with function 5 or higher)

Classification Results (errors in classification in bold)

Genus	Predicted Group Membership (Original Grouped Cases [Cross-Validated Grouped Cases])									
	R	Ap	C	Dr	An	M	Em	Eu	P	Di
Rhea (R)	12(12)	0 (0)	0 (0)	0 (0)	0 (0)	0 (0)	0 (0)	0 (0)	0 (0)	0 (0)
Apteryx (Ap)	0 (0)	10 (9)	0 (0)	0 (0)	0 (0)	0 **(1)**	0 (0)	0 (0)	0 (0)	0 (0)
Casuarius (C)	0 (0)	0 (0)	10 (10)	0 (0)	0 (0)	0 (0)	0 (0)	0 (0)	0 (0)	0 (0)
Dromaius (Dr)	0 (0)	0 (0)	0 (0)	10 (10)	0 (0)	0 (0)	0 (0)	0 (0)	0 (0)	0 (0)
Anomalopteryx (An)	0 (0)	0 (0)	0 (0)	0 (0)	17 (17)	0 (0)	0 (0)	0 (0)	0 (0)	**1 (1)**
Megalapteryx (M)	0 (0)	0 (0)	0 (0)	0 (0)	0 (0)	3 (3)	0 (0)	0 (0)	0 (0)	0 (0)
Emeus (Em)[a]	0 (0)	0 (0)	0 (0)	0 (0)	0 (0)	0 (0)	4 (3)	0 (0)	0 **(1)**	0 (0)
	0 (0)	0 (0)	0 (0)	0 (0)	0 (0)	0 (0)	3 (2)	0 (0)	**1 (1)**	0 **(1)**
Euryapteryx (Eu)[a]	0 (0)	0 (0)	0 (0)	0 (0)	0 (0)	0 (0)	0 **(2)**	7 (3)	**1 (3)**	0 (0)
	0 (0)	0 (0)	0 (0)	0 (0)	0 (0)	0 (0)	0 (0)	7 (5)	**1 (3)**	0 (0)
Pachyornis (P)[a]	0 (0)	0 (0)	0 (0)	0 (0)	0 (0)	0 (0)	0 **(1)**	**1 (1)**	8 (7)	**3 (3)**
	0 (0)	0 (0)	0 (0)	0 (0)	0 **(1)**	0 (0)	0 (0)	**1 (1)**	8 (7)	**3 (3)**
Dinornis (Di)[a]	0 (0)	0 (0)	0 (0)	0 (0)	0 (0)	0 (0)	0 **(1)**	0 (0)	0 (0)	27 (26)
	0 (0)	0 (0)	0 (0)	0 (0)	0 (0)	0 (0)	0 (0)	0 (0)	0 (0)	27 (27)

[a]Classification was done two ways. In the first method, the prior probability that a case belonged to a particular group was assumed to be equal for all groups. In the second method, the prior probability was weighted according to the observed proportions of cases in each group, which varied considerably across the various genera. Results for the second classification are reported only where they differ from those of the equal prior probability run.
Prior probabilities equal: 94.7% of original grouped cases were correctly classified; 87.1% of cross-validated grouped cases were correctly classified.
Prior probabilities weighted according to observed number of cases in each group: 93.9% of original grouped cases were correctly classified; 89.5% of cross-validated grouped cases were correctly classified.

Similarity in pedal proportions seems to be a useful, but not an infallible, proxy for phylogenetic propinquity.

NON-AVIAN THEROPODS As with birds, we ran two PCAs, one including unguals, and one without them (Tables 5.10, 5.11). As in previous PCAs, PC 1 accounts for the greatest part of data variance. Principal Component 2 in both versions of the PCA contrasts forms with a relatively long digit IV and long distal phalanges of digits II and III (positive), on the one hand, with forms, on the other, that have broad toes and relatively long proximal phalanges of digits II and III (negative; see the tables for loadings of specific phalanx measurements on the components). Principal Component 3 is also similar in the two PCA versions and reflects an interesting diagonal axis across the foot. Positive values of PC 3 are associated with relatively long non-ungual phalanges of digits II and III and also the first phalanx of digit IV, while negative values are associated with broad toes, long unguals, and long phalanges of digit IV beyond the first phalanx.

As with overall digit lengths, the more rococo morphologies are sported by small theropods (Fig. 5.16). *Deinonychus* and *Bambiraptor* show strongly positive values of PC 2. *Mononykus* is highly negative on PC 3, while *Nedcolbertia*, *Coelophysis*, *Tanycolagreus*, and *Chirostenotes* have highly positive values of PC 3. Large theropods again are conservative. Tyrannosaurid points surround those for *Allosaurus*, and *Dilophosaurus* is not far away.

The sample size of fairly complete large theropod phalangeal skeletons is small, even if analysis is restricted to the first two phalanges of digits II–IV (4 carnosaurs, 11 tyrannosauroids [specifically tyrannosaurids]). Most of the scaled phalangeal lengths and widths show no significant difference between carnosaurs and tyrannosauroids (Wilks's lambda test). Phalanx IV2, however, seems to be relatively larger in tyrannosauroids. This is readily apparent in a bivariate comparison of the lengths of phalanges IV1 and IV2 (Fig. 5.17; a bivariate comparison greatly increases the sample size of large theropod points).

Cluster analyses of phalanx lengths and widths scaled to a common III1 length (Fig. 5.18A) identify a cluster composed of *Allosaurus* and tyrannosaurids, but the tyrannosaurids do not cluster together apart from *Allosaurus*. Ornithomimids likewise form a distinct cluster, as do the two dromaeosaurids. Other smaller theropods link together in a manner not readily reconciled with their known phylogenetic relationships. If log-transformed phalangeal lengths and widths are scaled against the mean of log-transformed measurements (Fig. 5.18B), the topology is much the same, although now there is a bit less tendency for *Allosaurus* and the tyrannosaurids to link together apart from other theropods. If the two versions of scaled phalangeal dimensions are run with unguals excluded to increase the sample size, there is still a tendency for carnosaurs and tyrannosaurids, and for ornithomimids, to cluster together (Fig. 5.18C, D). If the analysis is restricted to carnosaurs and tyrannosaurids, the two groups do not completely separate from each other.

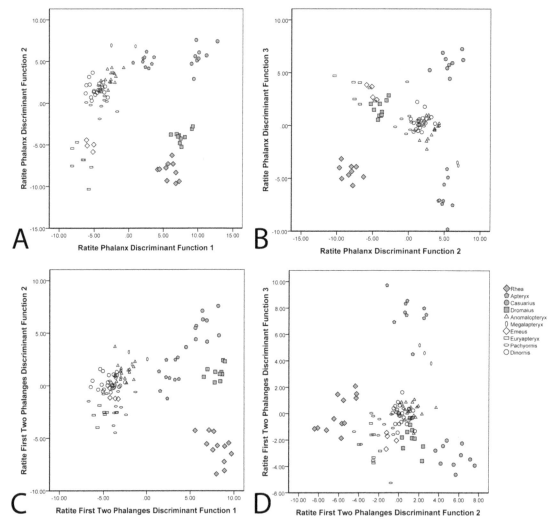

GROUND BIRDS, NON-AVIAN THEROPODS, AND ORNITHISCHIANS In our most inclusive analysis of foot shape, we compare phalangeal proportions in ground birds, NATs, and bipedal (or at least facultatively bipedal) ornithischian dinosaurs. The PCA (Table 5.12), after extracting the usual dominant first component, pulls out three additional shape components. Positive values of PC 2 are associated with relatively long phalanges from the middle portions of digits, while negative values correlate with relatively broad digits and (to a lesser extent) relatively long first phalanges and ungual phalanges (cf. Farlow and Chapman 1997). Positive values of PC 3 are mainly associated with relatively long proximal phalanges (especially of digit III), and negative values with relatively long distal phalanges and (to a lesser extent) broad digits. Positive values of PC 4 are associated primarily with relatively long unguals and a long digit III, and negative values with relatively long non-ungual phalanges of digit II.

There is remarkably little overlap among NATs, ground birds, and ornithischians when PC 3 is plotted against PC 2 (Fig. 5.19A). *Tenontosaurus*,

Leptoceratops, and the Proctor Lake ornithopod are well away from most other non-avian dinosaurs and birds, with highly negative values of PC 3, while *Genyornis* plots at the positive extreme of PC 3. Large theropods fall in a region of high positive values of PC 2 and intermediate values of PC 3; their nearest neighbors are ornithomimids, the moa *Dinornis*, and – oddly – the pheasant *Chrysolophus*. *Eoraptor*, *Conchoraptor*, and the basal euornithopod ?*Bugenasaura* are near kiwi (*Apteryx*), *Megalapteryx*, and *Aptornis*. *Nedcolbertia* plots among points for *Dinornis*, and near *Anomalopteryx* and *Cariama*. Hadrosaurids plot adjacent to *Iguanodon* and near points for *Casuarius*, *Pachyornis*, and *Euryapteryx*. Principal Component 4 (not shown) separates cassowaries (strongly positive) and *Genyornis* (strongly negative) from other birds and non-avian dinosaurs.

Canonical variate analysis (Table 5.13) yields astonishingly good separation of NATs, ground birds, and ornithischians (Fig. 5.19B). It is particularly noteworthy that all of the variable measurements selected by the forward stepwise procedure to create the functions are of proximal as opposed to distal phalanges. Unfortunately, simple characterization of the discriminating functions is not easy. Positive values of Discriminant Function 1 are associated with the lengths of III1, III2, II2, and IV2, and also the widths of II2 and IV2, while negative values are associated with the width of III2 and the length of IV1 (Table 5.13). Of these variables, the lengths of III1, IV2, and especially III2 show individual positive correlations with Discriminant Function 1, while the three phalanx widths show individual negative correlations with that function. Thus Discriminant Function 1 seems to be a hazy reflection of the same separation produced by PC 2 in the PCA (Table 5.12). It yields nearly complete separation between ornithischians, on the one hand, and NATs and birds, on the other (Fig. 5.19B).

Discriminant Function 2 is harder to interpret. It is strongly associated with the lengths of IV2, II2, and III1 and the widths of III2 and IV2 (positive direction) and the length of IV1 and the width of II2. Of these variables, the lengths of IV2 and II2 show respectable individual positive correlations with Discriminant Function 2, and the lengths of IV1 and III1 and the width of II2 show reasonable negative correlations. This second discriminant function is thus also reminiscent of PC 2 in the PCA. Non-avian theropods and ground birds separate fairly clearly along this function (Fig. 5.19B), with tyrannosaurids being especially different from birds along this axis.

In the CVA classification, ornithischians (Table 5.13) are almost never mistaken for birds (*Parksosaurus* is the genus in which such a mistaken classification generally occurs), and birds are almost never mistaken for ornithischians. Ground birds are seldom mistaken for non-avian theropods, but the latter are more frequently mistaken for birds. Non-avian theropods are never misclassified as ornithischians.

Bivariate plots (Fig. 5.20) illustrate the some of the discriminating variables with larger sample sizes. Although there is some overlap

5.13. Canonical variate analyses (CVAs) of phalanx lengths and widths of extant "struthioniforms" and moa. A, B) First three discriminant functions from a stepwise CVA using all scaled phalanx lengths (Table 5.8) and the scaled widths of the second phalanx of each digit. Positive values of Function 1 are associated particularly with long phalanges other than the first phalanges of each digit (particularly III2, III3, III4, II3, and the combined lengths of IV3 and IV4) and a broad III2. Note that Function 1 results in nearly complete separation of extant "struthioniforms" from moa. Positive values of Function 2 are especially associated with higher values of the combined length of phalanges IV3 and IV4 and the lengths of III1 and II3, and negative values especially with the lengths of II1, IV1, and III4. Positive values of Function 3 are associated with higher values of the lengths of most phalanges and the widths of phalanges II2 and III2. Negative values of Function 3 are associated with higher values of the lengths of II1 and III2. C, D) First three discriminant functions from a stepwise CVA using the scaled lengths of only the first two phalanges of digits II–IV and the scaled distal widths of the second phalanx of each digit (Table 5.9). Positive values of Function 1 are again most strongly associated with a long phalanx III2, and Function 1 continues to separate extant "struthioniforms" from moa. Positive values of Function 2 are associated with higher values of the lengths of phalanges III1 and IV2, and negative values of Function 2 with higher values of the length of phalanx II1 and IV1. Positive values of Function 3 are associated with higher values of the lengths of II1 and II2, and negative values of Function 3 with higher values of the lengths of phalanges IV1 and III1 and the width of phalanx III2.

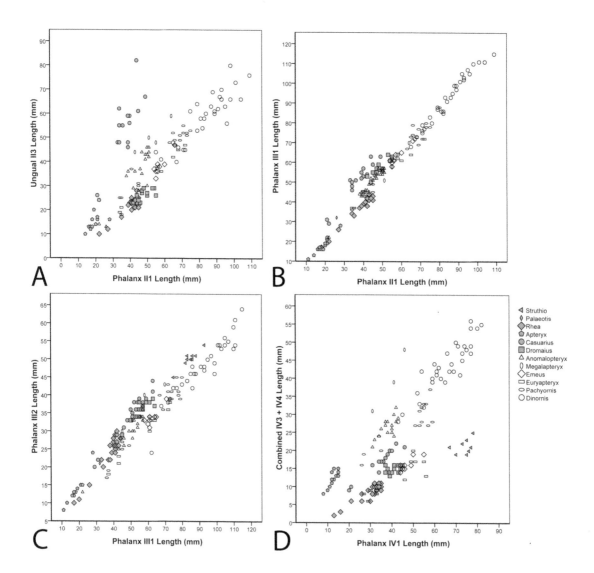

5.14. Bivariate comparisons of some of the phalangeal lengths identified as strongly contributing to the separation of genera in discriminant analyses of moa vs. "struthioniforms" (Fig. 5.13).

between groups, ornithischians (and especially iguanodonts and hadrosaurids) tend to have a relatively broader III2 for a given length of III2 than do ground birds, which in turn have a relatively broader III2 than do nonavian theropods. At large sizes the length of phalanx IV1 compared with that of III2 is greatest for ornithischians, intermediate for ground birds, and least for non-avian theropods. This comparison is less of a separator for smaller-bodied taxa, however.

Canonical variate analysis of NATs, ground birds, and ornithischians based on overall digit lengths and widths (Table 5.13) is not as effective in segregating the three categories as is CVA based on phalanx lengths and widths, but it nonetheless correctly assigns specimens to the three categories in almost 90 percent of cases. Discriminant Function 1 mainly reflects a contrast between the length of digit III and the width of phalanx III2 and so is readily reconciled with the CVA based on phalanx lengths and widths.

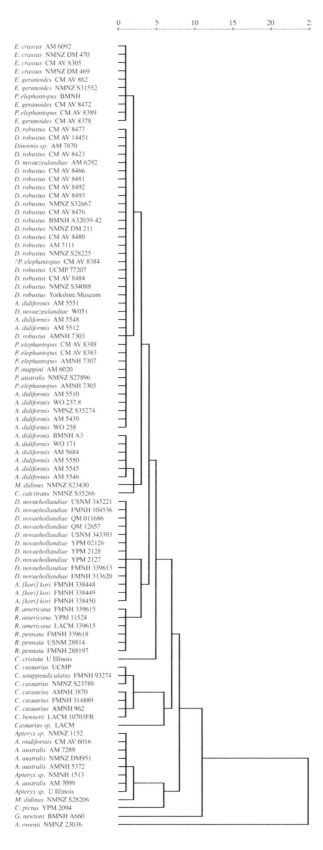

5.15. Cluster analysis of ground bird foot specimens with phalanx lengths and widths scaled to a common length of phalanx III1. Clusters were created using the between-groups linkage method, based on the squared Euclidean distance. Distances between clusters are expressed in terms of values between 0 and 25. A cluster analysis in which log-transformed phalanx lengths and widths are scaled by subtracting the mean of all log-transformed variables from each log-transformed variable (not shown) yields results fairly similar to those shown here.

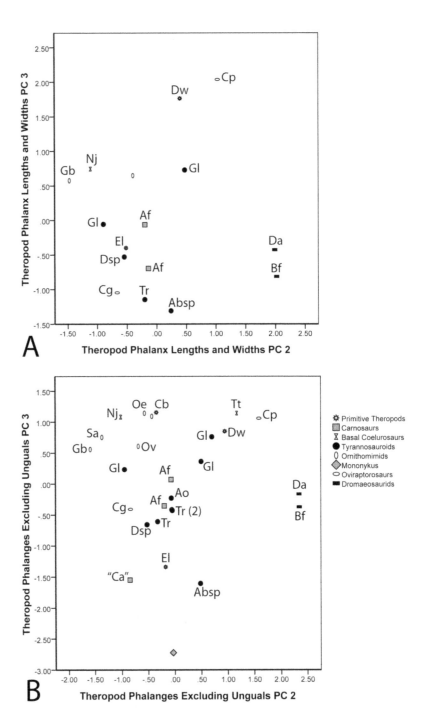

5.16. Principal components analysis of phalangeal lengths and widths in non-avian theropods, both including (A) and excluding (B) unguals (Tables 5.10, 5.11). Acronyms/ abbreviations labeling points as in Appendix 5.1. Principal Component 2 is a contrast between forms with a long digit IV and long distal phalanges of digits II and III (positive values) and forms with broad toes and relatively long proximal phalanges of digits II and III (negative values). Positive values of PC 3 relate to relatively long non-ungual phalanges of digits II and III and the first phalanx of digit IV, and negative values to broad toes, long unguals, and long digit IV phalanges beyond the first.

A cluster analysis of scaled phalanx lengths and widths that includes NATs, ground birds, and ornithischians creates a dendrogram that is too huge to reproduce here. However, its results are largely consistent with the already described PCAs and CVAs. Clusters that form early in the agglomeration schedule (indicating considerable similarity of specimens) include (*a*) most specimens of moa; (*b*) *Dromaius* + *Ardeotis*;

Parameter	PC 1 Loading	PC 2 Loading	PC 3 Loading
Phalanx II1 length	0.337 (0.970)	−0.067 (−0.192)	0.043 (0.125)
Phalanx II2 length	0.322 (0.984)	0.024 (0.073)	0.034 (0.105)
Phalanx II2 distal width	0.346 (0.970)	−0.072 (−0.201)	−0.034 (−0.095)
Phalanx II3 length	0.264 (0.938)	0.059 (0.210)	−0.047 (−0.167)
Phalanx III1 length	0.286 (0.993)	−0.012 (−0.041)	0.025 (0.087)
Phalanx III2 length	0.290 (0.983)	−0.033 (−0.110)	0.030 (0.101)
Phalanx III2 distal width	0.352 (0.978)	−0.070 (−0.195)	−0.020 (−0.056)
Phalanx III3 length	0.288 (0.982)	0.026 (0.090)	0.039 (0.132)
Phalanx III4 length	0.289 (0.987)	−0.002 (−0.007)	−0.021 (−0.070)
Phalanx IV1 length	0.290 (0.986)	0.031 (0.105)	0.025 (0.083)
Phalanx IV2 length	0.301 (0.976)	0.054 (0.174)	−0.009 (−0.029)
Phalanx IV2 distal width	0.362 (0.984)	−0.054 (−0.148)	−0.027 (−0.073)
Phalanx IV3 length	0.299 (0.961)	0.068 (0.217)	−0.014 (−0.045)
Phalanx IV4 length	0.271 (0.922)	0.102 (0.348)	0.005 (0.016)
Phalanx IV5 length	0.295 (0.990)	0.004 (0.015)	−0.024 (−0.081)
Eigenvalues (% variance)	1.419 (94.923)	0.042 (2.831)	0.012 (0.834)
Cumulative variance explained	94.923	97.754	98.588

Table 5.10. Principal components (PC) analysis (using a covariance matrix) of log-transformed measurements of phalangeal lengths and the distal widths of phalanges II2, III2, and IV2 in non-avian theropods ($N = 16$)

Note: Kaiser-Meyer-Olkin measure of sampling adequacy = 0.646; Bartlett's test of sphericity: $\chi^2 = 597.905$, $p < 0.001$.

Parameter	PC 1 Loading	PC 2 Loading	PC 3 Loading
Phalanx II1 length	0.347 (0.980)	−0.048 (−0.136)	0.039 (0.110)
Phalanx II2 length	0.342 (0.988)	0.034 (0.100)	0.015 (0.043)
Phalanx II2 distal width	0.372 (0.973)	−0.064 (−0.168)	−0.057 (−0.148)
Phalanx III1 length	0.313 (0.994)	−0.006 (−0.019)	0.029 (0.092)
Phalanx III2 length	0.314 (0.981)	−0.023 (−0.072)	0.051 (0.160)
Phalanx III2 distal width	0.375 (0.985)	−0.059 (−0.156)	−0.017 (−0.046)
Phalanx III3 length	0.304 (0.982)	0.020 (0.064)	0.047 (0.151)
Phalanx IV1 length	0.324 (0.990)	0.029 (0.088)	0.019 (0.058)
Phalanx IV2 length	0.335 (0.982)	0.053 (0.155)	−0.017 (−0.051)
Phalanx IV2 distal width	0.380 (0.986)	−0.043 (−0.112)	−0.037 (−0.096)
Phalanx IV3 length	0.319 (0.970)	0.067 (0.203)	−0.028 (−0.085)
Phalanx IV4 length	0.279 (0.942)	0.080 (0.270)	−0.028 (−0.095)
Eigenvalues (% variance)	1.347 (96.076)	0.028 (2.030)	0.015 (1.035)
Cumulative variance explained	96.076	98.106	99.141

Table 5.11. Principal components (PC) analysis (using a covariance matrix) of log-transformed measurements of phalangeal lengths (excluding the unguals) and the distal widths of phalanges II2, III2, and IV2 in non-avian theropods ($N = 27$)

Note: Kaiser-Meyer-Olkin measure of sampling adequacy = 0.899; Bartlett's test of sphericity: $\chi^2 = 967.819$, $p < 0.001$.

(c) *Rhea*; (d) most specimens of *Casuarius* (which do not join particularly early with clusters containing *Dromaius* apart from other birds); (e) most specimens of *Apteryx*; (f) most specimens of large theropods; (g) dromaeosaurids; (h) hadrosaurids + iguanodonts; and (i) *Tenontosaurus* + other small to medium-sized ornithischians. Non-avian theropods do not cluster together apart from birds; rather, particular NAT clusters link up with various bird clusters: ornithomimids and *Nedcolbertia* with the big moa cluster, *Dilophosaurus* and *Chirostenotes* with *Chrysolophus*, and large theropods with *Apteryx* and some specimens of *Anomalopteryx*

5.17. Comparison of the lengths of phalanges IV1 and IV2 in carnosaurs vs. tyrannosauroids (specifically tyrannosaurids). Tyrannosaurids have a relatively long phalanx IV2.

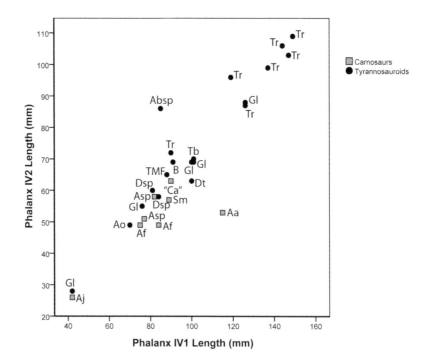

Table 5.12. Principal components (PC) analysis (using a covariance matrix) of log-transformed measurements of phalangeal lengths and the distal widths of phalanges II2, III2, and IV2 in non-avian theropods, large ground birds, and selected ornithischian dinosaurs (*N* = 143)

Parameter	PC 1 Loading	PC 2 Loading	PC 3 Loading	PC 4 Loading
Phalanx II1 length	0.248 (0.964)	−0.010 (−0.040)	0.054 (0.210)	−0.019 (−0.076)
Phalanx II2 length	0.241 (0.905)	0.083 (0.313)	−0.029 (−0.110)	−0.063 (−0.237)
Phalanx II2 distal width	0.335 (0.973)	−0.059 (−0.172)	−0.032 (−0.094)	−0.015 (−0.045)
Phalanx II3 length	0.250 (0.942)	−0.004 (−0.016)	−0.045 (−0.171)	0.070 (0.263)
Phalanx III1 length	0.230 (0.941)	0.002 (0.008)	0.078 (0.318)	0.003 (0.013)
Phalanx III2 length	0.167 (0.839)	0.057 (0.288)	0.080 (0.403)	0.032 (0.160)
Phalanx III2 distal width	0.343 (0.963)	−0.086 (−0.242)	0.004 (0.011)	−0.021 (−0.059)
Phalanx III3 length	0.184 (0.885)	0.084 (0.403)	0.011 (0.051)	0.024 (0.116)
Phalanx III4 length	0.247 (0.970)	−0.018 (−0.072)	−0.030 (−0.120)	0.031 (0.121)
Phalanx IV1 length	0.238 (0.953)	−0.016 (−0.065)	0.066 (0.262)	0.001 (0.005)
Phalanx IV2 length	0.231 (0.921)	0.082 (0.325)	0.012 (0.046)	−0.006 (−0.022)
Phalanx IV2 distal width	0.351 (0.979)	−0.064 (−0.178)	−0.013 (−0.037)	−0.023 (−0.064)
Combined lengths phalanges IV3 and IV4	0.231 (0.882)	0.098 (0.374)	−0.063 (−0.239)	−0.013 (−0.049)
Phalanx IV5 length	0.267 (0.975)	−0.012 (−0.044)	−0.039 (−0.141)	0.033 (0.122)
Eigenvalues (% variance)	0.946 (89.361)	0.049 (4.667)	0.030 (2.872)	0.014 (1.352)
Cumulative variance explained	89.361	94.029	96.901	98.253

Note: Kaiser-Meyer-Olkin measure of sampling adequacy = 0.923; Bartlett's test of sphericity: χ^2 = 5394.194, $p < 0.001$.

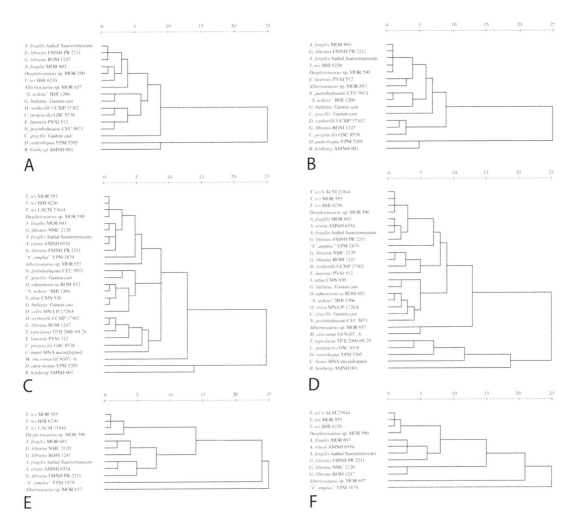

5.18. Cluster analyses of non-avian theropod foot specimens using scaled phalanx lengths and widths. Phalanx lengths and widths were scaled either to a common length of phalanx III1 (A, C, E) or by log-transforming each phalanx length and width, creating a scaling factor by computing the mean of the log-transformed phalanx lengths and widths and subtracting the scaling factor from each log-transformed phalanx length or width (B, D, F). Clusters were created either including lengths of unguals (A, B) or excluding unguals (C–F), the latter to increase sample sizes. Clusters were created using the between-groups linkage method, based on the squared Euclidean distance. A–D) All non-avian theropods in the sample. In all analyses carnosaurs and tyrannosaurids tend to cluster together apart from other non-avian theropods (particularly when all measurements are scaled to a common length of phalanx III1), but tyrannosaurids do not cluster together apart from carnosaurs. E–F) Carnosaurs and tyrannosaurids only.

and *Megalapteryx*. At a still higher (and statistically more problematic) linkage in the dendrogram, big ornithopods join the NAT and ground-bird cluster apart from *Tenontosaurus* and the smaller ornithischians. Thus the dendrogram does not do a particularly good job of reflecting phylogenetic propinquity.

Digit I does not figure in any of our analyses and is variably developed in ground birds, NATs, and bipedal ornithischians. In *Tenontosaurus* and most basal euornithopods, digit I is relatively long, and in some of these dinosaurs it should routinely have impressed in footprints. In large theropods and moa, digit I is present but so small that it would either not

5.19. Comparison of phalangeal proportions of non-avian theropods, ground birds, and selected ornithischian dinosaurs. A) Principal components analysis. Positive values of PC 2 are associated with relatively long phalanges from the middle portions of the digits, while negative values are associated with relatively broad phalanges and relatively long first (most proximal) phalanges. Principal Component 3 is a contrast between forms with relatively long proximal phalanges (positive) and forms with relatively long distal phalanges and relatively broad phalanges (negative). Individual points or groups of points associated with particular species or genera of birds and dinosaurs are labeled; acronyms/abbreviations as in Appendix 5.1. B) Canonical variate analysis. Ornithischians are almost completely segregated from non-avian theropods and ground birds, but there is some overlap between the latter two groups.

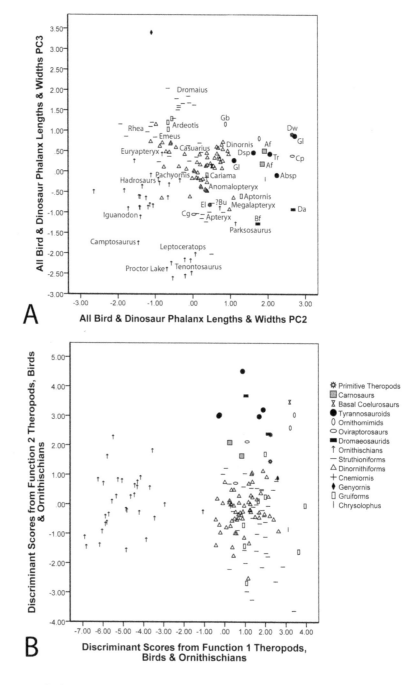

touch the ground or leave a very small impression. Digit I is entirely lost in most ornithomimosaurs, many ground birds, *Iguanodon*, and hadrosaurids. Although the relative development of the hallux is a character that could potentially assist in assigning some footprints to their makers, it seems to be of little value in determining which bipedal dinosaurs are most closely related.

Our results allow some generalizations about the extent to which foot shape is a reliable clue to group membership. The results are mixed. It is encouraging that feet of members of the same avian genus tend to be more like each other than other bird genera and that feet of the closely related moa *Emeus* and *Euryapteryx* are very similar in shape. It is equally encouraging that PCAs and CVAs are able to distinguish moa from most "struthioniforms" and ground birds, NATs, and ornithischians from each other.

On the one hand, it is also nice for ichnologists that we don't even have to have complete pedal skeletons in order to make a reasonable guess as to their systematic affinities. Tridactyl dinosaur footprints often provide information about digit lengths, but the lengths of individual phalanges can be interpreted only when tracks show good digital pad impressions. Our results indicate that, although an ability to interpret complete foot skeletons from tracks allows the greatest discrimination of trackmakers, the ability to interpret a few distinctive phalanges (and particularly the larger, more proximal phalanges), and even overall digit lengths, may often be enough to allow a fair chance at systematic interpretation.

That's the good news.

On the other hand, the moa *Megalapteryx*, as its name might suggest, has a foot shape about as much like that of *Apteryx* as like those of other moa. The bird whose foot is most consistently similar to that of the emu is the non-ratite kori bustard (*Ardeotis*), and not *Casuarius*. Nor should we forget the ostrich *Struthio*, whose didactyl foot is arguably more specialized than those of other ratites. If all we knew of ostriches and other big birds were their feet (or footprints), we doubt that anyone would suspect that *Casuarius* and *Struthio* are more closely related to *Dromaius* than is *Ardeotis*.

The more limited data for NATs tell a similar story. From the presently available information, one would be hard-pressed to detect a pattern in theropod foot shapes that approximates relationships discerned from cladistic analysis of the entire theropod skeleton. Indeed, the most striking result is the great similarity in overall foot shape among very large theropods, regardless of their relationships (cf. Farlow and Chapman 1997; Farlow 2001). It is possible, of course, that if we had sample sizes of large theropod specimens as large as those for extant and recently extinct ratites that we would be able to distinguish between feet of carnosaurs and tyrannosaurids as well as we can between "struthioniforms" and moa and to discriminate among genera of large theropods as well as we can do for large ground birds. But our data for NATs suggests that there is greater variability in foot shape among small and medium-sized NATs than among clades of large theropods. If so, our success in discriminating among foot skeletons of genera of ground birds may reflect the fact that these modern theropods are likewise much smaller, and so potentially more variable pedally, animals than the Mesozoic giants.

Phalanx Length and Width Treatment:

Variable[a]	Standardized Canonical Discrimination Coefficients (Correlations between Individual Variables and Discriminant Function)	
	Function 1	Function 2
III2 length	0.674 (0.566)*	0.013 (−0.103)
II2 length	0.218 (0.061)	0.308 (0.479)*
III1 length	1.520 (0.336)	0.310 (−0.359)*
III2 distal width	−1.746 (−0.399)*	0.382 (−0.175)
IV2 distal width	1.287 (−0.350)*	1.333 (−0.076)
II2 distal width	0.226 (−0.379)*	−1.475 (−0.259)
IV1 length	−0.576 (0.174)	−1.001 (−0.470)*
IV2 length	0.177 (0.189)	0.548 (0.513)*
Eigenvalue (% of variance)	6.427 (88.2)	0.858 (11.8)
Canonical correlation	0.930	0.680

Digit Length and Width Treatment:

Variable[a]	Function 1	Function 2
Digit III Length	2.450 (0.750)*	1.815 (−0.151)
IV2 Distal Width	1.466 (−0.465)*	−0.124 (0.095)
II2 Distal Width	0.636 (−0.452)	1.818 (−0.495)*
III2 Distal Width	−0.233 (−0.559)*	1.070 (0.361)
Eigenvalue (% of variance)	2.858 (90.0)	0.319 (10.0)
Canonical correlation	0.861	0.492

[a]Variables were entered in the table in the order in which they were entered in the forward stepwise procedure.
*Largest absolute correlation between each variable and any discriminant function.

The kinds of phenetic information commonly used to define tridactyl dinosaur ichnotaxa (footprint lengths and widths, digit lengths, phalangeal pad dimensions, interdigital angles, projection of digit III beyond the limits of digits II and IV) should be useful in making minimum estimates of the number of different kinds of trackmakers in local ichnofaunas (cf. Farlow and Pianka [2000] for varanid lizards). Footprint shapes may also provide clues to the kinds of local and even regional trackmakers, particularly if skeletal specimens are available for comparison. However, tridactyl footprint shapes are not completely reliable for assessing how closely related the trackmakers were. While it may be possible to say something about the degree of relatedness of trackmakers based on similarities of footprint shape within a particular fauna, it is not safe to assume that similar tridactyl footprints from different regions or times were made by closely related dinosaurs.

Our cautionary remarks about identifying tridactyl trackmakers seem particularly apropos for footprints of large non-avian theropods, but the news isn't entirely bleak. Farlow (2001:417–421) concluded that "pedal phalangeal skeletons of large ceratosaurs, allosaurs, and tyrannosaurs are indistinguishable." Our updated phalangeal shape analyses remain consistent with that result (apart from the possible difference between

Group	Treatment	Predicted Group Membership (Original Grouped Cases [Cross-Validated Grouped Cases])		
		Non-avian Theropods	Birds	Ornithischians
Non-avian Theropods	Phalanx Lengths and Widths	15 (15)	**1 (1)**	0 (0)
		13 (8)	**3 (8)**	0 (0)
	Digit Lengths and Widths	12 (11)	**4 (5)**	0 (0)
		6 (2)	10 (14)	0 (0)
Birds	Phalanx Lengths and Widths	**2 (6)**	105 (101)	0 (0)
		1 (0)	106 (107)	0 (0)
	Digit Lengths and Widths	14 (15)	93 (92)	0 (0)
		4 (0)	103 (107)	0 (0)
Ornithischians	Phalanx Lengths and Widths	0 (0)	**1 (1)**	31 (31)
		0 (0)	1 (1)	31 (31)
	Digit Lengths and Widths	0 (0)	**1 (2)**	31 (30)
		0 (0)	**3 (5)**	29 (27)

[a]Classification was done two ways. In the first method, the prior probability that a case belonged to a particular group was assumed to be equal for all groups. In the second method, the prior probability was weighted according to the observed proportions of cases in each group, which varied considerably across the three groups.
Prior probabilities equal: Phalanx lengths and widths treatment: 97.4% of original grouped cases correctly classified; 94.8% of cross-validated grouped cases correctly classified. Digit lengths and widths treatment: 87.7% of original cases correctly classified; 85.8% of cross-validated cases correctly classified.
Prior probabilities weighted according to observed number of cases in each group: Phalanx lengths and widths treatment: 96.8% of original grouped cases correctly identified; 94.2% of cross-validated grouped cases correctly identified. Digit lengths and widths treatment: 89.0% of original cases correctly classified; 87.7% of cross-validated cases correctly classified.

carnosaurs and tyrannosaurids in the relative lengths of phalanges IV1 and IV2; see Fig. 5.17), but our data on relative metatarsal lengths suggest that Farlow (2001) missed a pedal "forest" as he focused on the phalangeal "trees" therein. At least some tyrannosaurids differ from carnosaurs in having an elongate MT IV relative to MTs II and III (Figs. 5.4–5.5). This suggests that tyrannosaurid footprints might differ from those of carnosaurs in having a digit IV impression that projects farther forward (distally) from the back of the print than that of digit II and possibly a digit III impression that does not project as far beyond the tips of digits II and IV as in carnosaurs (Fig. 5.21).

McCrea et al. (2005) described tridactyl dinosaur footprints from the Late Cretaceous Dinosaur Park Formation of Alberta, one of which (TMP 81.34.1) is figured here (Fig. 5.1E). McCrea and his colleagues identified TMP 81.34.1 as a natural cast of a right footprint of a large ornithopod, most likely a hadrosaur. The print indeed shows features seen in many prints attributed to large ornithopods: a relatively short "free" length of digit III compared with the length of the "palm" of the print, a fairly broad digit III impression compared with its free length, and indentations along both sides of the print "heel." However, the ornithopod prints to which TMP 81.34.1 is most similar are Early and early Late Cretaceous ichnotaxa like *Amblydactylus* and *Caririchnium*. Unambiguous Late Cretaceous hadrosaur prints commonly show exaggeratedly stout digit impressions that have extremely short free lengths (Langston 1960; Currie

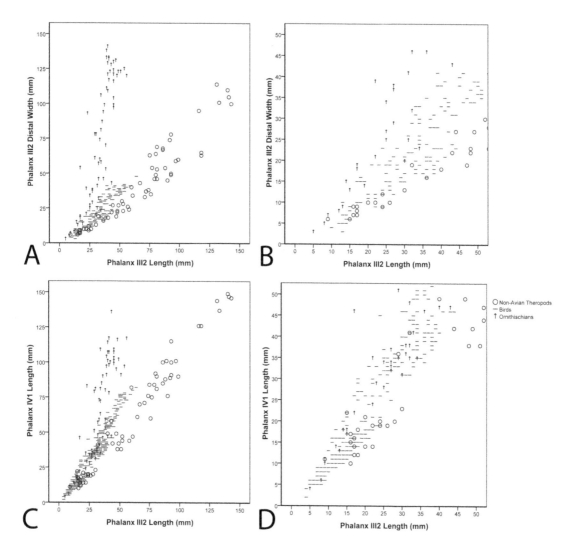

5.20. Bivariate separation of pedal proportions in non-avian theropods (NATs), ground birds, and ornithischian dinosaurs. A) Phalanx III2 distal width vs. length. B) Same as (A), but showing smaller forms. Ornithischians tend to have a relatively broader III2 than do ground birds or NATs. Non-avian theropods tend to have a narrower III2 than do birds, but there is considerable overlap between the two groups. C, D) Ornithischians tend to have a relatively long IV1 compared with III2 than do birds and NATs, especially at large sizes. Non-avian theropods tend to have a relatively shorter IV1 than do birds.

et al. 1991; Carpenter 1992), features that can readily be rationalized with hadrosaurid phalangeal skeletons.

With a total length of about 56 cm, TMP 81.34.1 is the right size to be a hadrosaur footprint. However, the digit impressions, stout as they are, do not show the extreme relative breadth (compared with their free lengths) we would expect in a hadrosaur track; the impressions of digits II and IV in particular seem too narrow, and the impression of digit IV looks like it may terminate in a narrow claw mark.

We therefore suggest that TMP 81.34.1 could in fact be a poorly preserved tyrannosaurid footprint. A second Dinosaur Park Formation print (TMP 93.36.282), comparable in size to TMP 81.34.1 but with poorer preservation of the distal portions of impressions of digits III and IV, may be an even better candidate for a tyrannosaurid print (McCrea et al. 2005) in that it has relatively narrower digit impressions. In both footprints the free length of the digit IV impression is noticeably longer than that of digit II. Carpenter (1992:fig. 5C) illustrated a footprint from the "Mesaverde"

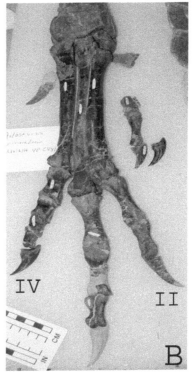
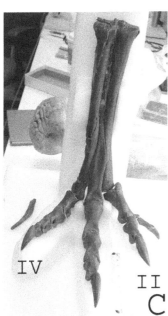
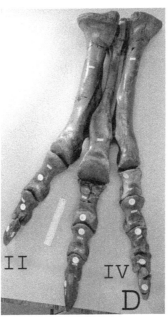

5.21. Foot skeletons of carnosaurs (A, B) and tyrannosaurids (C–E); digits II and IV labeled on each. A) Cast of *Allosaurus fragilis* (Aathal Sauriermuseum). B) *Allosaurus* sp. (UMNH VP C481). C) *Gorgosaurus libratus* (FMNH PR 2211). D) *Albertosaurus* sp. (MOR 657). E) *Daspletosaurus* sp. (MOR 590). Note the longer forward projection of digit IV than of digit II in the tyrannosaurids as compared with *Allosaurus*.

Group that he attributed to a tyrannosaurid and which likewise shows a longer free length of digit IV than of II. However, Manning et al. (2008) described a likely tyrannosaurid footprint from the Hell Creek Formation in which, contrary to our suggestion, the digit II impression projects slightly farther forward than does the digit IV impression. Additional specimens of carnosaur and tyrannosaurid foot skeletons, and especially of footprints potentially assignable to tyrannosaurids, are needed to establish how consistently our hypothesized correlation between an osteological and an ichnological character of large theropod footprints holds up.

Several authors (Olsen 1995; Farlow and Chapman 1997; Olsen et al. 1998; Wilson and Carrano 1999; Carrano and Wilson 2001; Farlow 2001; Wilson 2005) have suggested that a cladistic approach to identifying trackmakers might be superior to overall phenetic comparisons of the kind attempted in this paper. Indeed, the cladistic approach has proved successful in interpreting the identities of sauropod trackmakers and in tracing the temporal origin of key sauropodomorph synapomorphies (Wilson and Carrano, 1999; Wilson, 2005).

The list of synapomorphies potentially recognizable in trace fossils presented by Carrano and Wilson (2001:appendix 1) includes characters that might be used to identify theropod prints (e.g., tridactyl shape, presence of claw marks, features of the digit I impression that might register in resting traces). However, most of the synapomorphies listed by Carrano and Wilson (2001) for identifying different clades within theropods require that the trackmaker sit on the ground and make integument impressions; the only feature potentially associated with a normal footprint made by a NAT was the short, clawless impression of digit II expected for paravians.

We have identified one, and possibly two, pedal skeletal features that might register in trace fossils and therefore distinguish carnosaur from tyrannosaurid footprints (relative lengths of digit II and IV free lengths; relative lengths of phalanges IV1 and IV2). Coria et al. (2002:fig. 4) illustrated what might be a distinctive feature of the foot of *Aucasaurus:* proportionally shorter unguals (compared with other pedal phalanges) than in other large theropods. It is possible, then, that a cladistic analysis of pedal proportions in ground birds and NATs would provide a better matchup between theropod clades and footprint shapes than our phenetic analyses.

Functional Considerations

Phalangeal proportions of birds have been shown to differ between arboreal and terrestrial species (Hopson 2001; Zhou and Farlow 2001), indicating that phalangeal proportions can carry a strong functional signal. Some of the similarities in foot shape between non-avian dinosaurs and ground birds (Fig. 5.19) may be morphological coincidences of no particular functional significance, but some might be worth pursuing. It seems intuitively plausible, for example, that kiwi, some small NATs, and some small ornithopods might have engaged in similar motions of the foot during locomotion or foraging that could have resulted in similar phalangeal proportions.

Carnosaurs and tyrannosaurids tend to group together in this analysis, just as they did in all previous comparisons of digital or phalangeal proportions. We speculate that gigantism in large theropods constrained possibilities for variability in phalangeal shape, either directly or indirectly. Direct constraints could have involved common adaptations for weight support (related to withstanding impact against the ground? cf. Alexander et al. [1986]) that precluded great variability in phalangeal shape. However, this hypothesis is weakened by the very different pedal proportions from those of large theropods seen in *Iguanodon* and hadrosaurs (cf. Moreno et al. 2007), dinosaurs just as heavy as large theropods. Indirect constraints include the possibility that large theropods, because of the need to keep their balance during locomotion and other activities (cf. Farlow et al. 1995, 2000), or mechanical and energetic difficulties in engaging in unusually athletic activities (Hutchinson and Garcia 2002), would have been less likely to evolve novel ways of attacking prey that employed their feet than would smaller theropods (cf. Holtz 2003; Fastovsky and Smith 2004). The hypothesis that large theropods were constrained to less variable foot shapes than were smaller forms can be tested by the discovery of additional foot specimens of the big hunters, of as many different clades as possible (cf. Coria et al. 2002; Carrano 2007; Maganuco et al. 2008).

Acknowledgments

We thank curators in museums around the world for access to specimens in collections under their care, Richard McCrea for discussions about dinosaur tracks from the Dinosaur Park Formation, Yvonne Zubovic for statistical advice, and Matt Bonnan and Mike Parrish for critical comments that led to great improvements of this chapter. Jim Whitcraft assisted in the creation of illustrations. This research was supported by grants from the National Science Foundation to JOF.

Literature Cited

Alexander, R. McN., M. B. Bennett, and R. F. Ker. 1986. Mechanical properties and function of the paw pads of some mammals. *Journal of Zoology* 209:405–419.

Anderson, A. 1989. *Prodigious Birds: Moas and Moa-Hunting in Prehistoric New Zealand.* Cambridge University Press, Cambridge.

Baird, D. 1957. Triassic reptile footprint faunules from Milford, New Jersey. *Bulletin of the Museum of Comparative Zoology* 17:449–519.

Baker, A. J., L. J. Huynen, O. Haddrath, C. D. Millar, and D. M. Lambert. 2005. Reconstructing the tempo and mode of evolution in an extinct clade of birds with ancient DNA: the giant moas of New Zealand. *Proceedings of the National Academy of Science* USA 102:8257–8262.

Bunce, M., T. H. Worthy, T. Ford, W. Hoppitt, E. Willerslev, A. Drummond, and A. Cooper. 2003. Extreme reversed sexual size dimorphism in the extinct New Zealand moa *Dinornis. Nature* 425:172–175.

Bunce, M., T. H. Worthy, M. J. Phillips, R. N. Holdaway, E. Willerslev, J. Haile, B. Shapiro, R. P. Scofield, A. Drummond, P. J. J. Kamp, and A. Cooper. 2009. The evolution of the extinct ratite moa and New Zealand Neogene paleogeography.

Proceedings of the National Academy of Sciences USA 106:20646–20651.

Carpenter, K. 1992. Behavior of hadrosaurs as interpreted from footprints in the "Mesaverde" Group (Campanian) of Colorado, Utah, and Wyoming. *Contributions to Geology, University of Wyoming* 29:81–96.

Carr, T. D., and T. E. Williamson. 2010. *Bistahieversor sealeyi*, gen. et sp. nov., a new tyrannosauroid from New Mexico and the origin of deep snouts in Tyrannosauroidea. *Journal of Vertebrate Paleontology* 30:1–16.

Carrano, M. T. 2007. The appendicular skeleton of *Majungasaurus crenatissimus* (Theropoda: Abelisauridae) from the Late Cretaceous of Madagascar. *Journal of Vertebrate Paleontology* 27(2, suppl.):163–179.

Carrano, M. T., and J. A. Wilson. 2001. Taxon distributions and the tetrapod track record. *Paleobiology* 27:564–582.

Chiappe, L. M., and L. M. Witmer (eds.). 2002. *Mesozoic Birds: Above the Heads of Dinosaurs.* University of California Press, Berkeley, California.

Cooper, A. 1997. Studies of avian ancient DNA: from Jurassic Park to modern island extinctions; pp. 345–373 in D. P. Mindell (ed.), *Avian Molecular Evolution and Systematics.* Academic Press, San Diego, California.

Cooper, A., C. Lalueza-Fox, S. Anderson, A. Rambaut, J. Austin, and R. Ward. 2001. Complete mitochondrial genome sequences of two extinct moas clarify ratite evolution. *Nature* 409:704–707.

Cooper, A., C. Mourer-Chauviré, G. K. Chambers, A. von Haeseler, A. C. Wilson, and S. Pääbo. 1992. Independent origins of New Zealand moas and kiwis. *Proceedings of the National Academy of Sciences* USA 89:8741–8744.

Coria, R. A., L. M. Chiappe, and L. Dingus. 2002. A new close relative of *Carnotaurus sastrei* Bonaparte 1985 (Theropoda: Abelisauridae) from the Late Cretaceous of Patagonia. *Journal of Vertebrate Paleontology* 22:460–465.

Cracraft, J. 2001. Avian evolution, Gondwana biogeography and the Cretaceous-Tertiary mass extinction event. *Proceedings of the Royal Society of London* B 268:459–469.

Cracraft, J., F. K. Barker, M. Braun, J. Harshman, G. J. Dyke, J. Feinstein, S. Stanley, A. Cibois, P. Schikler, P. Beresford, J. García-Moreno, M. D.

Sorenson, T. Yuri, and D. P. Mindell. 2004. Phylogenetic relationships among modern birds (Neornithes): towards an avian tree of life; pp. 468–489 in J. Cracraft and M. J. Donoghue (eds.), *Assembling the Tree of Life.* Oxford University Press, Oxford.

Currie, P. J., D. Badamgarav, and E. B. Koppelhus. 2003. The first Late Cretaceous footprints from the Nemegt locality in the Gobi of Mongolia. *Ichnos* 10:1–12.

Currie, P. J., G. C. Nadon, and M. G. Lockley. 1991. Dinosaur footprints with skin impressions form the Cretaceous of Alberta and Colorado. *Canadian Journal of Earth Sciences* 28:102–115.

Currie, P. J., E. B. Koppelhus, M. A. Shugar, and J. L. Wright (eds.). 2004. *Feathered Dragons: Studies on the Transition from Dinosaurs to Birds.* Indiana University Press, Bloomington.

Dalla Vecchia, F. M., A. Tarlao, G. Tunis, and S. Venturini. 2000. New dinosaur track sites in the Albian (Early Cretaceous) of the Istrian Peninsula (Croatia). *Memorie di Scienze Geologiche* 52:193–292.

Dyke, G. J., and M. Van Tuinen. 2004. The evolutionary radiation of modern birds (Neornithes): reconciling molecules, morphology and the fossil record. *Zoological Journal of the Linnean Society* 141:153–177.

Ericson, P. G. P., C. L. Anderson, T. Britton, A. Elzanowski, U. S. Johansson, M. Källersjo, J. I. Ohlson, T. J. Parsons, D. Zuccon, and G. Mayr. 2006. Diversification of Neoaves: integration of molecular sequence data and fossils. *Biology Letters* 2:543–547.

Fain, M. G., and P. Houde. 2004. Parallel radiations in the primary clades of birds. *Evolution* 58:2558–2573.

Farlow, J. O. 2001. *Acrocanthosaurus* and the maker of Comanchean large-theropod footprints; pp. 408–427 in D. H. Tanke and K. Carpenter (eds.), *Mesozoic Vertebrate Life: New Research Inspired by the Paleontology of Philip J. Currie.* Indiana University Press, Bloomington.

Farlow, J. O., and R. E. Chapman. 1997. The scientific study of dinosaur footprints; pp. 519–553 in J. O. Farlow and M. K. Brett-Surman (eds.), *The Complete Dinosaur.* Indiana University Press, Bloomington.

Farlow, J. O., and P. M. Galton. 2003. Dinosaur trackways of Dinosaur State Park, Rocky Hill, Connecticut;

pp. 248–263 in P. M. Letourneau and P. E. Olsen (eds.), *The Great Rift Valleys of Pangea in Eastern North America,* vol. 2: *Sedimentology, Stratigraphy, and Paleontology.* Columbia University Press, New York.

Farlow, J. O., and M. G. Lockley. 1993. An osteometric approach to the identification of the makers of early Mesozoic tridactyl dinosaur footprints; pp. 123–131 in S. G. Lucas and M. Morales (eds.), *The Nonmarine Triassic. New Mexico Museum of Natural History and Science Bulletin,* no. 3. New Mexico Museum of Natural History and Science, Albuquerque.

Farlow, J. O., and E. R. Pianka. 2000. Body form and trackway pattern in Australian desert monitors (Squamata: Varanidae): comparing zoological and ichnological diversity. *Palaios* 15:235–247.

Farlow, J. O., J. B. Smith, and J. R. Robinson. 1995. Body mass, bone "strength indicator," and cursorial potential in *Tyrannosaurus rex. Journal of Vertebrate Paleontology* 15:713–725.

Farlow, J. O., S. M. Gatesy, T. R. Holtz, Jr., J. R. Hutchinson, and J. M. Robinson. 2000. Theropod locomotion. *American Zoologist* 40:640–663.

Farlow, J. O., W. Langston, Jr., E. E. Deschner, R. Solis, W. Ward, B. L. Kirkland, S. Hovorka, T. L. Reece, and J. Whitcraft. 2006. *Texas Giants: Dinosaurs of the Heritage Museum of the Texas Hill Country.* Heritage Museum of the Texas Hill Country, Canyon Lake, Texas, 105 pp.

Fastovsky, D. E., and J. B. Smith. 2004. Dinosaur paleoecology; pp. 614–626 in D. B. Weishampel, P. Dodson, and H. Osmólska (eds.), *The Dinosauria,* 2nd ed. University of California Press, Berkeley.

Feduccia, A. 1996. *The Origin and Evolution of Birds.* Yale University Press, New Haven, Connecticut.

García-Ramos, J. C., L. Piñuela, and J. Lires. 2004. *Guía del Jurásico de Asturias: Rutas por los Yacimientos de Huellas de Dinosaurios.* Zinco Comunicación, Gijóon, Asturias, Spain.

Gauthier, J., and L. F. Gall (eds.). 2001. *New Perspectives on the Origin and Early Evolution of Birds.* Yale Peabody Museum, New Haven, Connecticut.

Gierliński, G. 1995. *Śladami Polskich Dinozaurów.* Polska Oficyna Wydawnicza, Warsaw.

Glut, D. F. 1997. *Dinosaurs: The Ency-clopedia.* McFarland & Co., Jefferson, North Carolina.

Hackett, S. J., R. T. Kimball, S. Reddy, R. C. K. Bowie, E. L. Braun, M. J. Braun, J. L. Chojnowski, W. A. Cox, K.-L. Han, J. Harshman, C. J. Hud-dleston, B. D. Marks, K. J. Miglia, W. S. Moore, F. H. Sheldon, D. W. Steadman, C. C. Witt, and T. Yuri. 2008. A phylogenomic study of birds reveals their evolutionary history. *Science* 320:1763–1768.

Haddrath, O., and A. J. Baker. 2001. Complete mitochondrial DNA ge-nome sequences of extinct birds: rat-ite phylogenetics and the vicariance biogeography hypothesis. *Proceed-ings of the Royal Society of London* B 268:939–945.

Harshman, J., E. L. Braun, M. J. Braun, C. J. Huddleston, R. C. K. Bowie, J. L. Chojnowski, S. J. Hackett, K.-L. Han, R. T. Kimball, B. D. Marks, K. J. Miglia, W. S. Moore, S. Reddy, F. H. Sheldon, D. W. Steadman, S. J. Step-pan, C. C. Witt, and T. Yuri. 2008. Phylogenomic evidence for multiple losses of flight in ratite birds. *Pro-ceedings of the National Academy of Sciences* USA 105:13462–13467.

Holtz, T. R., Jr. 2003. Dinosaur preda-tion: evidence and ecomorphol-ogy; pp. 325–340 in P. H. Kelley, M. Kowalewski, and T. A. Hansen (eds.), *Predator-Prey Interactions in the Fossil Record.* Kluwer Academic/Plenum, New York.

Hopson, J. A. 2001. Ecomorphology of avian and nonavian theropod phalangeal proportions: implica-tions for the arboreal vs. terrestrial origin of bird flight; pp. 211–235 in J. Gauthier and L. F. Gall (eds.), *New Perspectives on the Origin and Early Evolution of Birds: Proceedings of the International Symposium in Honor of John H. Ostrom.* Peabody Museum of Natural History, New Haven, Connecticut.

Houde, P., A. Cooper, E. Leslie, A. E. Strand, and G. A. Montaño. 1997. Phylogeny and evolution of 12S rdNA in Gruiformes (Aves); pp. 121–158 in D. P. Mindell (ed.), *Avian Molecular Evolution and Systematics.* Academic Press, San Diego.

Hutchinson, J. R., and M. Garcia. 2002. *Tyrannosaurus* was not a fast runner. *Nature* 415:1018–1021.

Huynen, L., C. D. Millar, R. P. Sco-field, and D. M. Lambert. 2003. Nuclear DNA sequences detect species limits in ancient moa. *Nature* 425:175–178.

Johnsgard, P. A. 1991. *Bustards, Hemi-podes, and Sandgrouse: Birds of Dry Places.* Oxford University Press, Oxford.

Kvale, E. P., G. D. Johnson, D. L. Mick-elson, K. Keller, L. C. Furer, and A. A. Archer. 2001. Middle Jurassic (Bajo-cian and Bathonian) dinosaur mega-tracksites, Bighorn Basin, Wyoming, U.S.A. *Palaios* 16:233–254.

Langston, W., Jr. 1960. A hadrosaurian ichnite. *National Museum of Canada Natural History Papers,* no. 4:1–9.

Lee, K., J. Feinstein, and J. Cracraft. 1997. The phylogeny of ratite birds: resolving conflicts between molecular and morphological data sets; pp. 173–211 in D. P. Mindell (ed.), *Avian Molecular Evolution and Systematics.* Academic Press, San Diego.

Leonardi, G. 1989. Inventory and statistics of the South American dinosaurian ichnofauna and its paleobiological interpretation; pp. 165–178 in D. D. Gillette and M. G. Lockley (eds.), *Dinosaur Tracks and Traces.* Cambridge University Press, Cambridge.

Livezey, B. C., and R. L. Zusi. 2001. Higher-order phylogenetics of mod-ern Aves based on comparative anat-omy. *Netherlands Journal of Zoology* 51:179–205.

Livezey, B. C., and R. L. Zusi. 2007. Higher-order phylogeny of modern birds (Theropoda, Aves: Neornithes) based on comparative anatomy. II. Analysis and discussion. *Zoologi-cal Journal of the Linnean Society* 149:1–95.

Lockley, M. G. 1991. *Tracking Dino-saurs: A New Look at an Ancient World.* Cambridge University Press, Cambridge.

Lockley, M. G., and A. P. Hunt. 1995. *Dinosaur Tracks and Other Fossil Footprints of the Western United States.* Columbia University Press, New York.

Lockley, M. G., and C. Meyer. 2000. *Di-nosaur Tracks and Other Fossil Foot-prints of Europe.* Columbia University Press, New York.

Maganuco, S., A. Cau, and G. Pasini. 2008. New information on the abelisaurid pedal elements from the Late Cretaceous of NW Madagascar (Mahajanga Basin). *Atta della Soci-età Italiana di Scienze Naturali e del Museo Civico di Storia Naturale in Milano* 149:239–252.

Manning, P. L., C. Ott, and P. L. Falking-ham. 2008. A probable tyrannosau-rid track from the Hell Creek Forma-tion (Upper Cretaceous), Montana, United States. *Palaios* 23:645–647.

Martínez, R. N., P. C. Sereno, O. A. Alc-ober, C. E. Colombi, P. R. Renne, I. P. Montañez, and B. S. Currie. 2011. A basal dinosaur from the dawn of the dinosaur era in southwestern Pangaea. *Science* 331:206–210.

Mayr, G., A. Manegold, and U. S. Johansson. 2003. Monophyletic grounds within "higher land birds" – comparison of morphological and molecular data. *Journal of Zoo-logical Systematics and Evolutionary Research* 41:233–248.

McBrayer, L. D., and C. E. Corbin. 2007. Patterns of head shape variation in lizards: morphological correlates of foraging mode; pp. 271–301 in S. M. Reilly, L. D. McBrayer, and D. B. Miles (eds.), *Lizard Ecology: The Evo-lutionary Consequences of Foraging Mode.* Cambridge University Press, Cambridge.

McCrea, R. T., P. J. Currie, and S. G. Pemberton. 2005. Vertebrate ichnol-ogy; pp. 405–416 in P. J. Currie and E. B. Koppelhus (eds.), *Dinosaur Provincial Park: A Spectacular An-cient Ecosystem Revealed.* Indiana University Press, Bloomington.

Milner, A. R. C., and S. Z. Spears. 2007. Mesozoic and Cenozoic Paleoichnol-ogy of Southwestern Utah. 2007 Annual Meeting, Rocky Mountain Section, Geological Society of America (field trip). Utah Geological Association Publication no. 35. Utah Geological Association Publication, St. George.

Moratalla, J. J., J. Hernán, and S. Jimé-nez. 2003. Los Cayos dinosaur track-site: an overview on the Lower Creta-ceous ichno-diversity of the Cameros Basin (Cornago, La Rioja Province, Spain). *Ichnos* 10:229–240.

Moreno, K., M. T. Carrano, and R. Sny-der. 2007. Morphological changes in pedal phalanges through ornithopod dinosaur evolution: a biomechanical approach. *Journal of Morphology* 268:60–63.

Murray, P. F., and P. Vickers-Rich. 2004. *Magnificent Mihirungs: The Colos-sal Flightless Birds of the Australian Dreamtime.* Indiana University Press, Bloomington.

Olsen, P. E. 1995. A new approach for recognizing track makers. *Geological Society of America, Abstracts with Programs* 27:72.

Olsen, P. E., J. B. Smith, and N. G. Mc-Donald. 1998. Type material of the type species of the classic theropod footprint genera *Eubrontes, Anchisauripus,* and *Grallator* (Early Jurassic, Hartford and Deerfield Basins, Connecticut and Massachusetts, U.S.A.). *Journal of Vertebrate Paleontology* 18:586–601.

Padian, K. 2004. Basal Avialae; pp. 210–231 in D. B. Weishampel, P. Dodson, and H. Osmólska (eds.), *The Dinosauria,* 2nd ed. University of California Press. Berkeley.

Paul, G. S. 2006. Turning the old into the new: a separate genus for the gracile iguanodont from the Wealden of England, pp. 69–77 in K. Carpenter (ed.), *Horns and Beaks: Ceratopsian and Ornithopod Dinosaurs.* Indiana University Press, Bloomington.

Paul, G. S. 2008. A revised taxonomy of the iguanodont dinosaur genera and species. *Cretaceous Research* 29:192–216.

Pérez-Lorente, F. (ed.). 2003. *Dinosaurios y Otros Reptiles Mesozoicos en España.* Gobierno de La Rioja, Universidad de La Rioja, Spain.

Rainforth, E. C. 2007. Ichnotaxonomic updates from the Newark Supergroup; pp. 49–57 in E. C. Rainforth (ed.), *Contributions to the Paleontology of New Jersey (II): Field Guide and Proceedings, Geological Association of New Jersey 24th Annual Conference and Field Trip.* East Stroudsburg University, East Stroudsburg, Pennsylvania.

Schult, M. F., and J. O. Farlow. 1992. Vertebrate trace fossils; pp. 34–63 in C. G. Maples and R. R. West (eds.), *Trace Fossils,* Paleontological Society Short Course, no. 5, Knoxville, Tennessee.

Shapiro, B., D. Sibthorpe, A. Rambaut, J. Austin, G. M. Wragg, O. R. P. Bininda-Emonds, P. L. M. Lee, and A. Cooper. 2002. Flight of the dodo. *Science* 295:1683.

Sibley, C. G., and B. L. Monroe, Jr. 1990. *Distribution and Taxonomy of Birds of the World.* Yale University Press, New Haven, Connecticut.

Smith, J. B., and J. O. Farlow. 2003. Osteometric approaches to trackmaker assignment for the Newark Supergroup ichnogenera *Grallator, Anchisauripus,* and *Eubrontes;* pp. 273–292 in P. M. LeTourneau and P. E. Olsen (eds.), *The Great Rift Valleys of Pangea in Eastern North America,* vol. 2: *Sedimentology, Stratigraphy, and Paleontology.* Columbia University Press, New York.

Sullivan, C., D. W. E. Hone, T. D. Cope, L. Yang, and L. Jun. 2009. A new occurrence of small theropod tracks in the Houcheng (Tuchengzi) Formation of Hebei Province, China. *Vertebrata Palasiatica* 47:35–52.

Thulborn, T. 1990. *Dinosaur Tracks.* Chapman & Hall, London.

Weishampel, D. B., P. Dodson, and H. Osmólska (eds.). 2004. *The Dinosauria,* 2nd ed. University of California Press, Berkeley.

Wilson, J. A. 2005. Integrating ichnofossil and body fossil records to estimate locomotor posture and spatiotemporal distribution of early sauropod dinosaurs: a stratocladistic approach. *Paleobiology* 31:400–423.

Wilson, J. A., and M. T. Carrano. 1999. Titanosaurs and the origin of "wide-gauge" trackways: a biomechanical and systematic perspective on sauropod locomotion. *Paleobiology* 25:252–267.

Worthy, T. H. 2005. Rediscovery of the types of *Dinornis curtus* Owen and *Palapteryx geranoides* Owen, with a new synonymy (Aves: Dinornithiformes). *Tuhinga* 16:33–43.

Worthy, T. H., and R. N. Holdaway. 2002. *The Lost World of the Moa: Prehistoric Life of New Zealand.* Indiana University Press, Bloomington.

Zhou, Z., and J. O. Farlow. 2001. Flight capability and habits of *Confuciusornis;* pp. 237–254 in J. Gauthier and L. F. Gall (eds.), *New Perspectives on the Origin and Early Evolution of Birds: Proceedings of the International Symposium in Honor of John H. Ostrom.* Peabody Museum of Natural History, Yale University, New Haven, Connecticut.

Zou, H., and L. Niswander. 1996. Requirement for BMP signaling in interdigital apoptosis and scale formation. *Science* 272:738–741.

Non-avian Theropods

Basal theropods: *Eoraptor lunensis* (El; considered a basal sauropodomorph by Martínez et al. 2011), *Herrerasaurus ischigualastensis*, *Guiabasaurus candelariensis*

Basal neotheropods: *Coelophysis bauri* (Cb), *Dilophosaurus wetherilli* (Dw), *Velocisaurus unicus*

Ceratosaurs: *Ceratosaurus nasicornis* (Cn), *Majungasaurus crenatissimus* (Mc)

Note: In some graphs basal theropods, basal neotheropods, and ceratosaurs are combined into a category labeled as "primitive" theropods.

Spinosauroids: *Megalosaurus bucklandi* (Mb), *Piatnitzkysaurus floresi* (Pf), *Chilantaisaurus tashuikouensis* (Ct)

Carnosaurs: *Allosaurus fragilis* (Af), *Allosaurus* sp. (Asp, Aj), "*Camptosaurus [Camptonotus] amplus*" ("Ca"; possibly = A. *fragilis* [Glut, 1997]), *Acrocanthosaurus atokensis* (Aa), *Saurophaganax maximus* (Sm)

Basal coelurosaurs: *Compsognathus longipes*, *Nedcolbertia justinhofmanni* (Nj), *Ornitholestes hermanni*, *Nqwebasaurus thwazi*, *Juravenator starki*, *Tanycolagreus topwilsoni* (Tt), *Dilong paradoxus* (Note: In graphs of metatarsal lengths the last two taxa are coded as basal tyrannosauroids.)

Tyrannosauroids: *Appalachiosaurus montgomeriensis* (Am), *Alectrosaurus olseni* (Ao), *Albertosaurus sarcophagus* (As), *Albertosaurus* sp. (Absp), unnamed tyrannosaurid from the Two Medicine Formation of Montana (TMF), *Gorgosaurus libratus* (Gl), *Daspletosaurus torosus* (Dt), *Daspletosaurus* sp. from the Kirtland Formation of New Mexico (NMT) (now *Bistahieversor sealeyi*; Carr and Williamson 2010), other *Daspletosaurus* species, *Tarbosaurus bataar* (Tb), *Tyrannosaurus rex* (Tr, possibly including the Burpee tyrannosaurid "Jane," B)

Ornithomimosaurs: *Harpymimus okladnikovi*, *Garudimimus brevipes*, *Ornithomimus edmontonicus* (Oe), *Ornithomimus velox* (Ov), *Ornithomimus* sp., *Struthiomimus altus* (Sa), Hell Creek Formation ornithomimid ("*Struthiomimus sedens*"), *Archaeornithomimus asiaticus*, *Gallimimus bullatus* (Gb)

Oviraptorosaurs: *Avimimus portentosus*, *Chirostenotes pergracilis* (Cp), *Oviraptor philoceratops*, *Ingenia yanshini*, *Conchoraptor gracilis* (Cg), *Elmisaurus elegans*, *Elmisaurus rarus*

Troodontids: *Troodon formosus*, *Tochisaurus nemegtensis*

Dromaeosaurids: *Achillobator giganticus*, *Deinonychus antirrhopus* (Da), *Bambiraptor feinbergi* (Bf)

Alvarezsaurids: *Mononykus olecranus*

Appendix 5.1: Breakdown of avian and non-avian dinosaur taxa analyzed in this paper, with a key to abbreviations used to identify taxa on graphs

Birds

"Struthioniforms": *Dromaius novaehollandiae*, *Rhea americana*, *R.* (or *Pterocnemia*) *pennata*, *Casuarius casuarius*, *C. unappendiculatus*, *C. bennetti. Apteryx australis, A. owenii, Palaeotis weigelti*

Dinornithiforms: *Dinornis robustus, D. novaezealandiae, Emeus crassus, Euryapteryx curtus, Euryapteryx gravis* (formerly *E. geranoides* [Worthy, 2005; and now regarded as conspecific with *E. curtus*; Bunce et al. 2009]), *Anomalopteryx didiformis, Megalapteryx didinus, Pachyornis geranoides* (formerly *P. mappini* [Worthy, 2005]), *P. australis, P. elephantopus*

Galliforms: *Chrysolophus pictus*

Anseriforms: *Diatryma gigantea, Genyornis newtoni, Cnemiornis calcitrans*

Gruiforms: *Cariama cristata, Aptornis otidiformis, Ardeotis kori, Otis tarda, Eupodotis senegalensis, Neotis heuglinii*

Columbiforms: *Raphus cucullatus*

Ornithischians

Basal euornithopods: ?*Bugenasaura infernalis* (?Bu), *Parksosaurus warreni*, Proctor Lake ornithopod, "*Laosaurus minimus*"

Tenontosaurus tilletti, Tenontosaurus sp.

Camptosaurus dispar, Camptosaurus sp.

Iguanodon: *I. bernissartensis, I. atherfieldensis* (Paul [2006, 2008] assigns this second species to a new genus, *Mantellisaurus*.)

Hadrosaurids: *Brachylophosaurus canadensis, Edmontosaurus regalis, E. annectens, Lophorothon atopus, Corythosaurus casuarius, Lambeosaurus lambei, Hypacrosaurus altispinus, Saurolophus osborni*, unidentified juvenile lamebeosaurine

Basal ceratopsians: *Psittacosaurus meileyingensis, Montanoceratops* sp., *Leptoceratops gracilis*

Note from the authors: During the long interval between the time when this paper was written and the time when it was finally copyedited, there were a number of changes in nomenclature and/or classification of some of the taxa in the preceding list. The authors are aware of these changes, some of which are indicated above, but think the main points made by our paper are unaffected by them.

6.1. Left lateral view of endo-cranial cast of wild *Alligator mississippiensis*. Skull length 34.3 cm, estimated body length, 2.6 m. Scale bar equals 1 cm.

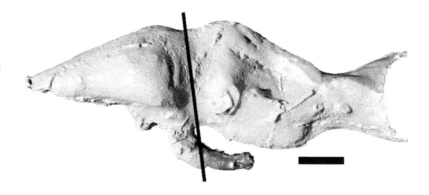

Relative Size of Brain and Cerebrum in Tyrannosaurid Dinosaurs: An Analysis Using Brain-Endocast Quantitative Relationships in Extant Alligators

Grant R. Hurlburt, Ryan C. Ridgely, and Lawrence M. Witmer

Abstract

Brain and cerebrum mass are estimated from endocasts of three tyrannosaurid taxa (*Tyrannosaurus rex*, *Gorgosaurus*, and *Nanotyrannus*) using morphological and quantitative brain-endocast relations in a size series of sexually mature alligators (*Alligator mississippiensis*). The alligator size series (N = 12) ranged from the smallest sexually mature size to the largest size commonly encountered. Alligator brain mass (MBr) increased regularly with increasing body mass, while the ratio of brain mass to endocast volume (MBr:EV) declined regularly from 67 percent to 32 percent. The ratio of cerebrum mass to cerebrocast was 38 percent in the largest alligators and regularly exceeded the MBr:EV ratio by 5.6 percent. For estimates from endocasts of non-avian dinosaurs of unknown sex, a MBr:EV ratio of 37 percent was used, the mean of the ratio of the largest male and female alligators. A corresponding 42 percent ratio was used for the cerebrum-cerebrocast ratio.

Relative brain size was measured as Encephalization Quotients (EQs) based on brain-body relations in extant non-avian reptiles (REQs) and birds (BEQs). *Tyrannosaurus rex* has the relatively largest brain of all adult non-avian dinosaurs, excepting certain small maniraptoriforms (*Troodon*, *Bambiraptor*, and *Ornithomimus*), which are well within the extant bird relative brain size range. The relative brain size of *T. rex* is within the range of extant non-avian reptiles and, at most, 2 standard deviations (SDs) above the mean of non-avian reptile log REQs, which are normally distributed. *Gorgosaurus* REQs overlapped the lower end of the *T. rex*. Log BEQs of all theropods, excepting small maniraptoriforms, were well below the range of extant birds. *Nanotyrannus* log REQs were anomalously high for an adult, but the difference between *Nanotyrannus* log REQs and *T. rex* values paralleled the difference between log REQs of the smallest subadult and largest alligators. *Nanotyrannus* cerebrum:brain ratios were also consistent with those of an older juvenile or youngest subadult. Cerebrocast:endocast ratios of the three *T. rex* endocasts ranged from 41.1 to 43.5 percent, and cerebrum mass:brain mass (MCb:MBr) ratios range from 47.5 to 49.53 percent, more than the lowest ratios for extant birds (44.6 percent) but very close to ratios (45.9–47.9 percent) typical of the smallest sexually mature alligators. In

Carcharodontosaurus saharicus, these ratios were 37.1 percent and 42.1 percent, respectively, the latter essentially identical to actual MCb:MBr ratios (40.76–42.91 percent) of the two largest alligators. Although the relative brain size of *Carcharodontosaurus* (SGM-Din 1), was approximately two thirds that of *T. rex*, the MCb:MBr ratio of the former was only 5.5–7.5 percent less than that of *T. rex*.

Introduction

Endocasts (endocranial casts) are natural or artificial casts made from the endocranial (or brain or cranial) cavity of vertebrates, or they exist as virtual endocasts produced by laser or X-ray computed tomography (CT scans). The external morphology of an endocast corresponds to the external surface of the dura mater, of which the surface topography reflects contained sinuses and underlying blood vessels.

Excepting species in which the brain filled the cranial cavity, the endocasts of non-avian dinosaurs strongly resemble those of crocodilians in general proportions and specific anatomical features. Among non-avian dinosaurs, the brain apparently filled the cranial cavity in the following taxa: pachycephalosaurs, small theropods, *Archaeopteryx*, and the hypsilophodont *Leaellynasaura* (Russell 1972; Hopson 1979; Nicholls and Russell 1981; Rich and Vickers-Rich 1988; Currie and Zhao 1993; Osmólska 2004; Evans 2005). This is inferred because the skull surface, and thus the corresponding endocast, reproduces the contours of the gross brain divisions (cerebrum, midbrain, cerebellum) and, in some cases, the blood vessels of the brain surface, as in extant birds (Iwaniuk and Nelson 2002), pterosaurs (Hopson 1979; Witmer et al. 2003), and most mammals (Hurlburt 1982, and references cited therein). In most other dinosaurs, the endocranial surface does not bear impressions of brain divisions or the cerebral (vs. dural) blood vessels, indicating that the brain either did not fill the endocranial cavity or, at most, contacted the endocranial surface at the lateral poles of the cerebrum, as in hadrosaurs (Evans 2005). These endocasts resemble those of extant crocodilians, in which the brain does not fill the cranial cavity (Hopson 1979). In the largest alligators, the only brain parts that contact the endocranial walls are the lateral poles of the cerebrum (Hurlburt, unpublished results). This resemblance makes endocasts of crocodilians excellent models for the brain-endocast relationship in most dinosaurs.

Brain-Endocast Relations in Extant Non-avian Reptiles

Brain volume traditionally has been estimated for non-avian reptiles using a brain-mass:endocast volume ratio (MBr:EV) of 0.5; that is, the brain occupies 50 percent of the brain cavity. This ratio is based on an observation of the *Sphenodon* brain (Dendy 1910) and the MBr:EV ratio in one *Iguana* specimen (Jerison 1973). Although a very rough approximation (and one re-evaluated here), it provided a productive starting point (Hopson 1977, 1980). Larsson et al. (2000) made an important contribution in analyzing relative size of the brain and cerebrum in several theropods and comparing them to non-avian reptiles and birds. Their approach made use of laser-scan data and analyzed brain and cerebrum size in a phylogenetic context. They tested, for the first time, the hypothesis of increasing size of the cerebrum relative to the rest of the brain in a phylogenetic context and also pioneered the study of brain division scaling in extinct taxa.

Relative Brain Size and Encephalization Quotients

Brain and body mass are highly correlated, as exemplified by the high correlation coefficients of equations in this paper. Relative brain size is the size of the brain compared to the size of the body, usually measured as body mass. It has been used to infer cognitive capacity and thermoregulatory mode in extinct vertebrates (Jerison 1973; Hopson 1977, 1980) and associated with complex cognitive behavior in birds and mammals (Lefebvre et al. 2002; Marino 2002). Because the specific gravity of the brain is approximately unity (Jerison 1973; Hurlburt 1982), brain size can be expressed as mass or volume. A commonly used measure of relative brain size is the Encephalization Quotient (EQ; Jerison 1973), which is the ratio of brain mass to a predicted brain mass, obtained from the brain-body equation of a reference group, such as non-avian reptiles or birds (Jerison 1973). Jerison (1973) was the first to state that dinosaurs had brains of the size expected for non-avian reptiles of their body mass. He supported this statement with graphical illustrations of brain-body relations in reptiles and dinosaurs, although four of his ten dinosaurs actually had LVEQs (Lower Vertebrate EQs; discussed below) less than those of the least encephalized reptile (Hurlburt 1996). His EQs were based on two equations, both with a slope (b) of 0.67, and with intercepts (a) of 0.007 and 0.07 for "lower" and "higher" vertebrates, respectively (Jerison 1973). He fitted the 0.67 slope "by eye" to the brain-body point scatter because it is the coefficient relating volume to surface area. He considered it a "theoretical" slope, but it is more properly termed a "hypothetical" slope. Hurlburt (1996) developed Reptile EQs (REQs), and Bird EQs (BEQs) using species-based brain-body equations. The reduced major axis regression (RMA) brain-body equation for non-avian reptiles by species ($N = 62$) is

$$\log \mathrm{MBr} = -1.810 + (0.553 \times \log \mathrm{MBd}), \tag{1}$$

where MBd is body mass, $r = 0.9616$, and the 95 percent confidence limits (CL) of b are 0.5150 and 0.5915. The RMA brain-body equation for birds by species ($N = 174$) is

$$\log \mathrm{MBr} = -0.930 + (0.590 \times \log \mathrm{MBd}), \tag{2}$$

where $r = 0.9355$ and the 95 percent CL of b are 0.5578 and 0.6213. These equations were based on much larger and more taxonomically comprehensive samples than those of Jerison (1973). The 95 percent confidence

**Institutional Abbre-
viations** AMNH, American
Museum of Natural History,
New York; BMNH, The Natural
History Museum, London; CM,
Carnegie Museum of Natural
History; CMNH, Cleveland
Museum of Natural History,
Cleveland; FMNH, Field
Museum of Natural History,
Chicago; NMC, National
Museums of Canada, Ottawa;
KUVP, Kansas University Natu-
ral History Museum, Lawrence;
ROM, Royal Ontario Museum,
Toronto; RTMP, Royal Tyrrell
Museum of Paleontology,
Drumheller, Alberta; SGM,
Ministere de l'Energie et des
Mines, Rabat, Morocco; UUVP,
University of Utah, Salt Lake
City.

limits (95 percent CL) of both equations exclude 0.67, constituting clear evidence that the 0.67 slope is inappropriate and falsifying Jerison's hypothesis. Moreover, the categories of birds (the monophyletic Aves) and the grade of non-avian reptiles are more appropriate than the categories of "lower vertebrates" (fish, amphibians, and reptiles) and "higher" vertebrates (birds and mammals) used by Jerison (1973). The bird slope is steeper than the reptile slope but within the 95 percent CL of the reptile slope. The bird intercept is approximately 10 times that of the non-avian reptile slope (Hurlburt 1996).

The corresponding EQ formulae (based on species-level equations) are

$$REQ = MBr/(0.0155 \times MBd0.553) \tag{3}$$

and

$$BEQ = MBr/(0.117 \times MBd0.590), \tag{4}$$

where $antilog_{10} -1.810 = 0.0155$; $antilog_{10} -0.930 = 0.117$. The RMA equation is appropriate for data with unequal variation in x and y variables, as is typical of brain and body data (Sokal and Rohlf 1981). From this point on, the term "non-avian reptiles" refers to extant non-avian reptiles.

The purposes of this study are (1) to determine three regression equations relating (*a*) log endocranial volume (EV), (*b*) log brain mass (MBd), and (*c*) REQ to TL (snout to tail-tip length) in a size series of alligators; (2) to determine ratios of brain-mass:endocast-volume (MBr:EV) and cerebrum mass:cerebrocast-volume (MCb:CbcV) in a size series of alligators; (3) to determine REQ and log REQ ranges of reptiles and BEQ and log BEQ ranges of birds; (4) to estimate endocranial and cerebrocast volume from actual or virtual endocasts of several large theropod dinosaurs and to estimate MBr and MCb from dinosaur endocasts using MBr:EV and MCb:CbcV ratios in alligators; (5) to compare relative brain size of dinosaurs with that of reptiles and birds, using REQs and BEQs; (6) to calculate ratios of cerebrum to total brain size in dinosaurs and compare these ratios to those of reptiles and birds; (7) to discuss methods of analyzing brain size, including methods of obtaining volumes from dinosaur endocasts; and (8) to test hypotheses regarding evolution of brain size.

Methods and Materials

The fossil specimens consisted of three taxa of the Late Cretaceous tyrannosaurids *Tyrannosaurus rex* (AMNH 5029, AMNH 5117, FMNH PR 2081), *Gorgosaurus libratus* (ROM 1247), and "*Nanotyrannus lancensis*" (CMNH 7541), as well as two allosauroids, the late Jurassic allosaurid *Allosaurus* (UUVP 294) and the late Cretaceous carcharodontosaurid *Carcharodontosaurus* (SGM-Din 1). In addition, EQs were calculated from endocast data for *Archaeopteryx* (BMNH 37001) and three small theropods: *Bambiraptor* (KUVP 129737), *Ornithomimus* (NMC 12228), and *Troodon* (RTMP 86.36.457 and RTMP 79.8.1). To provide a context for the theropod data, we also provide data for other dinosaur taxa. The relations of brain to endocast and of cerebrum to cerebrocast were determined from a size series of 12 sexually mature alligators (*Alligator mississippiensis*), of which half were wild and half were pen-raised domestic animals.

Volumetric Relations between Brain and Endocast and between Cerebrum and Cerebrocast in *Alligator mississippiensis*

In the alligator sample, MBd ranged from 11.3 to 276.9 kg, TL ranged from 1.613 to 3.810 m, and MBr ranged from 4.47 to 10.51 g (Hurlburt and Waldorf 2002). Although alligators have been known to reach TL slightly exceeding 4 m, generally there is little growth after 3.5 m TL in males and 2.6 m TL in females. Alligators can be sexually mature at 1.60 cm (Woodward et al. 1991). Thus, the sample ranged from the smallest sexually mature individual – that is, the smallest subadult – to the largest commonly encountered size. It thus constitutes a useful comparison sample for studying relations between relative brain size and ontogenetic age in extinct, non-avian archosaurs.

Brains were removed immediately postmortem in the alligators and weighed within 10–30 minutes, following removal of the olfactory tracts. Brain weight included pia mater but excluded the pituitary gland, dura mater, arachnoid, grossly visible blood vessels, and any dried blood, which occurred between the meninges in some specimens. Brains were dissected into gross divisions (cerebrum, cerebellum, optic lobes [= tectum], and brain stem, including diencephalons), which were then weighed. The first three divisions were cut off from the brain stem in a horizontal plane. Divisions were fixed in 10 percent formalin. The skulls were cleaned, and endocasts were made by applying successive layers of latex to the skull and calvarium. Because the specific gravity of brain tissue approximates unity (one), brain volume and mass are used interchangeably. Volumes of alligator endocasts were determined by suspending endocasts from an electronic balance, once in air and once immersed in water. The difference in the two masses equals the mass (g) of water displaced by the volume of the cast. Because the specific gravity of water is one, this mass equaled the volume in milliliters (Alexander 1985).

Limits and landmarks on alligator and dinosaur endocasts were chosen to correspond to the brain portion of the endocast. The limits were anteriorly, the point where the cerebrum narrows to meet the olfactory tract, and posteriorly, the stump of the hypoglossal nerve (XII). Endocast portions beyond these limits were removed, as were foramen fillings corresponding to nerves and blood vessels. Cerebrocast volumes were determined by suspending the cast with the water line at the posterior cerebrocast boundary line and again subtracting the wet mass from the dry mass.

In alligators, the cerebrocast boundary line was somewhat oblique, from rostrodorsal to caudoventral (Fig. 6.1). The line always fell just at the posterior contact of the cast of the infundibulum connecting to the pituitary. The cerebrocast (endocast portion corresponding to the

Other Terminology and Abbreviations 95 percent CL, 95 percent confidence limits; BEQ, Bird Encephalization Quotient based on species-level equations; CbcV, cerebrocast volume (ml); cerebrocast, endocast portion corresponding to cerebrum; DGI, Double Graphic Integration; EV, endocast volume (ml); HVEQ, Higher Vertebrate Encephalization Quotient; log, $\log_{10}$; LVEQ, Lower Vertebrate Encephalization Quotient; MBd, body mass (g); MBr, brain mass (g); MCb, cerebrum mass (g); REQ, Reptile Encephalization Quotient based on species-level equations; RMA, reduced major axis (regression equation); SD, standard deviation; TL, total length measured from snout to tail tip.

forebrain) includes the cerebrum and also unavoidably includes the portions corresponding to the diencephalon, optic chiasma, and optic tracts because the cerebrum lies dorsal and lateral to these brain components. Alligator cerebrum-cerebrocast ratios are the ratio of the cerebrum alone to the cerebrocast.

The MBr:EV ratio was determined for 12 alligators, and least squares regression equations were calculated with EV and MBr as dependent variables and TL as the independent variable. Reptile Encephalization Quotients of a sample including four additional specimens (total N=16) were calculated to describe the ontogenetic pattern relative to TL. These four were included to increase sample size although cerebrum data were unavailable for these specimens. Total length was used because some alligators were pen-raised and heavier for their length than wild alligators.

Relative Brain Size in Extant Non-avian Reptiles and Birds

Both non-avian reptile log REQs and bird log BEQs were normally distributed, unlike either non-avian reptile REQs or bird BEQs (all logarithms are $\log_{10}$ in this paper). Accordingly, dinosaur log REQs and log BEQs were compared to ranges of these parameters in non-avian reptiles and birds and analyzed in terms of z-scores (SD units) as appropriate (Sokal and Rohlf 1981). Additionally, the relationship between log REQs and TL in alligators was calculated (Hurlburt, unpublished data) to describe the ontogeny of REQ in a modern group.

Estimating Body Mass, Brain Mass, and Relative Brain and Cerebrum Size in Dinosaurs

Except for those of *Nanotyrannus* and *Bambiraptor*, dinosaur body mass (MBd) estimates were taken from the literature. In most cases, two estimates were used to cover a range of reasonable possible masses because no robust mechanism for MBd estimation has been universally accepted for extinct vertebrates (Hurlburt, 1996, 1999). The MBd of *Nanotyrannus lancensis* (CMNH 7541) was calculated from estimated femoral circumference. Femoral length was calculated from premaxilla-quadrate skull length (572 mm; Gilmore 1946) using Currie's (2003) least squares regression equations for tyrannosaurids (N = 26, r = 0.980) and tyrannosaurines (N = 14, r = 0.988), giving femoral lengths of 589.33 and 563.06 mm, respectively. Femoral circumference was estimated from femoral length, using the "All Theropods" RMA regression equation (n = 33, r = 0.9923) of Christiansen (1999), giving femoral circumferences of 192.72 and 182.84 mm, respectively. Body mass was calculated using the equation

$$W = 0.16\,Cf2.73, \tag{5}$$

where W is mass (g), and Cf is minimum femur circumference (mm) (Anderson et al. 1985). Adult MBd of *Bambiraptor* was calculated from Cf of an adult femur cast provided by David Burnham, and of *Ornithomimus* from *Cf* of ROM 852, an adult femur. Volumes of dinosaur endocasts and cerebrocasts were taken by one of three methods: (1) Double Graphic Integration (DGI; see Jerison 1973; Hurlburt 1999) of illustrations of endocasts of *Carcharodontosaurus saharicus* (SGM-Din 1; see Larsson 2001) and of *Tyrannosaurus rex* (AMNH 5029; see Hopson 1979), (2) volume calculation from virtual endocasts produced from three dimensional CT scans of the four tyrannosaurids, and (3) water displacement by the wet-dry method, as described above, of an endocast of *Allosaurus fragilis* (UUVP 294). Volumes derived from CT scans were generated using Amira 3.1.1 visualization software.

The posterior limit of the cerebrocast in dinosaurs was defined as a vertical line in the transverse plane just posterior to the bulge corresponding to the cerebrum. This line was usually just rostral to the rostral limit of the base of the cast of the trochlear nerve (IV) where apparent. Dinosaur MBr and MCb were obtained by applying the MBr:EV and MCb:CbcV ratios of the largest alligators and also by application of the widely used 50 percent ratio. The 50 percent ratio was used both for comparison to results of other studies and because it is the ratio of alligators halfway between the youngest and oldest sexually mature specimens. In addition, MBr and MCb values for *Nanotyrannus* were also obtained using MBr:EV and MCb:CbcV ratios typical of the youngest subadult alligators, owing to the hypothesis that it is a juvenile, as indicated by suture contacts, and as bone grain indicates juvenile status for *Nanotyrannus* (Carr 1999). No ratios are known for juvenile alligators. Calculated dinosaur log EQs were compared to ranges of log REQs of non-avian reptile species ($N = 62$) and of log BEQs of bird species ($N = 174$). Encephalization Quotients for other dinosaur species, including small theropods, were also used to provide a context for analysis of large theropods. Encephalization Quotients for small theropods and *Archaeopteryx* were calculated from endocast and MBd data in Russell (1972; see Nicholls and Russell 1981), Hopson (1977), Currie and Zhao (1993), Hurlburt (1996), Elzanowski (2002), Burnham (2004), and Dominguez Alonso et al. (2004) or as described in Methods. Encephalization Quotients were calculated for other dinosaurs (stegosaurs, ankylosaurs, ceratopsians, and sauropods), applying MBr:EV and MCb:CbcV ratios in alligators to EV. These data are from Hurlburt (1996) with two exceptions. The EV of *Stegosaurus* (64.18 ml) was obtained by DGI of the endocast of CM 106 figured in Galton (2001). The EV of *Iguanodon* was obtained by DGI of the brain cavity of BMNH R2501, an isolated endocranium (Andrews 1897; Norman 1986), after excluding an area corresponding to an extensive sinus complex (Norman and Weishampel 1990).

Brain:Endocast and Cerebrocast:Cerebrum Ratios in *Alligator mississippiensis*

Results and Discussion

The MBr:EV (brain mass to endocast volume) ratio decreased from 68 percent in the smallest to 32 percent in the largest alligators. Among the

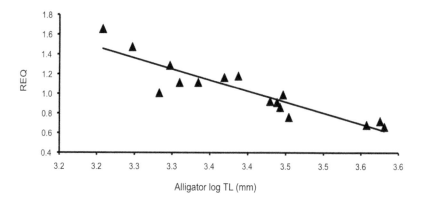

6.2. Reptile Encephalization Quotients (REQs) of alligators and log TL (total length) of *Alligator missippiensis*. Least squares regression equation: REQ = 8.56 + (−2.22 × log TL; $r = 0.907$).

three largest males, the TL range was 3610–3840 mm, the MBr range was 9.82–10.71 g, and the MBr:EV ratio range was 31.09–34.82 percent with a mean of 32.6 percent. The largest female (TL 284.5 cm, MBr 8 g) had an MBr:EV ratio of 42.04 percent. Alligator REQs regularly declined from 1.648 (log value = 0.217) in the smallest alligator (TL = 1613 mm) to 0.667 (log value =−0.176) in the largest alligator (TL = 3810 mm), in a mixed sample of wild and domestic alligators ($N = 16$; see Fig. 6.2). The associated least squares regression equation, REQ = 8.558 + (−2.215 × log TL), had 95 percent CL of (−2.804–1.627) enclosing the slope ($b = −2.2732$) of the smaller sample ($N = 12$).

Endocast volume increases faster than MBr relative to TL. In the least squares regression equation relating log EV (ml) to log TL (mm), the slope ($b = 1.811$; 95 percent CL 1.652, 1.978; $r = 0.992$) is statistically significantly larger than the slope of the equation relating log MBr (g) to TL (mm), where $b = 0.997$ (95 percent CL 0.890, 1.104; $r = 0.989$), although both variables increase with increasing body size (Hurlburt and Waldorf 2002). To take into account the pronounced sexual dimorphism of alligators, in which the greatest male size markedly exceeds that of females, MBr estimates from EV in mature dinosaurs and crocodilians should apply the largest male ratio (33 percent) for undoubted males, the male-female mean (37 percent) when specimen sex is unknown, and the largest female ratio (42 percent) for undoubted females when sexual dimorphism is known. The mean cerebrum:cerebrocast ratio of the largest males and females of 42 percent was applied to estimate cerebrum mass from dinosaur cerebrocasts.

Relative Brain Size in Extant Non-avian Reptiles and Birds

Reptile Encephalization Quotients of the 62 non-avian reptile species ranged from 0.402 to 2.404; the BEQs of 174 extant avian species ranged from 0.357 to 2.986. Figure 6.3 shows polygons that surround the non-avian reptile and bird brain-body data on which the EQ equations are based. Figure 6.4 shows the same polygons and unlabeled dinosaur brain-body data with the slopes for the Lower Vertebrate and Higher Vertebrate EQ (LVEQ and HVEQ) equations. The REQ slope, empirically

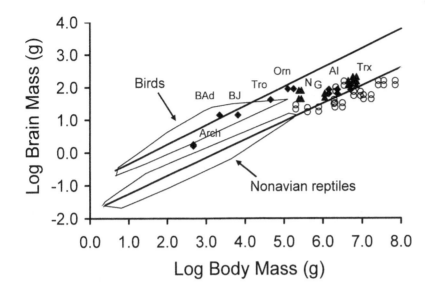

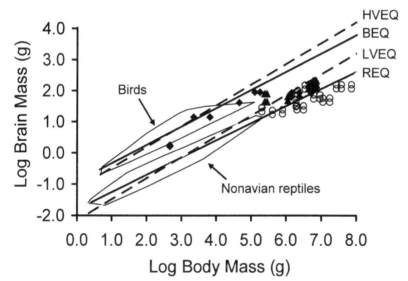

6.3. Log brain (MBr) and body mass (MBd) of dinosaurs, plotted with slopes of brain-body equations of non-avian reptile species (lower slope) and bird species (upper slope). Polygons surrounded brain-body point scatters of non-avian reptiles (N = 62) and birds (N = 174), as indicated. Legend: filled triangles, tyrannosaurids; filled diamonds, other theropods; hollow circles, other dinosaurs. Abbreviations: Al, *Allosaurus;* Arch, *Archaeopteryx,* BAd, *Bambiraptor,* estimated adult values; BJ, *Bambiraptor,* juvenile; G, *Gorgosaurus;* N, *Nanotyrannus;* Orn, *Ornithomimus;* Trx, *Tyrannosaurus rex;* Tro, *Troodon.*

6.4. Comparison of slopes of Lower Vertebrate Encephalization Quotient (LVEQ), Higher Vertebrate Encephalization Quotient (HVEQ), Reptile Encephalization Quotient (REQ), and Bird Encephalization Quotient (BEQ) equations. Polygons and dinosaur data as in Figure 6.3. Jerison (1973) chose the 0.67 slope of both the LVEQ and HVEQ because it is the coefficient of the volume to surface area relationship and fitted the intercepts (0.007 and 0.07) "by eye" to the Lower and Upper Vertebrate brain-body point scatters. The LVEQ slope passes above most dinosaur data points, and four of Jerison's ten dinosaur genera had LVEQs less than those of the least-encephalized reptiles. The REQ slope passes through the middle of the dinosaur distribution. Both the REQ and BEQ equations were empirically derived. Lower Vertebrates: fish, amphibians, and reptiles; Higher Vertebrates: birds and mammals.

derived from reptile brain-body data, passes through the middle of the dinosaur brain-body distribution, whereas the LVEQ slope (dashed), passes above most dinosaur brain-body points, and results in lower LVEQs for several dinosaurs than for the least-encephalized reptiles (Hurlburt 1996). Neither non-avian reptile REQs nor bird BEQs were normally distributed (Hurlburt 1996), but both non-avian reptile log REQs (range: −0.396–0.381) and bird log BEQs (range: −0.447–0.475) were normally distributed. Accordingly, dinosaur log EQs were compared to statistics of distribution (mean and SD) of reptile log REQs and bird log BEQs. The reptile log REQ distribution has mean −0.0087 and SD 0.1968; the mean + 2 SDs is −0.402 and 0.385, respectively. The bird log BEQs distribution has mean 0.0002 and SD 0.1815; the mean ± 2 SDs is −0.363 and 0.363, respectively (see Tables 6.1 and 6.2; Figs. 6.5 and 6.6).

6.5. Log$_{10}$ REQs of dinosaurs and 62 non-avian reptile species. Vertical dashed lines indicate non-avian reptile mean log REQ and mean ± 2 SDs. Non-avian reptile log REQ range, −0.318–0.237; mean, −0.0087; SD, 0.1968; mean ± 2 SDs, −0.403– 0.385. Open bars, reptiles; cross-hatched bars, dinosaurs; filled bars, reptile and dinosaur bars of equal height. Abbreviations: A and B indicate two MBd estimates (two MBrs for Arch.); Ad, adult; Carch, *Carcharodontosaurus,* Gorg, *Gorgosaurus;* Juv, juvenile; Nan, *Nanotyrannus;* Trx 2081 & 5117, *Tyrannosaurus rex* FMNH PR 2081 and AMNH 5117. Other abbreviations as in Figure 6.3.

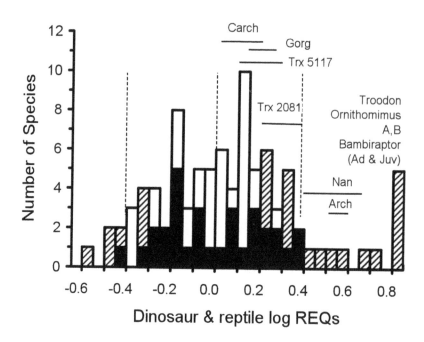

6.6. Log$_{10}$ BEQs of *Archaeopteyx,* dinosaurs, and 174 bird species. Bird log BEQ range, −0.447– 0.475; mean, 0.0002; SD, 0.1815; mean ± 2 SDs, −0.363–0.363. Extant bird species are underlined. Open bars, birds; cross-hatched bars, dinosaurs; filled bars, bird and dinosaur bars of same height. Dashed lines and abbreviations as in Figure 6.4.

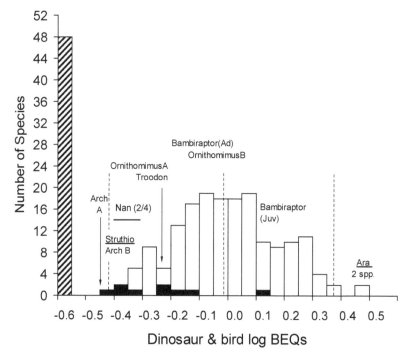

Body Mass, Endocast Volume, Brain Mass, and Cerebrum Mass in Dinosaurs

Tyrannosaurus is assigned a MBd range of 5000–7000 kg, from Holtz (1991) and Henderson (1999; modified from 7200 kg), respectively, unless other estimates for individual specimens were available. This MBd range encloses the estimates of 6250 kg for *T. rex* by Christiansen (1999), of 6650.9 kg by Seebacher (2001), and estimates by all researchers cited by Seebacher (2001), except Anderson et al. (1985). The estimate by Anderson et al. (1985) was more than 2.8 SDs below the mean of estimates by six researchers using various methods, more than twice the next greatest deviation (analysis of data in Seebacher 2001). The large FMNH PR 2081 specimen of *T. rex* had an estimated ontogenetic age of 28.9 years (Erickson et al. 2004). The *Gorgosaurus* MBd estimate (1105 kg) is calculated from *Cf* (mm) of RTMP 94.12.602 by Erickson et al. (2004), who suggests it is likely an underestimate (Erickson, pers. comm., 2005). Estimates for individual *T. rex* specimens of 5634 kg for FMNH PR 2081 and 4312 kg for AMNH 5117 kg are from Erickson et al. (2004) and Erickson (pers. comm., May 2006) and are based on the equation of Anderson et al. (1985). Given the discrepancy between the results of Anderson et al. (1985) and those of other researchers for *T. rex*, these are probable underestimates but permit comparisons to the other *T. rex* specimens. Anderson et al. (1985) produced their equation for bipedal dinosaurs by fitting a regression line with a slope of 2.73 to the MBd: *Cf* data point for a model and femur of *Troodon*. Their *T. rex* and *Allosaurus* MBd estimates were two thirds of Colbert's (1962) estimates from scale models, and their *Anatosaurus* MBd estimate exceeded the estimate from Colbert's (1962) model. The 2.73 slope was the exponent of an equation predicting MBd from the sum of femur and humerus circumferences of 33 mammal species, of which 29 massed under 500 kg and only two exceeded 1500 kg (Anderson et al. 1985). This exponent may notably underestimate MBd in large terrestrial amniotes (>1500 kg) because leg bones of the smaller amniotes have smaller duty factors (Alexander et al. 1979) and experience greater compressive forces during locomotion (Hurlburt 1996). The method may be more accurate for smaller theropods near the mass of *Troodon* (45 kg), such as *Bambiraptor* and *Nanotyrannus*.

Body mass estimates for *Allosaurus* of 2300 and 1400 kg were obtained from Colbert (1962) and Anderson et al. (1985), respectively. Body masses of 5000 and 7000 kg were used for *Carcharodontosaurus*, assuming its MBd to be in the same range as *T. rex*, following Larsson et al. (2000).

Table 6.1 lists endocast volumes for theropods. The DGI endocast volume (404 ml) estimate for AMNH 5029 is about 106 percent of the presumably more accurate CT estimate (381.76 ml) and is not used in the following analyses of *T. rex*. The over-estimate of endocast volume by the DGI method is likely due to erroneous inclusion of volumes of the concave regions dorsal to the brainstem and lateral to the optic lobes and cerebellum, but it is accurate when the subject analyzed is convex (Hurlburt 1999). The EV produced by Larsson et al. (2000) for AMNH 5029 appears to be an underestimate caused by methodological limitations and is not used in the following analyses. Higher values were obtained in this study for the *Allosaurus* EV than by Larsson et al. (2000), although both used water displacement. The estimates calculated in this study are used here because there was more control, but EQs resulting from both sets of estimates are similar. Double Graphic Integration produced higher

Table 6.1. Body mass, endocast volumes, associated brain volumes, REQs and log REQs of large alligators (ROM R8328 and R8333) and large theropod dinosaurs

Specimen	Meth	EV (ml)	MBr 37% (ml)	MBr 50% (ml)	MBd (t)	REQ 37%	REQ 50%	log REQ 37%	log REQ 50%
T 5117 (4.3)	CT	313.64	116.05	156.8	4.312	1.604	2.168	0.205	0.336
T 5117 (7)	CT	313.64	116.05	156.8	7.00	1.227	1.659	0.089	0.220
T 5029 (5)	CT	381.76	141.25	190.9	5.00	1.799	2.432	0.255	0.386
T 5029 (7)	CT	381.76	141.25	190.9	7.00	1.494	2.019	0.174	0.305
T 2081 (5.65)	CT	414.19	153.25	207.1	5.654	1.824	2.465	0.261	0.392
T 2081 (7)	CT	414.19	153.25	207.1	7.00	1.621	2.190	0.210	0.340
T 5029 (5)	DGI	404.0	149.48	202.0	5.00	1.904	2.573	0.280	0.410
G 1247	CT	128.93	47.70	64.47	1.11	1.400	1.892	0.146	0.277
N 7541 (0.24)	CT	111.18	41.14	55.59	0.24	2.812	3.800	0.449	0.580
N 7541 (0.28)	CT	111.18	41.14	55.59	0.28	2.597	3.510	0.414	0.545
N 7541 (0.24)	CT	111.18	67% = 74.491		0.24	67% = 5.092		67% = 0.707	
N 7541 (0.28)	CT	111.18	67% = 74.491		0.28	67% = 4.703		67% = 0.672	
C Din 1 (5)	DGI	263.68	97.56	131.8	5.00	1.243	1.680	0.094	0.225
C Din 1 (7)	DGI	263.68	97.56	131.8	7.00	1.032	1.394	0.014	0.144
C Din 1 (5)	Lsr	224	82.88	112	5.00	1.056	1.427	0.024	0.154
A 294 (1.4)	WaH	187.9	69.52	93.95	1.40	1.791	2.420	0.253	0.384
A 294 (2.3)	WaH	187.9	69.52	93.95	2.30	1.361	1.839	0.134	0.265
A 294 (1.4)	WaL	169.0	62.53	84.5	1.40	1.610	2.176	0.207	0.338
ROM R8328	WaH	27.34	10.50	—	0.238	0.721	—	0.060	—
ROM R8333	WaH	32.94	10.51	—	0.277	0.663	—	0.055	—

Note: REQ = MBr/(0.0155 × MBd$^{0.553}$), both MBr and MBd in grams. Non-avian reptile REQ range: 0.402–2.404. Nonavian reptile log REQ mean ± 2 SDs, −0.403–0.385. MBr, brain mass estimated from EV using alligator MBR:EV ratios. 37% or 50% indicates ratio used, but 67% used for *Nanotyrannus* (italicized in table). Two MBd estimates for most species (see Methods). Numbers in parentheses after specimens indicate MBd. Abbreviations: A 294, *Allosaurus* UUVP 294; ROM R8328 and R8333, data for ROM alligator specimens; C Din 1, *Carcharodontosaurus saharicus* SGM Din-1; CT, computed tomography scans; DGI, Double Graphic Integration; EV, endocast volume; G 1247, *Gorgosaurus* ROM 1247; Lsr, laser scan (Larsson et al. 2000); MBd, body mass; MBr, brain mass; Meth, method by which volumes were obtained from endocasts; MBd, body mass in metric tons; MBr, brain mass (% indicates ratio); N 7541, *Nanotyrannus lancensis* CMNH 7541; REQ, Reptile Encephalization Quotient; T 5117, *Tyrannosaurus rex* AMNH 5117; T 5029, *T. rex* AMNH 5029; T 2081, *T. rex* FMNH PR 2081; WaH, volumes by water displacement by Hurlburt; WaL, volumes by water displacement by Larsson.

values than the laser scan for *Carcharodontosaurus* and are perhaps less accurate for reasons given above, but EQs again are similar between values from laser scan and DGI methods.

Dinosaur MBr was estimated by the mean of the MBr:EV ratio of the largest males and female alligators (37 percent) because the sexes of individual dinosaur specimens is unknown. Similarly, dinosaur cerebrum mass was obtained by applying the mean ratio of the largest males and females (42 percent). The possibility that some dinosaur specimens were not full adults is dealt with by application of the 50 percent MBr:EV ratio. All dinosaurs were treated as adults, including *Allosaurus* UUVP 294, which J. Madsen (pers. comm., June 2005) considered to be an adult despite Rogers (1999) regarding it as a subadult. Brain mass was considered to equal EV in small theropods and *Archaeopteryx*.

Table 6.1 provides estimates of body mass (MBd), endocast volume (EV), estimated brain mass (MBr), and methods by which EV was obtained. Figure 6.3 plots brain body data of large theropods, small theropods and other dinosaurs (stegosaurs, ankylosaurs, ceratopsians, and

Table 6.2. BEQs and log BEQs of theropod dinosaurs

Specimen	Meth	EV (ml)	MBd (t)	BEQ (37%)	BEQ (50%)	log BEQ (37%)	log BEQ (50%)
T 5117 (4.3)	CT	313.64	4.312	0.121	0.163	−0.918	−0.787
T 5117 (7)	CT	313.64	7.00	0.091	0.123	−1.042	−0.911
T 2081 (5.65)	CT	414.19	5.654	0.136	0.184	−0.867	−0.736
T 2081 (7)	CT	414.19	7.00	0.120	0.162	−0.921	−0.791
T 5029 (5)	CT	381.76	5.00	0.135	0.182	−0.871	−0.740
T 5029 (7)	CT	381.76	7.00	0.110	0.149	−0.957	−0.826
T 5029 (5)	DGI	404.0	5.00	0.143	0.193	−0.846	−0.715
T 5029 (7)	DGI	404.0	7.00	0.117	0.158	−0.932	−0.801
G 1247	CT	128.93	1.11	0.111	0.150	−0.955	−0.824
N 7541 (0.24)	CT	118.18	0.24	0.236	0.318	−0.628	−0.497
N 7541 (0.28)	CT	118.18	0.28	0.216	0.292	−0.665	−0.534
N 7541 (0.24)	CT	118.18	0.24	67% = 0.427		67% = −0.370	
N7541 (0.28)	CT	118.18	0.28	67% = 0.392		67% = −0.407	
A 294 (1.4)	WaH	187.9	1.40	0.141	0.190	−0.852	−0.722
A 294 (2.3)	CT	"	2.30	0.105	0.142	−0.979	−0.849
C Din 1 (5)	DGI	263.68	5.00	0.093	0.126	−1.031	−0.901
C Din 1 (7)	DGI	"	7.00	0.076	0.103	−1.118	−0.987
ROM R8328	WaH	10.50	0.238	0.060	—	−1.219	—
ROM R8333	WaH	10.51	0.277	0.055	—	−1.258	—

Note: BEQ = MBr/(0.117 × MBd$^{0.590}$); both MBr and MBd in grams. Encephalization Quotients are estimated from MBd and estimated MBr in Table 6.1. Note smallest subadult ratio (67%) used for *Nanotyrannus* in 50% column (italicized in table). Bird BEQ range: 0.357–2.986. Bird log BEQ mean ± 2 SDs, −0.363–0.363. Abbreviations: BEQ, Bird Encephalization Quotient. Other abbreviations as in Table 6.1.

Species	MBd (g)	EV (ml)	REQ	BEQ	log REQ	log BEQ
Ornithomimus A	175000	87.9	7.145	0.606	0.854	−0.218
Ornithomimus B	125000	87.9	8.606	0.739	0.935	−0.132
Troodon	45000	41.0	7.067	0.630	0.849	−0.201
Bambiraptor Juv	2240	14.00	12.680	1.263	1.103	0.101
Bambiraptor Ad	6581.96	14.0	6.986	0.669	0.844	−0.175
Archaeopteryx A	468	1.60	3.445	0.363	0.537	−0.440
Archaeopteryx B	468	1.76	3.789	0.400	0.579	−0.398

Table 6.3. Body mass, endocast volume, EQs and log EQs for three Late Cretaceous small theropods and *Archaeopteryx*

Note: Endocast data from *Archaeopteryx* (BMNH 37001), *Bambiraptor* (KUVP 129737), *Ornithomimus* (NMC 12228), and *Troodon* (RTMP 86.36.457 and RTMP 79.8.1). Data from Hurlburt (1996), except for *Bambiraptor* and *Archaeopteryx* (see text). MBr = EV since the brain filled the cranial cavity. Abbreviations: Ad, adult; Juv, juvenile. Other abbreviations as in Tables 6.1 and 6.2.

sauropods) with slopes and polygons that surround brain-body data of reptile (N = 62) and bird (N = 174) species.

Relative Brain Size in Dinosaurs

Tables 6.1 and 6.2 provide REQs, BEQs, log REQs, and log BEQs for all large theropods analyzed, and Table 6.3 gives data for three Late Cretaceous small theropods and *Archaeopteryx*. Figures 6.5 and 6.6 are histograms of log REQs and log BEQs of large theropods, small theropods, and other dinosaurs. Encephalization Quotients based on 37 percent MBr:EV ratios are more consistent with the analytical method, but EQs from 50 percent ratios are provided for reasons given above. Comparisons among species are made using raw (i.e., not log-transformed) EQ data, which are more easily comprehended; comparisons to reptile and bird distributions are made with log EQs, which are normally distributed, unlike raw EQs.

TYRANNOSAURIDAE *Tyrannosaurus rex* has the largest relative brain size of any dinosaur, other than some small theropods. Reptile Encephalization Quotients of *T. rex* range from 1.2 to 1.82 (37 percent ratio) and 1.66 to 2.47 (50 percent ratio). The highest *T. rex* log REQ (50 percent ratio) is no more than 2 SDs above the mean of reptile REQs, with one exception, from a 50 percent MBr:EV ratio for FMNH PR 2081, an unlikely ratio because this is the ontogenetically oldest and most mature *T. rex* in the sample (Table 6.1; Fig. 6.5). A high log REQ from a 50 percent MBr:EV ratio for the DGI EV value is discounted because DGI probably overestimates total EV. Log BEQs of *T. rex* are almost 4 SDs below the mean bird log BEQ and well below the lowest bird log BEQ (−0.447; see Table 6.2 and Fig. 6.6). It appears that the body size sequence of *T. rex* specimens increases through AMNH 5117, AMNH 5029, and FMNH PR 2081, and the pattern of EV increasing with body size is typical of alligators, as is continuing increase in body size with ontogenetic age (Table 6.1).

Gorgosaurus (ROM 1247) is less encephalized than *T. rex*, and only the REQ (1.89) derived from a 50 percent MBr:EV ratio reaches the lower end (37 percent ratio) of the REQ range of *T. rex* (Table 6.2; Fig. 6.5), while a 67 percent MBr:EV ratio produces an REQ of 2.54. ROM 1247 is clearly a subadult (Carr 1999). If *Gorgosaurus* and *T. rex* follow a similar growth trajectory, these REQs are lower than would be expected of a subadult *T. rex* because ontogenetically younger alligators have larger relative brain sizes than adults.

Reptile Encephalization Quotients of *Nanotyrannus* (e.g., 2.812–2.597 for a 37 percent MBr:EV ratio) clearly exceed the REQ range of *Tyrannosaurus rex*, even for a 50 percent MBR:EV ratio for *T. rex* (Table 6.1). Log REQs (37 percent ratio: 0.414–0.449) are more than 2 SDs above the reptile log REQ mean (mean + 2 SDs = −0.403–0.385). Larger MBR:EV ratios produce even higher REQs for *Nanotyrannus* (Table 6.1). Expressed as SDs of reptile log REQs (z-scores), *Nanotyrannus* log REQs are 0.86 SDs above *T. rex* values whether comparing 37 percent or 50 percent MBr:EV ratios and are as much as 2.99 SDs above the reptile log REQ mean (Sokal and Rohlf 1981). When calculated from a 67 percent MBr:EV ratio, *Nanotyrannus* log REQs are 3.64–3.64 SDs above the reptile mean and about 1.5–2.0 SDs above the highest *T. rex* values (z-scores = 1.47 for 37 percent ratio, 2.13 for 50 percent ratio; see Table 6.1). If *Nanotyrannus* is a juvenile or young subadult of *T. rex* or a similar tyrannosaurid, the difference between its log REQs and those of an adult tyrannosaurid such as *T. rex* approximates the difference (2.0 SDs) between log REQs of the smallest subadult and largest adult alligators in the comparison sample (Tables 6.1, 6.2; Figs. 6.2, 6.3). *Nanotyrannus* log REQs are consistent with those of a young subadult or older juvenile tyrannosaurid. Conversely, even the smallest *Nanotyrannus* MBr estimate produces REQs much larger than those of any adult dinosaur whose brain does not fill the cranial cavity; we consider this to be unlikely and therefore inconsistent with adult ontogenetic status. *Nanotyrannus* log BEQs are more than 2 SDs below the bird log BEQ mean, even with the large 67 percent MBR:EV ratio (Table 6.2; Fig. 6.6).

CARCHARODONTOSAURUS AND ALLOSAURUS Data for the two allosauroids indicate that *Carcharodontosaurus* had an REQ range of 1.032–1.240, less than that of *Allosaurus* UUVP 294 (1.361–1.791) and about two thirds that of *Tyrannosaurus rex* (Table 6.1; Figs. 6.3, 6.5). Neither dinosaur enters the bird BEQ range (Table 6.2; Fig. 6.6). However, the spread of *Allosaurus* MBd estimates is wide, and REQs from the larger MBd estimate approximate those of *Carcharodontosaurus*.

SMALL THEROPODS AND ARCHAEOPTERYX *Archaeopteryx* Brain mass ranged from 1.6 to 1.76 ml, exceeding an estimate of 1.47 ml from DGI of the figure of *Archaeopteryx* in Bühler (1985; also see Hurlburt 1996). Body mass was 468 g (Elzanowski 2002). The log REQ range (0.537–0.579) is more than 2.5 but slightly less than 3 SDs above the mean reptile log REQ, and its brain-body points overlapped the lower edge of the bird brain-body polygon (Figs. 6.3, 6.5). The log BEQs of *Archaeopteryx* (−0.365–−0.415) are within the bird log BEQ range, overlapping values for *Struthio* (Figs. 6.3, 6.6), although slightly more than 2 SDs below the bird log BEQ range. These results falsify the hypothesis that *Archaeopteryx* lies between the reptile and bird relative brain-body distributions (Larsson et al. 2000).

Cerebrocast:Endocast Volume and Cerebrum:Brain Mass Ratios in Theropod Dinosaurs

Larsson et al. (2000) suggested that relative cerebrum size increased in coelurosaurian dinosaurs, the lineage leading to and including birds, relative to allosauroids, a lineage including *Carcharodontosaurus*, a large theropod approximately equal in MBd to *Tyrannosaurus rex*. They compared the ratio of cerebrocast to EV, considering it equivalent to the MCb:MBr ratio. To test this hypothesis, we computed the same ratios using MCb estimates from applying the alligator MCb:CbcV ratio to CT scans of dinosaur endocasts. We also combine the result of a laser scan for EV with DGI of the cerebrocast of *Carcharodontosaurus* because DGI is fairly accurate for convex solids but less so for entire endocasts, as discussed above.

Cerebrocast volume:endocast volume ratios from CT scans of the three *Tyrannosaurus rex* and *Gorgosaurus* specimens ranged from 41.1 to 43.5 percent, and MCb:MBr ratios ranged from 47.5 to 49.53 percent (Tables 6.4 and 6.5). While MCb:MBr ratios estimated for *T. rex* enter the lower end of ratios typical of birds, they are very close to ratios (45.9–47.9 percent) typical of the smallest sexually mature alligators (Table 6.5).

Table 6.4. Endocast (EV) and cerebrocast (CbcV) volumes, associated brain and cerebrum mass (MCb), and associated CbcV:EV and MCb:MBr ratios of large alligators and theropod dinosaurs

Specimen	Meth	Brain part of EV (ml)	CbcV (ml)	EV less CbcV (ml)	MBr = 37% EV (ml)	MCb = 42% CbcV (ml)	CbcV: EV Ratio (%)	MBr: MCb Ratio (ml)
T 5117	CT	313.64	131.4	182.3	116.05	55.17	41.88	47.54
T 2081	CT	414.19	170.2	244.0	153.25	71.48	41.09	46.65
T 5029	CT	381.76	165.9	215.8	141.25	69.69	43.47	49.34
T 5029	DGI	404.0	147.8	256.2	149.48	62.07	36.58	41.53
T 5029	Lsr	343	111.8	231.2	126.91	46.96	32.59	37.00
G 1247	CT	128.93	56.0	72.9	47.70	23.52	43.43	49.30
N 7541 (Ad)	CT	111.18	64.73	46.45	41.14	27.19	58.22	66.09
N 7541 (Sub)	CT	111.18	64.73	46.45	74.49	46.61	58.22	62.57
C Din 1	DGI	263.68	83.1	180.6	97.56	34.90	31.51	35.77
C Din 1	Lsr,DGI	224	83.1	140.9	82.88	34.90	37.10	42.11
C Din 1	Lsr	224	53.7	170.3	—	—	23.97	24.00
A 294	WaH	187.9	101.7	86.2	69.52	42.71	54.12	61.44
A 294	WaL	169.0	46.7	122.3	—	—	27.63	27.63
ROM R8328 4	WaH	27.34	10.9	6.2	10.50	4.28	39.87	40.76
ROM R8333	WaH	32.94	11.3	6.0	10.51	4.51	34.15	42.91

Note: Theropod dinosaur volumes were calculated using MBr:EV (37%) and MCb:CbcV (42%) ratios of largest adult alligators. *Nanotyrannus* (N 7541) EV and CbcV were estimated using ratios from smallest subadult alligators, which were MBr:EV, 67%, and MCb:CbcV, 72%. Both Larsson et al. (2000) and the present study obtained *Allosaurus* volumes by water displacement. For *Carcharodontosaurus*, one CbcV: EV and one MCb:MBr ratio were obtained by combining CbcV from DGI with EV from a laser scan (Larsson et al. 2000). Abbreviations: Ad, adult; CbcV, cerebrocast volume; Sub, sub-adult. Other abbreviations as in Tables 6.1 and 6.2.

Table 6.5. Cerebrocast: endocast volume (CbcV:EV) and cerebrum:brain mass (MCb:MBr) ratios of dinosaurs, alligators, nonavian reptiles, and birds. *Ameiva* data from Platel (1979).

Specimen	Meth	CbcV:EV Ratio (%)	MCb:MbrRatio (%)
T. rex (N = 3)	CT scans	41.1–43.5	46.6–49.3
G 1247		43.3	49.3
N 7541 (Ad)		58.2	66.1
C Din 1 (5)	DGI	31.5	35.8
C Din 1 (7)	Lsr-DGI	37.1	42.1
A 294		54.1	61.4
Two smallest alligators		37.9–43.8	44.8–47.9
Two largest alligators		34.2–39.9	40.8–42.9
Reptiles: Mn ratio = 33.52. Actual range in column 3			23.57–43.56
Reptiles: Mn ratio ± 2 SDs			25.6–41.5
Ameiva and largest alligator *Alligator*			31.4, 40.8
Birds: Mn ratio = 63.7. Actual range in column 3			44.6–82.3
Birds: Mn ratio ± 2 SDs			47.4–80.0

Note: MCb:MBr ratios were calculated from cast volumes for all fossil specimens. Only MCb:MBr ratios are provided for extant species other than alligators. For the two smallest alligators, TL = 1613 mm and 1985 mm; for the two largest alligators, TL = 3759 mm and 3810 mm. Abbreviations: Mn, mean; Rep, Reptile; TL, snout to tail tip length. Other abbreviations as in Tables 6.1 and 6.2.

Because it cannot be determined whether a larger MCb:MBr ratio arises from a larger MCb or a decline in one or more of the other brain divisions, this result can be taken to indicate an avian-like condition, but tyrannosaurid MCb:MBr ratios are high. These ratios do not reflect the significant difference in brain size relative to body size between *T. rex* and birds, and thus that bird cerebrum size is relatively larger than in *T. rex*. The same applies to differences in relative cerebrum size between *Carcharodontosaurus* and *T. rex*. Ratios are useful in comparisons between related taxa of similar relative brain size.

Nanotyrannus has a CbcV:EV ratio of 58.2 percent. Its MCb:MBr ratios are 66.1 percent, using adult brain:endocast ratios, and 62.6 percent, using youngest subadult brain:cast ratios (Tables 6.4, 6.5). These CbcV:EV and MCb:MBr ratios are 15 percent or more higher than for other tyrannosaurids and than for alligators and other reptiles (Tables 6.4 and 6.5). While these ratios resemble those of the *Allosaurus* endocast, the endocasts are dissimilar in appearance, whereas the *Nanotyrannus* endocast resembles those of other tyrannosaurids. Because higher MCb:MBr ratios are typical of ontogenetically younger alligators, these data support the hypothesis that *Nanotyrannus* is a young subadult or juvenile, as do REQ data.

Larsson et al. (2000) proposed that cerebral volume is 100 percent greater in *T. rex* than in *Carcharodontosaurus*, accounting in part for the larger relative brain size of *T. rex*. Cerebrum mass:brain mass ratios of *Carcharodontosaurus* are as high as 42.1 percent when combining laser-scanned EV and DGI of CbcV, comfortably in the upper reptile range and only 5–7 percent less than those of *T. rex*. This rejects the hypothesis that cerebral volume is 100 percent greater in *T. rex* than in *Carcharodontosaurus* (Larsson et al. 2000). A quite high MCb:MBr ratio (61.4 percent) was obtained for *Allosaurus*. This may be due to experimental error or to relative size differences of other brain components.

Summary

The purpose of this study was assessment of the relative brain and relative cerebrum size of tyrannosaurid dinosaurs (*Tyrannosaurus rex*, *Gorgosaurus*, and *Nanotyrannus*) and comparison of these data to results for allosauroid dinosaurs (*Allosaurus* and *Carcharodontosaurus*). To measure relative brain size, EQs (Encephalization Quotients) were calculated using brain-body data for extant non-avian reptile species (N = 62) and extant bird species (N = 174). We compared dinosaur log EQs to the ranges of reptile log REQs and bird log BEQs because these samples were normally distributed, unlike either reptile REQs or bird BEQs. To estimate brain mass (MBr) from dinosaur endocast volume (EV), the MBr:EV ratio was determined in a size series, ranging from the smallest sexually mature to the largest commonly encountered size, of *Alligator mississippiensis*, an examplar of the extant archosaurian clade Crocodylia. The mean of MBr:EV ratios of the largest male and female was 37 percent, and the ratio was 67 percent in the smallest sexually mature alligators. Dinosaur MBr and MCb (cerebrum mass) were estimated from virtual endocasts produced from CT scans and also from laser scans and double graphic integration. Brain mass was estimated from EV in dinosaurs using the adult ratio and in *Nanotyrannus*, a possible juvenile, also using the youngest subadult ratio. Estimates were also made using the traditional 50 percent MBr:EV ratio, for comparison to previous studies and because this is the MBr:EV ratio in midrange subadult alligators, appropriate for the *Gorgosaurus* specimen. Relative brain sizes of small theropods and a wide sample of other dinosaurs were also determined to provide a context for evaluation of these large theropods. The cerebrum mass:brain mass (MCb:MBr) range was compared among dinosaurs and between dinosaurs and each of reptiles and birds.

Conclusion

This is the first study to use empirically based brain:endocast (MBr:EV) and cerebrum:cerebrocast (MCb:CbcV) ratios, derived from extant alligators, to estimate dinosaur relative brain and relative cerebrum size. It is also the first to measure dinosaur relative brain size by Reptile Encephalization Quotients (REQs) and Bird Encephalization Quotients (BEQs). Both MBr and EV increase with body size in alligators, but the MBr rate is significantly less, so that the MBr:EV ratio in alligators declines with increasing body size, as does REQ. Other than small theropods, which are well within the relative brain size range of extant birds, *Tyrannosaurus rex* has the largest relative brain size of any dinosaur but is within the relative brain size of extant reptiles and within 2 SDs of the mean of reptile log REQs. It is well below the relative brain size range of extant birds. *Gorgosaurus* plots at the lower end of log REQs of *T. rex*. The log REQs of *Nanotyrannus lancensis* are anomalously high for an adult but consistent with a juvenile or very young subadult age. The difference between its log REQs and those of an adult *T. rex* paralleled the difference between the youngest subadult and the oldest adult alligators of the comparison sample, when measured as log reptile REQ SD units. *Nanotyrannus* MCb:MBr ratios were also consistent with an older juvenile or young subadult ontogenetic age. *Carcharodontosaurus* has an REQ about two thirds that of *T. rex* and showed no increase in relative brain size compared to the late Jurassic *Allosaurus*, supporting a hypothesis of a trend of larger relative brain size in coelurosaurian compared to allosauroid dinosaurs. All three late Cretaceous small theropods (*Bambiraptor*, *Troodon*, and *Ornithomimus*) plotted well within in the bird log BEQ range and well above the reptile log REQ range. The relative brain size range of *Archaeopteryx* overlapped the lower edge of the bird log BEQ range and exceeded the reptile REQ range. Both tyrannosaurids and allosauroids had cerebrum mass:brain mass (MCb:MBr) ratios in the high end of the reptile range, and *T. rex* entered the low end of the bird MCb:MBr range. These values were also similar to those of subadult alligators. The MCb:MBr ratio of *Carcharodontosaurus* was less than 7 percent below that of *T. rex*, falsifying a hypothesis that a larger cerebrum accounted for the larger brain of *T. rex*.

Acknowledgments

We thank R. Elsey of the Rockefeller Wildlife Refuge in Louisiana, Allan Woodward of the Florida Fish and Wildlife Conservation Commission, and Bubba Stratton, an independent alligator control agent in Florida, for facilitating acquisition of alligators.

Literature Cited

Alexander, R. McN. 1985. Mechanics of posture and gait of some large dinosaurs. *Zoological Journal of the Linnean Society* 83:1–25.

Alexander, R. McN., G. M. O. Maloiy, B Hunter, A. S. Jayes, and J. Nturibi. 1979. Mechanical stresses in fast locomotion of buffalo (*Syncerus caffer*) and elephant (*Loxodonta africana*). *Journal of Zoology* 189:135–144.

Anderson, J. F., A. Hall-Martin, and D. A. Russell. 1985. Long-bone circumference and weight in mammals, birds, and dinosaurs. *Journal of Zoology,* London 207:53–61.

Andrews, C. W. 1897. Note on the cast of the brain of *Iguanodon. Annals and Magazine of Natural History.* 115:585–591.

Bühler, P. 1985. On the morphology of the skull of *Archaeopteryx*; pp. 135–140 in M. K. Hecht, J. H. Ostrom, G. Viohl, and P. Wellnhofer (eds.), *The Beginnings of Birds.* Freunde der Jura-Museums, Eichstatt.

Burnham, D. A. 2004. New information on *Bambiraptor feinbergi* (Theropoda: Dromaeosauridae) from the Late Cretaceous of Montana; pp. 67–111 in P. J. Currie, E. B. Koppelhus, M. A. Shugar, and J. L. Shugar (eds.), *Feathered Dragons: Studies on the Transition from Dinosaurs to Birds.* Indiana University Press, Bloomington.

Carr, T. D. 1999. Craniofacial ontogeny in Tyrannosauridae (Dinosauria, Theropoda). *Journal of Vertebrate Paleontology* 19:497–520.

Christiansen, P. 1999. Long bone scaling and limb posture in non-avian theropods: evidence for differential allometry. *Journal of Vertebrate Paleontology* 19:666–680.

Colbert, E. H. 1962.The weights of dinosaurs. *American Museum Novitates,* no. 2076.

Currie, P. J. 2003. Allometric growth in tyrannosaurids (Dinosauria, Theropoda) from the Upper Cretaceous of North America and Asia. *Canadian Journal of Earth Sciences* 40:651–665.

Currie, P., and X. Zhao. 1993. A new troodontid (Dinosauria, Theropoda) braincase from the Dinosaur Park Formation (Campanian) of Alberta. *Canadian Journal of Earth Sciences* 30:2231–2247.

Dendy, A. 1910. On the structure, development and morphological interpretation of pineal organs and adjacent parts of the brain in the Tuatara (*Sphenodon punctatus*). *Philosophical Transactions of the Royal Society,* ser. B, 201:227–331.

Dominguez Alonso, P. D., A. C. Milner, R. A. Ketcham, M.J. Cookson, and T. B. Rowe. 2004. The avian nature of the brain and inner ear of *Archaeopteryx. Nature* 430:666–669.

Elzanowski, A. 2002. Archaeopterygidae (Upper Jurassic of Germany); pp. 129–159 in I. Chiappe and L. M. Witmer (eds.), *Mesozoic Birds: Above the Heads of Dinosaurs* University of California Press, Berkeley.

Erickson, G. M., P. J. Makovicky, P. J. Currie, M. A. Norell, S. A. Yerby, and C. A. Brochu. 2004. Gigantism and comparative life-history parameters of tyrannosaurid dinosaurs. *Nature* 430:772–779.

Evans, D. C. 2005. New evidence on brain–endocranial cavity relationships in ornithischian dinosaurs. *Acta Palaeontologica Polonica* 50:617–622.

Galton, P. M. 2001. Endocranial casts of the plated dinosaur *Stegosaurus* (Upper Jurassic, Western USA): a complete undistorted cast and the original specimens of Othniel Charles Marsh; pp. 103–129 in K. Carpenter (ed.), *The Armored Dinosaurs.* Indiana University Press, Bloomington.

Gilmore, C. W. 1946. A new carnivorous dinosaur from the Lance Formation of Montana. *Smithsonian Miscellaneous Collections* 106:1–19.

Henderson, D. B. 1999. Estimating the masses and centers of mass of extinct animals by 3-D slicing. *Paleobiology* 25:88–106.

Holtz, T. R. 1991. Limb proportions and mass estimations in the Theropoda. *Journal of Vertebrate Paleontology* 11:35A.

Hopson, J.A. 1977. Relative brain size and behaviour in archosaurian reptiles. *Annual Review of Ecology and Systematics* 8:429–48.

Hopson, J. A. 1979. Paleoneurology; pp. 39–146 in C. Gans, R. G. Northcutt, and P. S. Ulinski (eds.), *Biology of the Reptilia,* vol. 9: *Neurology A.* Academic Press, New York.

Hopson, J. A. 1980. Relative brain size in dinosaurs: implications for dinosaurian endothermy; pp. 287–310 in R. D. K. Thomas and E. C. Olson (eds.), *A Cold Look at the Warm-Blooded Dinosaurs.* American Association for the Advancement of Science, Washington, D.C.

Hurlburt, G. R. 1982. Comparisons of brains and endocranial casts in the domestic dog *Canis familiaris* Linnaeus 1758. M.Sc. thesis, University of Toronto, Toronto, Ontario.

Hurlburt, G. R. 1996. Relative brain size in recent and fossil amniotes: determination and interpretation. Ph.D. dissertation, University of Toronto, Toronto, Ontario, 250 pp.

Hurlburt, G. R. 1999. Comparison of body mass estimation techniques, using recent reptiles and the pelycosaur *Edaphosaurus boanerges. Journal of Vertebrate Paleontology* 19:338–350.

Hurlburt, G. R., and L. Waldorf. 2002. Endocast volume and brain mass in a size series of alligators. *Journal of Vertebrate Paleontology* 23(3, suppl.):69A.

Iwaniuk, A. N., and J. E. Nelson. 2002. Can endocranial volume be used as an estimate of brain size in birds? *Canadian Journal of Zoology* 80:16–23.

Jerison, H. J. 1973. *Evolution of the Brain and Intelligence.* New York: Academic Press.

Larsson, H. C. E. 2001. Endocranial anatomy of *Carcharodontosaurus saharicus* (Theropoda: Allosauroidea) and its implications for theropod brain evolution; pp. 19–33 in D. H. Tanke and K. Carpenter (eds.), *Mesozoic Vertebrate Life.* Indiana University Press, Bloomington.

Larsson, H. C. E., P. C. Sereno, and J. A. Wilson. 2000. Forebrain enlargement among nonavian theropod dinosaurs. *Journal of Vertebrate Paleontology* 20:615–618.

Lefebvre, L., N. Nicolakakis, and D. Boire. 2002. Tools and brains in birds. *Behaviour* 139:939–973.

Marino, L. 2002. Convergence in complex cognitive abilities in cetaceans and primates. *Brain, Behavior and Evolution* 59:21–32.

Nicholls, E. L., and A. P. Russell 1981. A new specimen of *Struthiomimus altus* from Alberta, with comments on the classificatory characters of Upper Cretaceous ornithopods. *Canadian Journal of Earth Sciences* 18:518–526.

Norman, D. B. 1986. On the anatomy of *Iguanodon atherfieldensis* (Ornithischia: Ornithopoda). *Bulletin de l'Institut Royal des Sciences Naturelles de Belgique* 56: 281–372.

Norman, D. B., and D. B. Weishampel. 1990. Iguanodontidae and related ornithopods; pp. 510–533, in D. B. Weishampel, P. Dodson, and H. Osmólska, eds, *The Dinosauria.* University of California Press, Berkeley.

Osmólska, H. 2004. Evidence on relation of brain to endocranial cavity in oviraptorid dinosaurs. *Acta Paleontologica Polonica* 49:321–324.

Platel, R. 1976. Analyse volumetrique comparee des principales subdivisions encephaliques chez les reptiles Sauriens [Lizards]. *Journal für Hirnforschung* 17:513–537.

Platel, R. 1979. Brain weight–body weight relationships; pp. 147–171 in C. Gans, R. G. Northcutt, and P. S. Ulinski (eds.), *Biology of the Reptilia,* vol. 9: *Neurology A.* Academic Press, New York.

Rich, T. H., and P. Vickers-Rich. 1988. A juvenile dinosaur brain from Australia. *National Geographic Research* 4:148.

Rogers, S. W. 1999. *Allosaurus,* crocodiles, and birds: evolutionary clues from spiral computed tomography of an endocast. *Anatomical Record* 257:162–173.

Russell, D. A. 1972. Ostrich dinosaurs from the late Cretaceous of Western Canada. *Canadian Journal of Earth Sciences* 9:375–402.

Seebacher, F. 2001. A new method to calculate allometric length-mass relationships of dinosaurs. *Journal of Vertebrate Paleontology* 21:51–60.

Sokal, R. B., and F. J. Rohlf. 1981. *Biometry.* W. H. Freeman, New York.

Witmer, L. M., S. Chatterjee, J. Franzosa, and T. Rowe. 2003. Neuroanatomy of flying reptiles and implications for flight, posture and behaviour. *Nature* 425:950–953.

Woodward, A., C. T. Moore, and M. F. Delaney. 1991. Experimental Alligator Harvest: Final Report. Study no. 7567. Bureau of Wildlife Research, Florida Game and Fresh Water Fish Commission, Tallahassee, 118pp.

7.1. Tyler Keillor's flesh model of Jane.

Jane, in the Flesh: The State of Life-Reconstruction in Paleoart

7

Tyler Keillor

Tyler Keillor

My goal in creating a flesh reconstruction of an extinct animal is to provide the museum visitor with a sense of what the real live animal was all about. I don't want to give the exhibit viewer a cliché, a toy, a Hollywood prop, or something that's been seen in every kid's dinosaur book. I want the observer to see a restoration that is unique, that shows a creature, frozen in time, that endured various life processes, and that might challenge preconceived notions about the animal and elicit questions or thought. A reconstruction requires not just artistry and imagination but also the input of the latest scientific opinions and comparative observations of extant animals. A life reconstruction is, by nature, highly speculative, and being so is of less value scientifically than artistically (as an exhibit piece for the layperson). Nevertheless, a rigorously executed reconstruction may, through its very creation, yield new insight into paleontological questions and so can be a working model and an aid to scientific understanding. I'll let the task of bringing the Burpee Museum's juvenile tyrannosaur "Jane" (BMR P2002.4.1) back to "life" provide a glimpse into the behind-the-scenes aspects of paleoart (the depiction of ancient beasts; see Fig. 7.1). In this reconstruction, in particular, observations of extant reptiles yielded new insights into the external appearance of Jane's oral margin.

Introduction

I was fortunate enough to have been selected to complete the reconstruction of Jane's skull in the summer of 2004. Unlike the Field Museum's "Sue" (FMNH PR 2081), the skull of Jane was not preserved in one piece; rather, it was disarticulated and missing perhaps 40 percent of the cranium and jaws. By the time I became involved, the fossil skull bones had been fully prepared, molded, and cast by the Burpee Museum's team, and the cast parts were glued together to start building the skull model. To guide my progress as I sculpted the missing skull anatomy, the Burpee arranged for a panel of paleontologists to review my work. In addition to Jane's lead investigator, Michael Henderson, I consulted with Thomas Carr, Philip Currie, and Michael Parrish. The Burpee's chief preparator and collections manager, Scott Williams, served as facilitator to the process as well.

The skull reconstruction was fairly straightforward, keeping in mind that it involved restoring an incomplete skull of a 66 million-year-old dinosaur (Fig. 7.1). For bones that were missing from one side of the skull,

Restoring Jane's skull

Institutional Abbreviations BHI, Black Hills Institute; BMR, Burpee Museum of Natural History; CMNH, Cleveland Museum of Natural History, Cleveland, Ohio; FMNH, Field Museum of Natural History, Chicago; MNN, Musee National du Niger; RCI, Research Casting International; SGM, Ministere de l'Energie et des Mines, Rabat, Morocco.

I sculpted the mirror image of the bones that existed for the opposite side. It can be difficult to create a reversed bone without a guide, so I made the job easier with a trip to Kinko's (a photocopying shop). By printing out a reversed photocopy of the disarticulated casts (laid directly on the copier glass), I created an image of exactly what any missing bone should look like from different angles. A cast of the Cleveland Museum's tyrannosaur skull (CMNH 7541), as well as various casts from other tyrannosaur specimens, served as guides for sculpting anatomy that was missing completely. I used epoxy putty (Apoxie Sculpt) to sculpt the bones and more putty to hold them together as the skull took form. The putty starts soft and clay-like, then becomes firmer over several hours until it is hard enough to file or sand. Having been restored with epoxy, the finished skull model was strong and durable for handling and shipping.

The length of the preserved teeth is one aspect of a skull that proves to be a variable from specimen to specimen. Not only do teeth show different lengths as a result of the tooth-replacement process in the living animal, but postmortem decomposition loosens teeth, allowing them to slide partially or completely out of their sockets. In Jane, none of the premaxillary or anteriormost dentary teeth were preserved in their sockets. Most of the teeth were also absent from the entire left dentary. By contrast, both maxillae, as well as the right dentary, retained most of their teeth in place. The majority of the teeth that had fallen out were recovered with the specimen. In all likelihood, the animal died with a full mouthful of teeth, some of which subsequently fell out after death. The fact that some teeth fell out while others did not raises a question: Were the in situ teeth in their life positions, or had they slid partly out of their sockets as well, but not as far as the disarticulated teeth? For the purposes of the skull reconstruction, I did not set any of the disarticulated teeth back into their sockets so that they would be longer than the longest socketed teeth. This arrangement looks acceptable for the fossil skull model; for the flesh model, the question remains if any of the teeth are unnaturally long because of taphonomic distortion.

After a round of reviews and revisions, the final skull model was approved and shipped to Research Casting International (RCI) for molding and duplication. I created an articulated, removable unit from the palatal bones; this made the mold-making process easier. Each jaw ramus was left detached from its mate, again to aid in mold making. For me, an unforeseen, yet beneficial, part of the mold-making process was that the skull was cut lengthwise into a left and right half at RCI. Upon the return of the bisected skull model from RCI at the end of mold making, I realized it would provide me with a great opportunity. The divided skull allowed me to see, in cross section, the articulation of the jaw and its relation to the skull and the pathway of inhaled air through the skull from nostrils to palate, all of which would provide food for thought for the flesh model.

The ideal situation for the construction of a flesh model is to sculpt over a skull. By using a skull as the armature for a reconstruction, the proportions and size of the model are assured to be accurate. Since Jane's skull was not terribly deformed, and since experts had scrutinized the missing pieces that had been reconstructed, the final skull proved to be a very precise framework. But with the bisected skull in hand, I didn't want to rush into gluing it back together and covering it with clay; first, there was research to be done.

One feature of a reconstructed face that has garnered a lot of discussion recently is the lips, or absence thereof, particularly in theropods like *Tyrannosaurus rex* (Ford 1997). I find it useful to be aware of the styles of other paleoartists and their reasons for restoring dinosaurs in a given manner. Therefore, I knew before starting Jane's reconstruction that there are perhaps three variations on the theme of lips: there are crocodile-like reconstructions with no lips; there are reconstructions with partial lip flaps; and there are reconstructions with full lips that seal the mouth shut. There is a scientific rationale for the use of each style of lip.

In the "no-lip" reconstruction, the teeth erupt out of the margins of the jaw and skull very much like an alligator's. Typically, if the jaw is depicted closed, the maxillary teeth overhang to a level near or sometimes past the bottom of the dentary. If the jaws are open, the armament of huge upper and lower teeth is on full display. There is an acknowledged "wow factor" when it comes to showing the teeth of a gigantic predator, and some artists note that it is one of the things clients expect to see in their artwork. Usually, a large scale is depicted at the base of each tooth and the rest of the head and jaws have a similar scaly look, but nothing obstructs the view of the teeth.

The extant phylogenetic bracket for dinosaurs (birds and crocodilians; see Witmer 1995) is used as support for the no-lip style. Since birds and crocodilians don't have lips, then the least amount of inference is needed to restore dinosaurs lipless, as well. Recent work by Lawrence Witmer (2001) and his team have made mainstream news of the topic of soft-tissue features on dinosaurs. Papp and Witmer (1998) concluded that dinosaurs shouldn't have fleshy, movable lips. Similarly, "*Tyrannosaurus*, like any archosaur, would have lacked fleshy lips, and the upper teeth would normally be exposed when the jaws were closed," as Chris Brochu wrote in critique of a Douglas Henderson tyrannosaur painting (Brochu 2000:44). It is not uncommon for a theropod skull to be found with the jaws tightly closed, such that the dentary teeth are not visible at all. If a skull were fleshed out with this as the closed pose, an artist would be hard pressed to find a living animal to base the lip area on, other than a crocodilian. Indeed, an alligator's jaw can be put into a jaw-closed pose very similar to that seen in some theropod specimens (CMNH 7541, for example).

The Partial Lip Flap Reconstruction

The "partial lip flap" reconstruction may be the most commonly used approach. The maxillary teeth are still visible when the jaw is closed, but more of the teeth are covered by some manner of soft tissue. This "lip flap" may look a bit like a lizard lip, a thin band of scaly tissue that is fairly tightly appressed to the skull. The lower jaw has a similar band of lip tissue. Sometimes the upper and lower lips partly overlap the teeth, and sometimes they are pulled back from the teeth to reveal gum tissue.

Paul (1988) advocated simple lips of this sort in his book *Predatory Dinosaurs of the World*, a work that is routinely referenced by many paleoartists. An often-cited reason for the lip flaps is the need to protect the teeth and the mouth from drying out. Presumably, the lip flaps covered the base of the teeth and sealed when the mouth was closed to conserve moisture.

The Lipped Reconstruction

The least common type of depiction is the "lipped" reconstruction. Fewer artists take this route, as the teeth are completely covered and the "wow factor" of teeth is lost. The lips seal the mouth in the closed-mouth pose, so no maxillary teeth are visible poking out into view. To make room for the lips, this style requires that the closed-jaw pose be slightly less clenched than certain fossilized skulls demonstrate. If the mouth is shown open, the maxillary and dentary teeth may still be completely covered by the upper and lower lips, or some limited amount of tooth tips may be exposed.

The lipped reconstruction solves the problem of the mouth drying out by keeping moisture sealed in. Abler (2000) suggested that a full, sealed-lip style would allow for a wet oral environment that septic bacteria could have flourished in. Additionally, "sniffing" would be facilitated with lips that could seal, allowing inhaled air to fully pass through olfactory chambers in the snout instead of passing in through the mouth (Currie, pers. comm., February 2005).

How would Jane fit into this spectrum of reconstructions? I realized that one of the key questions that the amount of restored oral soft tissue hinges on, literally, is to know how far the living animal could have closed its mouth. As straightforward as that seems, one can observe that the closed-mouth pose is a variable in every artist's work, and one that directly affects the outcome of the soft tissue in the life study.

Jaw Closure

As mentioned above, fossilized theropod skulls have been found with the jaws tightly closed. While some artists have used this as the living animal's closed-mouth pose, I offer another interpretation. The jaws that are tightly clenched may show a postmortem deformation, akin to the "death curve" seen in the axial columns of many vertebrates under certain conditions. As tissues desiccated and shrank in the dead animal, the massive jaw closing muscles may have shortened and pulled the jaw tightly closed, more so than it would have in life. Punctures in the palate of Sue occurred after death, when the jaw's dentary teeth were closed further than they had been in life (Brochu 2003). In skulls that are preserved right-side up and resting on their jaws, overlying sediment compaction after burial could further crush the jaws closed in dorsoventral compression (Bakker et al. 1988).

The shape of the medial maxilla bears what some have considered a clue to the jaw's closed pose. It has been suggested that depressions ventral to the palatal shelf of the medial maxilla coincide with dentary tooth positions (Currie 2003). Peter Larson (pers. comm., February 2005) demonstrated for me how the toothed dentary of the *Tyrannosaurus rex* known as "Stan" (BHI 3033) could be made to line up to the depressions on its disarticulated medial maxilla. In skulls that are preserved "crushed closed," the dentary teeth seem to rest in these depressions. But the medial maxillary depressions are features that are not exclusive to tyrannosaurs alone.

For a diverse comparison, I looked at the medial maxilla of the relatively deep-skulled theropods *Carcharodontosaurus saharicus* (SGM-Din 1) and *Rugops primus* (MNN IGU1). These animals also show depressions ventral to the palatal shelf on the medial side of the maxilla. The great dorsal-ventral depth of these maxillae would result in an even more severe closed-mouth pose, if one were to interpret the maxillary depressions as dentary tooth rests. In a conservative jaw restoration (closed into the maxillary depressions), little of the animals' anterior lower jaw would be visible, and nearly the entire premaxillary and maxillary tooth row would be hanging from the overbite like icicles from a roof. This seems unlikely, especially given that the teeth in *Carcharodontosaurus* and *Rugops* are not as robust as those in *Tyrannosaurus rex* and would seem to be more vulnerable to damage.

Using the maxillary depressions as a landmark for the dentary teeth raises several questions. Functionally, why would the animal shut its mouth into this extreme closed-position? When the jaw is in this orientation, the opposing teeth have crossed well past the point of puncturing or slicing the food item. The animal's biting action would have already completely severed a piece of flesh from the prey; the morsel would be free in the mouth well before the jaws reached their fully closed position. Closing the mouth further would not produce better results since the functional range of the teeth has been past. Additionally, the pose would require a greater range of movement for the articular jaw joint, as well as greater jaw muscle lengths, to produce an identical gape in an animal that didn't close its jaw as far. Further, unlike any extant terrestrial taxa, the dry external integument of the dentary would perpetually sit against the wet surface of the medial maxilla in the extreme closed position (assuming the animal had salivary and oral glands, as do extant taxa).

A contrasting interpretation of the maxillary depressions frees the jaw from the need to close so far. The roots of the functional and the replacement teeth are long and occupy much of the maxilla. In Jane, the maxillary depressions can be seen as the areas between the convexities caused by the maxillary tooth-socket walls, and in some cases there appear to be additional collapsing of the medial maxilla directly over tooth roots. The regular symmetry of these depressions, therefore, seems to reflect the maxillary tooth/inter-tooth positions rather than indicating rests for dentary teeth. Similarly, Lamanna et al. (2002) describe alternating depressions and ridges on the medial maxilla of an abelisaurid as corresponding to tooth roots and replacement-teeth positions.

Additionally, while it is possible to align a dentary tooth into a disarticulated medial maxillary depression, the justification for lining up the teeth and depressions is made more difficult when one is dealing with the full skull. As I inspected Jane's skull model, the cranium could indeed be posed resting on top of the jaws, resulting in an extreme jaw-closed pose similar to that of the Cleveland Museum specimen (CMNH 7541). In this pose, the cross-sectional view showed that there is bone-to-bone contact between the quadrate and articular, the jugal and ectopterygoid come close to touching the surangular, and there is bone-to-bone contact between the dentary teeth and the palatal bones and medial maxillae as well. There is no room for soft tissue at the jaw joint, little room for tissue along the posterior third of the jaw where large jaw muscles would have needed room to exist and operate, and no room for soft tissue on the palate. However, since soft tissue is not taken into account in this version of the closed-jaw pose, it is advantageous to look at alternate jaw-closed configurations. Examination of extant taxa sheds light on the topic of jaw closure and soft tissues for comparison.

Birds

Birds will not be considered in great depth here. While both the dinosaur/bird connection and the topic of feathers have been making news in paleontology (Currie 2004), avian anatomy as applied to tyrannosaur head reconstructions is somewhat limited. Although birds form half of the dinosaurian extant phylogenetic bracket, many of their specialized features are not applicable to tyrannosaurs. No extant bird has teeth; the premaxilla, maxilla, and dentary are modified into the familiar, if variable, beak. The tough, keratinous growth of bills and crests leaves impressions on the bones of the skull, indicating their presence (Alexander 1994). These marks on bird skulls may be similar to imprinted areas of tyrannosaur skulls, such as nasal, lacrimal, and postorbital rugosities that could have been "crested" (Currie 1997); however, the details of the oral margin are very different. A flexible, lip-like structure, known as the commisure, gape, or rictus, forms the margin of the bird mouth behind the beak (Proctor and Lynch 1993). The beak itself exhibits a wide variety of forms to suit different functions. Within the mouth, the internal nares often appear as a single slit-like opening in the middle of the soft tissue of the palate. Further investigation into how the commisure and ridges of tissue inside the mouth help seal the oral cavity will aid in future reconstructions.

Being archosaurs, crocodilians often serve artists as extant analogs in tyrannosaurid soft-tissue reconstructions. However, there are several key components of crocodilian anatomy that may limit their effectiveness as theropod analogs. While very few skin impressions have been found for tyrannosaurs (Carpenter 1997), it is interesting to note that none of them resemble the soft-tissue scale patterns of crocodilians. More important, the adaptation to an aquatic lifestyle has resulted in specializations of the skull that must be considered contextually.

The pose of the closed jaw in crocodilian skulls superficially resembles that in tyrannosaur skulls. Alligators in particular have an overhanging maxillary tooth row that is reminiscent of that seen in tyrannosaurs. It follows that since crocodiles and alligators don't have lips, and their jaws close in a manner similar to those of certain tyrannosaur fossils, then artists may feel justified in giving the dinosaur a crocodilian soft-tissue pattern. If one were to take the preserved closed-jaw pose of a tyrannosaur and try to restore it with lips instead, one would end up with a jowly pouch of flesh hanging on the dentary (to sheath the maxillary teeth; see Ford 1997). This unlikely soft-tissue configuration makes the crocodilian model seem more plausible.

However, a closer look at the skulls of crocodilians reveals differences that affect comparisons with tyrannosaurs. Crocodiles have an interdigitating opposing tooth configuration (Edmund 1969) that is quite different from that seen in tyrannosaurs. The dentary teeth of tyrannosaurs occlude medial to the maxillary teeth, although there is evidence for occasional tooth-to-tooth contact (Schubert and Ungar 2005). In alligators, the maxillary teeth overhang the dentary, and the dentary teeth fit into distinct concavities in the secondary palate (Edmund 1969). The maxilla itself, however, does not overhang the dentary (as is the case in certain tyrannosaur closed-jaw specimens). The dorsoventrally flat crocodilian snout limits the degree to which the jaw can close. In tyrannosaurs, the relative dorsoventral depth of the snout permits the jaw to be preserved in a crushed-closed position that is more extreme than that seen in crocs. Crocodilians don't have lips, and it is plausible that their absence is an adaptation to an aquatic lifestyle. Terrestrial vertebrates must constantly protect their bodies from dehydration. The sealed lips of any lizard form a barrier to keep the moist oral cavity from drying by evaporation. In contrast, by spending all of its time in or near water, the crocodile's water supply is a constant part of its environment. In severe drought, when the water supply altogether evaporates, resident crocodiles will die from dehydration (Kiley et al. 1995). Crocodiles are noticeably absent from arid and desert environments (with the exception of isolated "desert" crocs that live in seasonal watering holes and then retreat to deep, cool, moist, underground burrows to wait out the dry season (Shine et al. 2001). The lizard lip of other reptiles is replaced in crocodiles by a thin veneer of skin tightly adhering to the skull. The specialized skin of the mouth area is rich in nerve endings, which provide sensory input for prey acquisition in murky water (Soares 2002).

Along with a lack of lips, another feature tied into an aquatic lifestyle is the unique crocodilian palatal structure. The roof of the mouth in all extant crocodilians is marked by the bony secondary palate, which separates the nasal airways from the oral cavity. The opening of the internal nares is at the back of the mouth, where a muscular valve at the back of the tongue and throat can close the airway off from the mouth. This feature allows a crocodilian to continue to breathe through its nostrils while the rest of its body is submerged, a major adaptation for underwater prey capture (Levy 1991). Even when closed, the crocodile's mouth cannot shut sufficiently to seal water out. The ability to separate the airway from the mouth has allowed the crocodilians to fully exploit their habitat. The caudal position of internal nares, coupled with the muscular valve to separate airflow from the oral region, have allowed the crocodilians to thrive in water without lips (Busbey 1994). Breathing and olfaction are separated from the permeable mouth with a bony structure that is absent in tyrannosaurs. The internal nares in tyrannosaurs perforate the palate about midway back from the snout tip (Ruben et al. 1997), as opposed to the posterior position in crocodilians.

Without a crocodilian-style bony secondary palate, how did the tyrannosaur prevent dehydration and accomplish olfaction? In the no-lip crocodilian style and the partial lip-flap style of tyrannosaur reconstructions, an extreme closed-jaw position is necessary to seal the oral cavity. What structure is actually sealing against what other structure in this model? Is the dry, external integument of the dentary sealing against the moist gum tissue of the medial maxilla? What happened to the seal during maxillary tooth replacement, when a gap would have been present along the tooth row? Again, in crocodilians a sealed oral cavity is not necessary. Even in alligators, the external surface of the dentary is not sealed against the medial maxilla. Alternatively, is it possible that the tyrannosaurs had a fleshy secondary palate and a muscular tongue valve (to allow olfaction and nose breathing without a sealed mouth) and an oral cavity devoid of moisture (to avoid evaporative dehydration without a sealed mouth)?

Did the extinct, terrestrial crocodiles have lips to prevent dehydration? Was the position of the internal nares caudal or rostral? If crocodilians are to be used as exclusive analogs for terrestrial dinosaurs, perhaps we must look at and understand the anatomy of extinct terrestrial crocodilians, such as *Simosuchus* and *Baurusuchus*, before making conclusions based on extant taxa alone. These comparative observations will be explored in future restorations.

Lizards

If the amphibious adaptations in crocodilians hinder terrestrial tyrannosaur comparisons, then perhaps the largest extant terrestrial reptiles might make good surrogates. The Komodo dragon (*Varanus komodoensis*), other large monitors, and some other lizards have similarities to tyrannosaurs that may be superficial; however, their shared dry terrestrial environment makes comparisons useful.

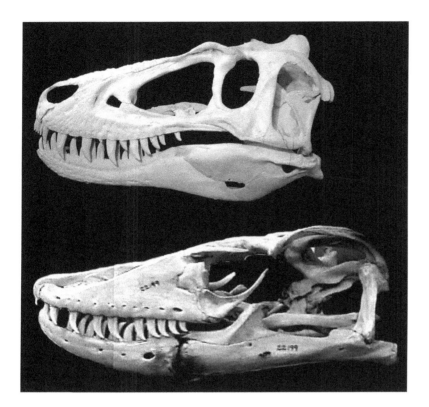

7.2. A Komodo dragon skull (*bottom*) is compared to Jane's skull (*top*). The skulls are scaled such that the longest maxillary teeth are about the same size. The position of jaw closure in Jane matches the Komodo.

The skull of a Komodo dragon is striking for its long, sharp, curved, serrated teeth. The teeth reflect the carnivorous diet of the lizard and serve it well in removing chunks of flesh from its quarry (Auffenberg 1981). While the socket arrangement of lizard teeth is different from those of dinosaurs, they do bear a resemblance in form and function to tyrannosaur teeth. The skull's dental arcade in both the lizard and the tyrannosaur is broadly U-shaped, while the jaw's dental arcade in both is narrower and V-shaped. In both, the dentary teeth close medial to the maxillary teeth (Figure 7.2).

Komodo dragons have rows of maxillary and dentary foramina, parallel to the tooth rows, that resemble a tyrannosaur's as well. The foramina on the lizard provide innervation and blood supply to the soft tissues of gums and lips (Bakker 1986). When the Komodo is in closed-jaw pose, the large teeth are not visible at all behind the lips, which seal together. The mandible has a thick buildup of soft tissue along the oral rim. This tissue allows the lips to seal, by making the lower jaw's narrow outline wider, to match the skull's broader outline.

Non-respiratory buccal oscillation is used by lizards and many other extant taxa for olfaction (Brainerd 2001). During this behavior, sealed lips allow inhaled air to pass through nasal chambers, instead of flowing in through the mouth. The lips also seal to prevent the moist mouth from drying by evaporation. To thermoregulate, however, lizards can gape and employ gular flutter to cool down by controlled evaporation.

In open-mouth pose, few if any of the gaping lizard's teeth are apparent. The upper lip hangs down to the level of the maxillary tooth tips. In medial view, the maxillary teeth can be seen, embedded in thick gums, medial to the upper lip. Thick gums cover the dentary teeth, as well. The lower lip margin is quite low on the jaw and does not cover the lower teeth at all. Since the upper lip matches the length of the maxillary teeth, the lower lip must be low enough on the jaw to permit opposing teeth to close functionally before the lips seal at the jaw's closed point (pers. obs.).

The closed-jaw pose in a Komodo results in sealed lips externally. The skull, meanwhile, exhibits a pose in which the opposing teeth have crossed, but the jaw is not closed as far as in crocodilians or tyrannosaur fossils. Computed tomographic views (available through DigiMorph. org, 2012) of preserved (but not skeletonized) monitor and other lizards in closed-mouth pose show that the jaws appear to be slightly open, even while the lips are sealed. The teeth in the jaw-closed pose are functionally engaged, and there is still room for soft tissues around the mouth (gums, muscles, lips). Yet a skeletonized specimen's jaws can be made to close further than the living animal would have closed its mouth. One can imagine how postmortem deformation during fossilization might preserve a Komodo skull in a crushed-closed pose similar to a tyrannosaur's. In this situation, would we reconstruct that Komodo specimen with exposed, overhanging maxillary teeth? It is interesting to note that the giant extinct monitor from Australia, *Megalania*, has been reconstructed by some artists as showing teeth in a dinosauresque style (Hallett 2004) and by others with lips covering teeth in a Komodo style (Knight 1985).

To compare Jane's skull and teeth to those of lizards, I set up the skull model in lateral view and photographed it. I posed the jaw closed, but only to the degree that the maxillary teeth were lined up with the foramina of the dentary. As noted earlier, the longest maxillary teeth may be longer than the living animal possessed, if the teeth shifted out of the sockets slightly after death. If the longest teeth were socketed slightly deeper and therefore shorter, the jaw could then be closed somewhat further than I could achieve with the skull as reconstructed. The jaw-closed pose I used is the same as that seen in the lizard scans from DigiMorph (with some "gap" between dentary and maxilla). Using the Adobe Photoshop image-editing program, I then scaled the image of Jane to that of several lizard skulls (images of an earless monitor lizard, *Lanthanotus borneensis*, and a Gila monster, *Heloderma suspectum*, both from DigiMorph, and an image of a Komodo dragon, *Varanus komodoensis*, FMNH 22199B, from the Field Museum, were used), so that the longest maxillary teeth were the same length among all the species. The skulls and teeth in all were very comparable in proportion when resized. Since the lizards' long sharp teeth are fully sheathed by lips, and since Jane's teeth were not proportionately much different in size from the lizards, it seems possible for Jane's teeth to have been sheathed by lips, as well.

The premaxillary and anteriormost dentary teeth in the Komodo and the tyrannosaur are as not as robustly built, or as deeply rooted, as the larger maxillary teeth along the sides of the snout. Because they are shorter, they are out of the way when the larger teeth are puncturing and pulling. It has been proposed that the premaxillary teeth in tyrannosaurs would nip meat off from tight quarters, based on tooth-marked bones (Erickson 2000). In the tyrannosaur skull that is posed in the lizard-style closed-jaw pose, the premaxillary teeth are not engaged with the anteriormost dentary teeth; there is a gap between opposing tooth tips. This would not render them useless, however. A bone could still be scraped clean, without being bitten through by the teeth at the front of the snout. If closing the premaxillary and anteriormost dentary teeth in a functional way without a gap between opposing tooth tips was imperative, then the longest maxillary and dentary teeth would need to be set deeper into their sockets, to prevent the skull from approaching the "crushed-closed" pose.

The closed-jaw pose of an animal provides the frame upon which to flesh out its skull. If the sealed lips like those in lizards are decided upon for a closed-mouth pose, then the open-mouth pose should not show alligator-style exposed teeth, and vice versa. Some artists have created closed- and open-mouthed individuals within the same scene. The open-mouthed individuals may have upper and lower teeth exposed. The closed-jaw individuals may have sealed lips with no teeth showing. This pattern does not exist in the world today, except in mammals where muscular lips allow full teeth baring as well as lip sealing. Most artists consider the mammalian model for lips too far outside of the extant phylogenetic bracket to be applied to a dinosaur reconstruction. As I sculpted the three-dimensional reconstruction of Jane, I could not rationalize a way to both have the teeth fully exposed while open mouthed yet have a sealed oral cavity while closed jaw. Based on my observations, I created a model that falls in line more with a terrestrial lizard than an aquatic crocodilian.

Jane's Details

For my Jane sculpture, tyrannosaur skin impressions were combined with skin patterns observed in monitors to create the look and form of much of the soft tissues. The lateral margin of the lower jaw is fleshed out to be as wide as the lateral margin of the upper lip. This would have allowed the lips to seal if the mouth were closed and is based on my observations of Komodo dragons from the Field Museum's Herpetology Department. The longest maxillary teeth are visible protruding from under the upper lip; this was a compromise I arrived at to keep some continuity between my sculpture and Michael Skrepnick's official Burpee Museum Jane portrait. The bases of the upper and lower teeth are covered by thick gum tissue: less than a monitor has, but consistent with the depth of marks on the teeth that may represent gum margins (Currie, pers. comm., February 2005). The lower lip edge is positioned well below the lower teeth and gum line. This would allow the jaw, if closed, to have both sealed lips and functionally occluded opposing teeth. The scale pattern is very fine overall, consistent with skin impression information provided to me by

Philip Currie and Peter Larson. For the lip margins and back of the neck, I created larger scales, following a monitor pattern.

Other features of my reconstruction may warrant brief explanation:

There is a sliver of a nictitating membrane in the inner corner of the eyes. The look of this structure is based directly on close-up footage of emus that I captured at a wildlife park. Nictitating membranes in dinosaurs are speculative, but they are widespread in the extant phylogenetic bracket and beyond (Owen 1866).

The top of the snout has a ridge of layered keratinous growth. The lacrimal, postorbital, and jugal prominences also have keratinous plates. These areas of Jane's skull show rugosities, which have been interpreted as supporting hard, crest-like growths in other theropods (Paul 2000).

The oral cavity is made to look very wet, representing a well-lubricated environment that is protected from evaporation by the lips. There are strands of saliva spanning between the jaw and palate as the mouth opens. This amount of saliva is consistent with that seen in Komodos.

I've depicted bits of decaying meat between the bases of several upper and lower teeth. It is conceivable that, as flesh was ripped from a carcass, bits could have become lodged in various crevices. As Abler (2000) pointed out, meat fibers would have fed oral bacteria, which could have given the animal a septic bite.

The position of the nostrils is based directly on the features of the premaxillae. Following Witmer's (2001) findings, the external nares are rostral and ventral in the bony nostril openings. The lips I've given Jane (which cover the premaxillary teeth entirely) give the snout the appearance of extra dorsoventral depth. This creates the illusion that the nostrils are too far dorsal within the openings; they are not.

I've added flakes of peeling integument to the skin in various places. All reptiles and members of the extant phylogenetic bracket molt or shed in various ways. My goal with this detail was to express that life processes were continually occurring. The patches of flaking hide could be interpreted as healing abrasions and scars or as molting skin. Either way, the keen observer may be reminded of something seen on a living animal and thereby see Jane as a living animal as well (as opposed to a "perfect" statue).

The Jane Experiment

I started my sculpture with the skull mounted to a stand with a 30° gape, but I realized later that for an ideal reconstruction (representing a fully functional animal that would have been able to live in a terrestrial environment) I should have started sculpting the model in closed-jaw pose and then opened the mouth to the desired angle. Admittedly, by showing a tyrannosaur in open-mouth pose, the artist is not responsible for what would happen to that animal when it closed its jaws. Would there be a seal to allow olfaction or to prevent evaporation? A sculpture is the only

7.3. The Jane experiment. (*Top to bottom:*) Jane's skull cast articulated in extreme closed-jaw pose; Todd Marshall's restoration based on the photo; Donna Braginetz's restoration based on the photo; Luis Rey's restoration based on the photo.

Images used with artists' permission; images copyright the respective artists.

way to fully explore the three-dimensional functional implications of a chosen soft-tissue course. Short of such a sculpture, an illustration that starts from the true closed-jaw pose is the next-best thing.

Surprisingly, I could not find a scientific consensus regarding the extent to which a live tyrannosaur might have closed its jaws. Based on my observations, I photographed Jane's skull in lateral view, in a closed-jaw pose consistent with large lizards (in which the opposing teeth cross but the maxillary teeth do not pass the line of the foramina on the dentary). This photograph became the starting point for the "Jane Experiment."

I thought it would be very interesting to send the photo to paleoartists and let them restore the flesh head as they liked—with one big rule: the pose of the jaw in the photo should be considered the extreme jaw-closed pose of the living animal. This would surely be unique because the majority of closed-jaw reconstructions seem to be based upon skulls preserved in a crushed-closed pose. As a matter of fact, the photo reference I had earlier sent to Michael Skrepnick for his official Burpee Museum portrait of Jane (a separate project from my experiment) consisted entirely of photos showing the skull model resting on the jaws in crushed-closed pose. Skrepnick *did* correct for this by allowing more room for soft tissue than the photos showed; however, the amount that he opened the jaws did not go as far as I subjectively chose for the Jane experiment later.

The paleoartist participants in the Jane experiment (who completed renderings at the time of this writing) include Donna Braginetz, Todd Marshall, and Luis Rey (see Fig. 7.3). As an ongoing exercise in the paleoart community, I'm continuing the Jane experiment with David Krentz, Michael Skrepnick, and others. Without a mummified tyrannosaur head for reference, the dinosaur artists will forever be investigating solutions to the problem of life reconstructions.

Of her experimental illustration, Donna Braginetz wrote me,

> If this is the maximum closure of the jaw, the resulting small gaps between the upper and lower teeth suggest fleshy lips that would provide a better seal to prevent the inner mouth tissues from drying. This presumes, however, that this animal "needed" to keep its tongue and gums moist. Crocodiles don't have a tight mouth closure, and parrots have a particularly dry mouth interior. An additional problem: When the mouth is closed, is there enough room for the lower jaw *plus* a lower lip to fit inside the upper? Or should the lower lip be wide and loose enough to enclose the upper teeth? My solution was to provide Jane with enough upper lip to cover the premaxillary teeth and enough lower lip to just meet the upper, while allowing the longest maxillary teeth to protrude.(Pers. comm., September 12, 2005)

Luis Rey's impressions of the experiment illuminate the artistic process. Upon seeing the skull photo I provided, Rey wrote me,

> I find that using your photograph I have had to go to extremes to find how to fill the gaps and have an efficient closure of the mouth, especially considering that the animal is supposed not to have had "lips" but more

a "crocodile or bird grin" and should have efficiently locked the jaws. I found the solution in actually adding some sort of lips and having the upper teeth barely visible protruding from the upper jaw on top of the lower. This is by no means a final word on this, but I'm trying it and see how it looks."(Pers. comm., March 14, 2005)

After some critical feedback from paleontologist colleagues of his, Rey wrote me that "they really didn't like the addition of lips and think that the model should shut the jaws naturally without gaps and much more than what the skull (photo) is showing" (pers. comm., August 15, 2005).

Braginetz's and Rey's experiences point out two major ideas. One, the degree of jaw closure in the live animal is an unknown, but there is a bias toward using crushed-closed specimens as models. Two, the artist will innately work to create a life portrait that solves problems, such as the gap between the jaw and skull. The use of a variety of extant taxa, both within and outside the extant phylogenetic bracket, can provide a pool of reference for solving life-reconstruction problems.

Japanese Paleoartists

It is interesting to note that the life reconstructions of Western paleoartists are in some ways different from those of Japanese paleoartists. Several widely published, contemporary Japanese paleoartists (Seiji Yamamoto, Takashi Oda, Mineo Shiraishi, among others) routinely depict tyrannosaurs with full lips that seal the teeth within the closed mouth (Manabe 2000). Do they know something that we don't, or do we know something that they don't? Or is this just a cultural or stylistic choice? In my opinion, many of the Japanese restorations look very lifelike and realistic. This may be because the lips, which give the dinosaurs a very lizard-like appearance, make them seem much more familiar and less monstrous. Their reasons for giving theropods lips are well summed up as a solution to evaporation in a terrestrial environment, by way of comparison to extant terrestrial reptiles (T. Oda, pers. comm., March 20, 2005). These artists note that clients may push to have teeth exposed but that baring the teeth is the less likely functional choice in a restoration (M. Shiraishi, pers. comm., March 8, 2005). There seems to be a human imperative to see the teeth, however, regardless of whether the anatomy dictates that they should be sheathed or not.

Summary

My work on Jane's reconstruction required addressing several conceptual problems. The degree of jaw closure seen in preserved tyrannosaur skulls may not always represent the degree to which the living animal would have closed its mouth. This raises the question of how far the animal could have closed its mouth. My observations of large, toothy, terrestrial reptiles forged my impression that the jaw closed only so far that the opposing maxillary and dentary teeth crossed in a functional way. The crocodilians' utility as an extant analog for tyrannosaur reconstructions is restricted by its aquatic adaptations. Separating the nasal airflow from air that comes in through the mouth enhances olfaction, and a feature that allows this (without a bony crocodilian palate) is reptilian lips. The oral cavity, presumably moist from oral glands, would avoid evaporation and the risk of dehydration with lips to seal the mouth closed.

In surveying the state of life reconstruction in paleoart, I'm reminded of early wildlife illustrators' attempts at depicting gorillas 100–150 years ago. Known only anecdotally and from trophy skulls at that time, gorillas were often depicted as rampaging, canine-baring beasts (Townsend 1890:28–29). Lacking an understanding of the living animal's anatomy and behavior, the artists instead fixated on the teeth, honing in on a primal human fear. Are today's museum exhibits similarly sensationalizing dinosaurs in the face of uncertainty? The tyrannosaur was an animal that had to deal with life in a terrestrial environment, and depictions must take this, as well as less-scrutinized aspects of its anatomy, into account.

In 1917, Henry Fairfield Osborn wrote,

> A serious restoration is to be regarded as a trial hypothesis, in which is expressed all the existing knowledge of the subject. It by no means discredits a restoration that, after a lapse of years, new knowledge may radically modify the scientist's conceptions and make a new restoration necessary. The best that the serious restorer succeeds in doing is to give an idea of the external proportions, based on the arrangement of muscles in adaptation to certain habits, and the general mode of life. Accepting it as such, a truly scientific and artistic restoration of an extinct animal and its environment is of very great value as an interpretation, and is, therefore, helpful to both the man of science and to the layman.(Osborn 1917:inside front cover)

While my sculpture of Jane is by no means the last word in tyrannosaur reconstructions, I hope to have at least presented some features that are not commonly seen (but are based on extant taxa) and to spur additional dialog and investigation.

Acknowledgments

Special thanks to John Lanzendorf, for opening doors to the world of paleontology and paleoart for me. Thanks to Paul Sereno, for providing myriad challenges and opportunities in my work as preparator and paleoartist. Thanks to Steve Brusatte for review of an earlier draft of this paper. Thanks to Scott Williams, for inviting me to participate in the symposium that spawned this book, and for believing in the value of reconstructions in a museum. Thanks to Donna Braginetz, Thomas Carr, Philip Currie, Michael Henderson, Peter Larson, Tetsuto Miyashita, Takashi Oda, Michael Parrish, Luis Rey, Mineo Shiraishi, and Michael Skrepnick for helpful comments during my work with Jane. Finally, a special thanks to my wife Kari, for her loving support and paleopatience.

Abler, W. L. 2000. The teeth of the tyrannosaurs; pp. 276–278 in G. P. Paul (ed.), *The Scientific American Book of Dinosaurs*. St. Martin's Press, New York.

Alexander, R. McN. 1994. *Bones: The Unity of Form and Function*. Macmillan, New York.

Auffenberg, W. 1981. *The Behavioral Ecology of the Komodo Monitor*. University Presses of Florida, Gainesville.

Bakker, R. T. 1986. *The Dinosaur Heresies*. Kensington, New York.

Bakker, R. T., M. Williams, and P. J. Currie. 1988. *Nanotyrannus*, a new genus of pygmy tyrannosaur, from the latest Cretaceous of Montana. *Hunteria* 1(5):1–30.

Brainerd, E. 2001. Buccal oscillation behavior in amniotes. *Journal of Vertebrate Paleontology* 21(3, suppl.):35A.

Brochu, C. A. 2000. In J. J. Lanzendorf, *Dinosaur Imagery: The Science of Lost Worlds and Jurassic Art: The Lanzendorf Collection*, edited by C. R. Crumley, 44. San Diego: Academic Press.

Brochu, C. A. 2003. Osteology of *Tyrannosaurus rex*: insights from a nearly complete skeleton and high-resolution computed tomographic analysis of the skull. *Memoirs of the Society of Vertebrate Paleontology* 7:1–138.

Busbey, A. B. 1994. The structural consequences of skull flattening in crocodilians; pp. 173–192 in J. J. Thomason (ed.), *Functional Morphology in Vertebrate Paleontology*. Cambridge University Press, Cambridge.

Carpenter, K. 1997. Tyrannisauridae; pp. 766–768 in P. J. Currie and K. Padian (eds.), *Encyclopedia of Dinosaurs*. Academic Press, Orlando.

Currie, P. J. 1997. Theropods; pp. 216–233 in J. O. Farlow and M. K. Brett-Surman (eds.), *The Complete Dinosaur*. Indiana University Press, Bloomington.

Currie, P. J. 2003. Cranial anatomy of tyrannosaurid dinosaurs from the Late Cretaceous of Alberta, Canada. *Acta Paleontologica Polonica* 48(2):191–226.

Currie, P. J. 2004. *Feathered Dragons: Studies on the Transition from Dinosaurs to Birds*. Indiana University Press, Bloomington.

Digimorph.org. 2012. *Digital Morphology: A National Science Foundation Digital Library at the University of Texas at Austin*. http://www.digimorph.org/.

Edmund, A. 1969. Dentition; pp. 117–200 in C. Gans, A. Bellairs, T. Parsons (eds.), *Biology of the Reptilia*, vol. 1: *Morphology A*. Academic Press, London and New York.

Erickson, G. M. 2000. Breathing life into *Tyrannosaurus rex*; pp. 267–275 in G. P. Paul (ed.), *The Scientific American Book of Dinosaurs*. St. Martin's Press, New York.

Ford, T. 1997. Did theropods have lizard lips? Southwest Paleontological Symposium–Proceedings. 1997:65–78.

Hallett, M. 2004. Jacket art for R. Molnar. 2004. *Dragons in the Dust*. Indiana University Press, Bloomington.

Kiley, R., C. Hughes, and D. Hughes. 1995. Last feast of the crocodiles. *National Geographic Television*. VHS video. National Geographic Video, Culver City, California.

Knight, F. 1985. Illustration; p. 153 in Megalania prisca, the Giant Goanna by T. H. Rich and P. V. Rich; in P. V. Rich, G. F. van Tets, and F. Knight, *Kadimakara: Extinct Vertebrates of Australia*. Pioneer Design Studio, Victoria.

Lamanna, M. C., R. D. Martinez, and J. B. Smith. 2002. A definitive abelisaurid theropod dinosaur from the early late cretaceous of Patagonia. *Journal of Vertebrate Paleontology* 22(1):58–69.

Levy, C. 1991. *Crocodiles and Alligators*. Chartwell Books, New Jersey.

Manabe, M. 2000. *Encyclopedia of Dinosaurs*. Gakken, Tokyo.

Osborn, H. F. 1917. Extinct animal forms (introductory statement). *Mentor* 5(13):inside front cover.

Owen, R. 1866. *On the Anatomy of Vertebrates*, vols. 1–2. Longmans, Green, London.

Papp, M. J., and L. M. Witmer. 1998. Cheeks, beaks, or freaks: a critical appraisal of buccal soft-tissue anatomy in ornithischian dinosaurs. *Journal of Vertebrate Paleontology* 18(3, suppl.): 69A.

Paul, G. S. 1988. *Predatory Dinosaurs of the World*. Simon & Schuster, New York.

Paul, G. S. 2000. *The Scientific American Book of Dinosaurs*. St. Martin's Press, New York.

Proctor, N. S., and P. J. Lynch. 1993. *Manual of Ornithology: Avian Structure and Function*. Yale University Press, New Haven, Connecticut.

Ruben, J., A. Leitch, W. Hillenius, N. Geist, and T. Jones. 1997. New insights into the metabolic physiology of dinosaurs; pp. 505–517 in J. O. Farlow and M. K. Brett-Surman (eds.), *The Complete Dinosaur*. Indiana University Press, Bloomington.

Schubert, B. W., and P. S. Ungar. 2005. Wear facets and enamel spalling in tyrannosaurid dinosaurs. *Acta Palaeontologica Polonica* 50(1):93–99.

Shine, T., W. Bohme, H. Nickel, D. F. Thies, and T. Wilms. 2001. Rediscovery of relict populations of the Nile crocodile. *Oryx* 35(3):260–262.

Soares, D. 2002. An ancient sensory organ in crocodilians. *Nature* 417:241–242.

Townsend, G. 1890. *Friend and Foe from Field and Forest.* L. P. Miller & Co., Chicago.

Witmer, L. M. 1995. The extant phylogenetic bracket and the importance of reconstructing soft tissue in fossils; pp. 19–33 in J. J. Thomason (ed.), *Functional Morphology in Vertebrate Paleontology.* Cambridge University Press, Cambridge.

Witmer, L. M. 2001. Nostril Position in Dinosaurs and Other Vertebrates and Its Significance for Nasal Function. *Science* 293:850–853.

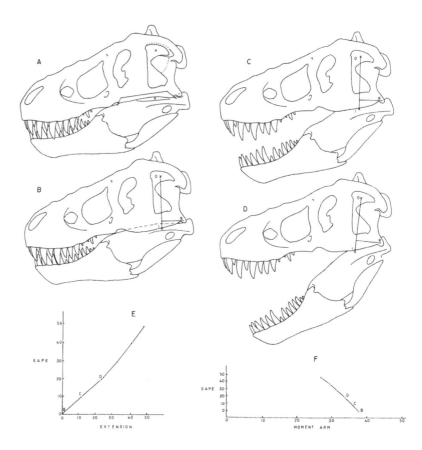

8.1. Illustration of the methodology for estimating the lever arm and muscle extension of the jaw muscles of *Tyrannosaurus rex*, using the *M. adductor mandibulae externus superficialis et medialis* as an example. A) First the areas of origin (dashed) and insertion of the muscles are determined (or postulated). Then a "center" of each area is estimated. This "center" is chosen subjectively, taking into account the angle of the surface with respect to the plane of the projection, the relative proportion of fibers originating from this area, and so forth. In this example, the "center" (indicated by "x") is set in the middle of the upper lobe of the infratemporal fenestra. The apparently large area of origin of the squamosal-quadratojugal flange probably contributed relatively few fibers, as the orientation of the muscle was parallel to the surface of the flange. However, the area of origin along the dorsal margin of the infratemporal fenestra is only small in projection since it is situated nearly perpendicular to the plane of projection. Hence, this area presumably contributed more fibers to the muscle than the squamosal-quadratojugal flange, although that area appears to be the greater. B) After the "centers" have been chosen, a line is constructed from the origin "center" to the insertion "center." The length of this line is taken as a measure of the length of the muscle fibers. The perpendicular distance from this line (OI) to the center of rotation of the quadrate condyles (r) is the lever arm for this muscle at this gape. The measurements were made for angles of 0°–50° of gape, with zero being taken as that gape for which the tips of several of the dentary teeth reach the ventral margin of the maxilla. C) The same for a gape of 10°. D) The same for a gape of 20°. E) Graph of extension vs. gape. The extension plotted here is the extension for zero gape subtracted from the extension for a given gape, so that the extension for zero gape is zero. This graph therefore represents the relation of gape to the length of the muscle extended beyond its (presumed rest) length with the mouth closed. The units of the abscissa are centimeters times 0.2: the reason for this unconventional unit is that the measurements were made on a one-fifth scale projection of the skull. The points on this graph labeled B, C, and D are likewise derived from the extensions illustrated in parts (B), (C), and (D), respectively. F) Graph of the lever arm versus gape. This graph is constructed from the lengths of the lever arms as shown in (B), (C), and (D). The abscissa represents the lever arm, and the ordinate the gape. The units of the axes of this graph are the same as those of the previous graph. The points on the graph labeled B, C, and D are derived from the lever arms illustrated in parts (B), (C), and (D), respectively.

A Comparative Analysis of Reconstructed Jaw Musculature and Mechanics of Some Large Theropods

8

Ralph E. Molnar

Abstract

Tyrannosaurus rex was compared, in terms of estimated lever arm and extension of the jaw adductors, with *Daspletosaurus torosus*, *Nanotyrannus lancensis*, *Allosaurus fragilis*, and *Ceratosaurus nasicornis*. *Daspletosaurus torosus* agrees reasonably closely with *T. rex* in these parameters, indicating that no great differences in feeding adaptation were apparent from this data. *Nanotyrannus* differs more from *T. rex*; these differences appear to be related to the relatively lower skull of *N. lancensis* and suggest that *N. lancensis* had a less powerful bite than the other tyrannosaurids examined.

   The muscular arrangement in *Allosaurus fragilis* is basically similar to that of *Tyrannosaurus rex*. The adductors of *A. fragilis* had generally lower lever arms, suggesting a weaker bite but possibly indicating a greater gape. The lever arms of *Ceratosaurus nasicornis* are reasonably similar to those of the tyrannosaurids, suggesting that both had stronger bites than *A. fragilis*.

Introduction

This work aims to infer from the reconstruction of the jaw musculature of *Tyrannosaurus rex* how these muscles may have acted and draw some implications from these inferences for the behavior of *T. rex* and to compare the positions and inferred actions of the muscles with those of other large theropods (*Allosaurus fragilis*, *Ceratosaurus nasicornis*, *Daspletosaurus torosus* and *Nanotyrannus lancensis*). The goal is to determine if differences of cranial structure between the selected taxa relate to differences in the inferred forms of the jaw muscles. This study originally formed part of a Ph.D. dissertation at the University of California, Los Angeles, in 1973. The many subsequent discoveries of theropod material, including that of *T. rex* since then, have not materially affected the conclusions reached here. The dissertation also included comparison with *Gorgosaurus libratus*. The analysis was based on Russell's (1970) figure, in turn based on FMNH PR 308, with the muscle placement assessed directly from a specimen, AMNH 5336. However, FMNH PR308 is now identified as *Daspletosaurus* sp. (Currie 2005), and AMNH 5336 is *G. libratus* (Currie 2005), so the analysis of "*Gorgosaurus libratus*" in the dissertation is unreliable.

Institutional Abbreviations AMNH, American Museum of Natural History, New York; CMNH, Cleveland Museum of Natural History, Cleveland, Ohio; DINO, Dinosaur National Monument, Vernal, Utah; FMNH, Field Museum of Natural History, Chicago; NMC, National Museum of Canada, Ottawa; USNM, National Museum of Natural History, Smithsonian Institution, Washington, D.C.

Level of Deduction	Basis of Deduction	Subject of Deduction
First	Observation	Form and position of muscles and of muscle attachment areas in modern analogs
Second	Inference from modern specimens	Muscle attachment areas in *Tyrannosaurus rex*
Third	Inference from first and second levels	Form and position of muscles in *Tyrannosaurus rex*
Fourth	Inference from third level	Properties of bite in *Tyrannosaurus rex*

Table 8.1. Levels of logical deduction

Reconstructing the musculature of *Tyrannosaurus rex* is an exercise in anatomy and logic of limited interest in itself. More interesting is taking the reconstruction a step further to infer properties and behaviors of the once-living organism. However, this carries with it the penalty of decreased confidence. Reconstructing musculature involves a certain degree of assumption (cf. Molnar 2008), and carrying this further into function adds to the burden of assumption. The conclusions presented here are based on several levels of analysis (Table 8.1): first, the observation of fossils; second, judicious interpretation from modern organisms, which is the first order of inference; third, deductions and extrapolations from the data used to reconstruct the musculature (the second order of inference); and fourth, further deductions and extrapolations regarding the life of *T. rex* (the third order of inference). The second- and third-order inferences—inferences based at least in part on other inferences—cannot be held with the same degree of confidence as those of first order, which themselves must be given less confidence than the observations on which they are based. Thus this is the most speculative of the series of essays (Molnar 1991, 2000, 2008) on the cranial structure of *T. rex*.

The reconstruction of the musculature (based largely on descriptions of crocodilians in the literature (particularly Iordansky 1964) and dissections of *Alligator mississipiensis* and *Paleosuchus trigonatus* and terminology used are given in Molnar (2008). The small amount of potential movement, if any, inferred between the cranial elements (Molnar 1991) justifies the approximation of an akinetic skull in interpreting the actions of the muscles.

Motions of bones produced by the actions of muscles are physical, rather than biological, effects, and thus the interpretation of the muscular actions is here treated as an exercise in elementary mechanics. The mandibles are taken to rotate rigidly about the craniomandibular joints, without any spreading of the mandibles as they opened or any motion inferred at either an intramandibular joint or the symphyseal contact. This is clearly an approximation, as the mandibles did spread laterally as they opened (Molnar 1991), there may well have been an intramandibular joint, and motion may also have been permitted at the symphysis (cf. Molnar 1991). Mandibular spreading and symphyseal motion appear to have been small and hence are omitted in the approximation here.

Muscle abbreviations
*M. add. mand. ext. prof.,
M. adductor mandibulae
externus profundus; M. add.
mand. ext. sup. med., M. adductor mandibulae externus
superficialis et medialis;
M. add. mand. post., M. adductor mandibulae posterior;
M. dep. mand., M. depressor
mandibulae; M. pteryg. dors.,
M. pterygoideus dorsalis.*

Given that the opening of the jaw is here assumed to be a simple rotation, its rotational or angular acceleration is the result of an imposed torque just as linear acceleration is the result of an imposed force. A torque (also known as "moment of force," or just "moment") is the product of a force multiplied by the perpendicular distance from the line of action of this force to the relevant axis of rotation. This perpendicular distance is termed the "lever arm" (or "moment arm") of the force. Since the jaw is analogous to a lever with the fulcrum at the craniomandibular joint, the force generated at any point along the mandible multiplied by its lever arm is equal to the sum of the products of the forces exerted by the adductors, each multiplied by its respective lever arm. If the forces generated by the muscles are equal, the contribution of each to the torque will depend only upon the lever arm. Since the forces generated by a muscle of an extinct organism cannot be directly determined, one way to approach an understanding of its contribution to the torque is by considering its lever arm. This, however, gives only a rough approximation without some idea of the relative strengths of the muscles. This complication is to some extent ameliorated because, with the exception of *Tyrannosaurus rex* and *Nanotyrannus lancensis*, the relative sizes (and hence, presumably the strengths) of the muscles seem to have been similar in the forms examined.

The vertical distance to which the jaw may be depressed is largely determined by two factors: the amount of rotation possible at the jaw joint (in turn determined by the forms of the joint surfaces of the involved elements and by the ligaments and joint capsule present at the joint), and the amount by which the jaw adductors can extend beyond their length when the mouth is closed. Although ligaments and a joint capsule were doubtless present in *Tyrannosaurus rex*, the amount by which they restricted movement cannot be directly determined. Since the medial condyle of the quadrate slides clear of the articular glenoid when the jaw is opened to about 45°, it would seem likely that ligaments would have restricted the jaw depression to less than 45°. The measurements were made for angles of 0°–50° of gape (beyond the likely maximum), with zero being taken as that gape for which the tips of the dentary teeth reach the ventral maxillary margin.

Fulton (1955) reported that skeletal muscle may be reversibly stretched to at least 15 percent of its equilibrium length (i.e., the length at which its resting tension is zero). In humans, at least some of the fibers of the appendicular musculature can be contracted to 57 percent of their fully extended length (Haines 1934), which amounts to an extension to 132.6 percent of their fully contracted length (Parrington 1955). In order for the canines to clear the mandible, one of the jaw adductors of the gorgonopsid *Leontocephalus intactus* (Kemp 1969) may have extended to about 250 percent of its fully contracted length. In view of these differences in measured or estimated extension (cf. also Gans and Bock 1965), it is uncertain what percentage extension may be reasonably used as maximal in determining the greatest extent of jaw opening. Hence, only relative percentage extensions are used here, and the maximum opening of the mouth is assumed to have been determined by osseous and ligamentous constraints. Measuring percentage extension permits some comparison of the degrees to which the mouth may be opened.

Although the name "*Nanotyrannus lancensis*" is used here, this does not indicate confidence that *N. lancensis* is a valid species: it may be, as proposed by Carr (1999), the juvenile of *Tyrannosaurus rex*. However, as demonstrated at "The Origin, Systematics, and Paleobiology of Tyrannosauridae," the conference held in Rockford, Illinois, on September 16–18, 2005, upon which this volume is based, uncertainties about its status remain.

Function of Jaw Musculature

Methods

The lever arms and lengths of the various jaw muscles were estimated from a physical model of a lateral projection of the skull and jaws of AMNH 5027 reproduced as plate 1 of Osborn (1912) at one-fifth natural size. The jaw was separately mounted and hinged so that it could be freely rotated relative to the skull. Details are given in Figure 8.1. Multiplying by five to correct for the scale gave the lever-arm estimate. Measurements were made for angles of depression at intervals of 10° to a maximum of 50°. As described in the previous section, this should cover the entire range of jaw openings.

The length of each muscle at each stage of jaw depression was estimated by measuring the distance between the inferred centers of the areas of origin and insertion with the model jaw rotated by the requisite angle. Multiplying by five to correct for the scale gave the estimate of muscle length. This length was then divided by the length of the muscle with the jaw closed and multiplied by 100 to give the percentage extension. The uncertainty in the measurements was 1 mm so that the values are accurate to two figures.

Results

M. ADDUCTOR MANDIBULAE EXTERNUS SUPERFICIALIS ET MEDIALIS In both *Alligator* and *Paleosuchus* this is a parallel-fibered muscle sheet just medial to the infratemporal fenestra and posterior ramus of the jugal. In *Tyrannosaurus rex*, this muscle was probably also thin with parallel fibers since the attachment areas are elongate anteroposteriorly but thin transversely (Molnar 2008). It had the second lowest lever arm over 0°–35° of jaw depression (Fig. 8.2). With the mouth shut, this muscle was the second shortest of the jaw adductors but had the second greatest rate of increase of extension with increasing depression of the jaw (Fig. 8.3). This is consistent with the observation of Gans and Bock (1965) that parallel-fibered muscles are capable of the greatest extensions and the inference that this was a parallel-fibered muscle in *T. rex*. In the transverse

Tyrannosaurus rex

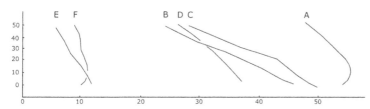

Nanotyrannus lancensis

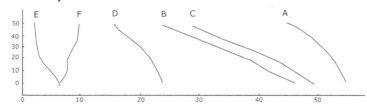

Daspletosaurus torosus

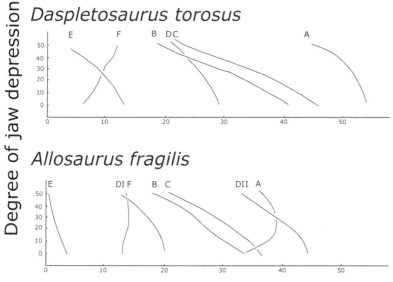

Allosaurus fragilis

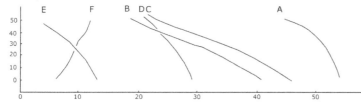

Ceratosaurus nasicornis

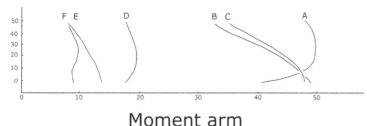

Moment arm

Degree of jaw depression

8.2. Graphs of the lever arm (abscissa) versus gape (ordinate) for the theropods studied here. The muscles represented in each graph are A) *M. pterygoideus dorsalis;* B) *M. adductor mandibulae externus profundus;* C) *M. pseudotemporalis;* D) *M. adductor mandibulae externus superficialis et medialis;* DI) *M. adductor mandibulae externus superficialis;* DII) *M. adductor mandibulae externus medialis;* E) *M. adductor mandibulae posterior;* and F) *M. depressor mandibulae.* The gape is in degrees, and the lever arm in fifths of centimeters.

8.3. Graphs of percentage extension vs. gape (by muscle) for the jaw muscles of the various theropods studied. The taxa represented are A) *Allosaurus fragilis*; A1) *M. adductor mandibulae externus superficialis*; A2) *M. adductor mandibulae externus medialis*; B) *Nanotyrannus lancensis*; C) *Daspletosaurus torosus*; D) *Tyrannosaurus rex*; and E) *Ceratosaurus nasicornis*. The units are the same as in Figure 8.2.

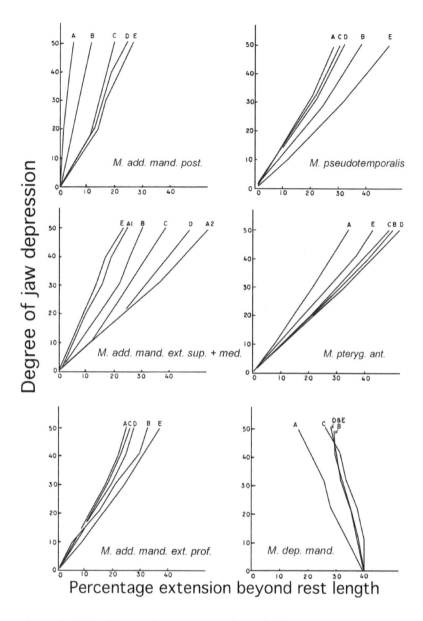

plane, the *M. add. mand. ext. sup. med.* would have exerted an almost vertical force on the mandible (Fig. 8.4).

M. ADDUCTOR MANDIBULAE EXTERNUS PROFUNDUS This is a parallel-fibered muscle of relatively small cross section but relatively large lever arm in both *Alligator* and *Paleosuchus*. In *Tyrannosaurus rex*, the relatively much larger area of origin of this muscle suggests that it was relatively larger than in the crocodilians. There is no indication of a zwischensehne in *T. rex* (as exists in crocodilians), so the muscle presumably inserted directly onto the mandible (Molnar 2008).

The lever arm of the *M. add. mand. ext. prof.* was moderate over all angles of depression (Fig. 8.2). This (with the *M. pseudotemporalis*) was the longest of the adductors. Portions of the M. *pterygoideus dorsalis* were longer. However, the distance from the center of the area of origin (taken as the center of the antorbital fenestra) to the center of the area of insertion of the M. *pteryg. dors.* was less. The maximal percentage of possible extension of the M. *add. mand. ext. prof.* was low, less than 130 percent of the muscle's length with the mouth closed (Fig. 8.3). In the transverse plane this muscle was inclined at about 15° to the medial of vertical (Fig. 8.4).

M. *ADDUCTOR MANDIBULAE POSTERIOR* This muscle is complexly pinnate in both *Alligator* and *Paleosuchus*, containing several tendon sheets (the A-, B-, and ls-tendons of Iordansky, 1964) that attach to the quadrate and laterosphenoid on prominent ridges. No such ridges are found in *Tyrannosaurus rex*, so presumably the tendons were absent, and the M. *add. mand. post.* was not complexly pinnate.

This muscle has the lowest lever arm of the jaw adductors (Fig. 8.2) and also a low maximum percentage extension (ca. 130 percent at 50° of jaw depression; see Fig. 8.3). The line of action of the M. *add. mand. post.* was directed nearly vertically in the transverse plane (Fig. 8.4).

M. *PSEUDOTEMPORALIS* Like the M. *add. mand. ext. prof.*, the pseudotemporal in *Alligator* and *Paleosuchus* also has parallel fibers, but a unipinnate insertion, and has a small cross section and moderate lever arm. Again, in *Tyrannosaurus rex*, the area of origin of this muscle appears to have been relatively much larger than in the crocodilians.

This muscle had the second greatest lever arm over all angles of jaw depression (Fig. 8.2). It had the greatest length when the jaw was closed (just greater than that of the M. *add. mand. ext. prof.*) and a relatively low percentage extension (to just over 130 percent). Similar to the M. *add. mand. ext. prof.*, this muscle was also inclined to about 15° medial of vertical in the frontal plane.

M. *PTERYGOIDEUS DORSALIS* This is the largest and presumably most powerful of the jaw adductors in both *Alligator* and *Paleosuchus*. It is parallel-fibered but unipinnate at its insertion. In *T. rex* this was large but is not clearly the largest adductor, as both the M. *add. mand. ext. prof.* and the M. *pseudotemporalis* are relatively larger than in crocodilians and thus would have approached, or perhaps surpassed, the anterior pterygoid in overall size in *Tyrannosaurus rex*. It is assumed here, as in Molnar (2008), that the anterior pterygoid took origin from the margins of the antorbital fenestra, rather than the alternative view of Witmer (1997) that this region was occupied by a paranasal sinus. The area of origin in *T. rex* is elongate anteroposteriorly but thin transversely. Hence this muscle was apparently more sheetlike in *T. rex* than in the living crocodilians, and there is no evidence that it was pinnate. It would seem best to assume

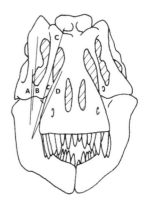

8.4. Skull of *Tyrannosaurus rex* in anterior view, with the lines of action of some of the jaw adductors indicated. A) M. adductor mandibulae posterior; B) M. adductor mandibulae externus superficialis et medialis; C) M. adductor mandibulae externus profundus and M. pseudotemporalis; and D) M. pterygoideus dorsalis.

that it was about equal in volume to the M. *pseudotemporalis* and the M. *add. mand. ext. prof.*

The M. *pteryg. dors.* had the greatest lever arm of the adductors; it did not decrease monotonically with increasing jaw opening, as for all of the other adductors, but reached its maximum at an angle of depression of about 10°, decreasing monotonically with both increasing and decreasing depression of the jaws (Fig. 8.2). This muscle was relatively long and had the greatest percentage extension of the adductors, about 150 percent for an angle of depression of 50° (Fig. 8.3). The M. *pteryg. dors.* would exert a force directed at about 25° medial of the vertical (in the transverse plane) on the mandible (Fig. 8.4).

M. *PTERYGOIDEUS VENTRALIS* AND M. *INTRAMANDIBULARIS* These two muscles will not be discussed here as there is no anatomical evidence for the location of either in *Tyrannosaurus rex* (Molnar 2008), although some inferences will be drawn regarding their function.

M. *DEPRESSOR MANDIBULAE* In *Alligator* and *Paleosuchus*, this partially unipinnate muscle arises from a short tendon attached to the exoccipital and also has a fleshy origin from the posterior face of the parietal. The insertion is also fleshy. The area of origin is similar in *Tyrannosaurus rex* to those of the crocodilians examined. No evidence for any tendons (and hence pinnation) in this muscle exists in *T. rex*.

The lever arm of the M. *dep. mand.* in *Tyrannosaurus rex* was relatively low for all of the angles of depression examined and did not obviously vary with the angle of depression (Fig. 8.2).

Discussion

The adductor musculature exerted a dorsally directed force on the mandibles, resulting either in closing the mouth or transmitting the force via the mandible (and teeth) to the prey. The ventrally directed reactions exerted by these muscles on the skull have already been analysed in Molnar (2000). There would also have been subsidiary effects of the adductors on the lower jaws, such as a tendency to rotate the posterior portion of the mandible about its longitudinal axis, holding the articular region appressed to the quadrate, maintaining the contact at the symphysis, and so forth. Determining the contribution of each of the muscles to each of these actions is difficult even with living creatures, as is the integration of individual contributions to obtain an idea of the resulting action as a whole. The attempt will be made, however, for it may safely be assumed that the jaw muscles did not act independently.

M. *ADDUCTOR MANDIBULAE EXTERNUS SUPERFICIALIS ET MEDIA-LIS* This muscle had a relatively small cross section, a relatively low lever arm, and a relatively large percentage extension and thus seems to have acted as a simple, relatively weak adductor of the jaw.

Acting alone, this muscle would have tended to disarticulate the jaw caudally. This tendency would have increased with increasing depression of the mandible, for as the jaw rotated the dorsoposterior component of the muscular force would have come to be nearly parallel to the long axis of the mandible. The component perpendicular to the long axis of the mandible, the smallest of those of any of the adductors, that would have tended to appress the jaw to the quadrate, would correspondingly have decreased. The force exerted was directed toward the back of the jaw at all angles considered. Since this muscle would have tended to pull the mandible posteriorly when the mouth was open, it presumably acted as a synergist with the *M. pteryg. dors.*, which would have tended to disarticulate the mandible anteriorly. This would have countered the strong anterior component exerted by that muscle, hence reducing the risk of dislocating the jaw joint.

M. ADDUCTOR MANDIBULAE EXTERNUS PROFUNDUS AND *M. PSEUDO-TEMPORALIS* These muscles share many properties and are thus treated together. Both had moderate to large lever arms and cross sections: the lever arm of the *M. pseudotemporalis* was greater than that of the *M. add. mand. ext. prof.*, whereas the cross-sectional area of the latter was probably greater. In both, the lever arm decreased with depression of the mandibles, the percentage extension was relatively low, and the line of action inclined at about 15° medial of vertical in the frontal plane when the jaw was shut. Both were probably strong adductors of the lower jaw. They show only a slight tendency to disarticulate the mandible since they acted in a nearly vertical direction in sagittal projection.

M. ADDUCTOR MANDIBULAE POSTERIOR The *M. add. mand. post.* had a relatively small cross section, the minimum lever arm, and a relatively low percentage extension. The very low lever arm of this muscle relative to the quadrate condyle implies a correspondingly low leverage in closing the mandibles. However, a small lever arm about the jaw joint implies a large one about a point of application of resistance at the front of the lower jaws. Hence, in *Tyrannosaurus rex*, as in crocodilians (Iordansky 1964), this muscle probably acted as an anti-luxation muscle, preventing the disarticulation of the jaws by resistance forces generated by the prey.

Originating from the pterygoid process of the quadrate and inserting onto the mandible, this muscle had a line of action approximately parallel to the long axis of the body of the quadrate. Pauwels (1948) showed that a muscle in this relationship to an asymmetrically loaded column can act to reduce the strain in that element (see also Young 1957:chap. 5). Hence this muscle may have acted to reduce stress in the body of the quadrate: its function then could have had less to do with closing the jaws than with maintaining the craniomandibular articular system.

M. PTERYGOIDEUS DORSALIS This would have acted to adduct the jaws, being most efficient when the jaws were opened at an angle of 10°

rather than when they were nearly closed, as with the other adductors. Its relatively large cross section, maximum lever arm, and the greatest percentage extension would have made the M. *pteryg. dors.* one of the main adductors.

It would have exerted a medial pull on each mandible (Fig. 8.4) as well as an anterior force at all angles of depression. This anterior component is relatively large, and the M. *pteryg. dors.* could have counteracted the posteriorly directed component of the M. *add. mand. ext. sup. med.*, which would have tended to disarticulate the cranio-mandibular joint.

M. *PTERYGOIDEUS VENTRALIS* If the M. *pterygoideus dorsalis* originated from the plate formed by the pterygoid process of the quadrate and the quadrate process of the pterygoid (and inserted onto the medial face of the surangular), it would have occupied a position much like that of the M. *add. mand. post.*, and its action might be expected to have been similar. If this muscle originated from the ectopterygoid (as suggested by Ostrom 1969), then it would have exerted a strongly anteriorly directed force upon the mandible, as it would have been orientated nearly parallel to the M. *pteryg. dors.* In either case, it would have exerted a more medially directed force on the mandible than any of the other adductors.

M. *INTRAMANDIBULARIS* The M. *intramandibularis* would have acted to pull posteriorly on the dentary. This would have provided a centripetal force on the mandible as it closed, as well as resisting the anteriorly directed forces exerted on the mandible by the prey.

M. *DEPRESSOR MANDIBULAE* Unlike most other theropods, tyrannosaurids do not have a retroarticular process, the depressor having inserted immediately behind, and almost below, the jaw joint. The force was directed nearly vertically on the mandible such that its mechanical efficiency in opening the mouth must have been very low because of its short lever arm. However, a terrestrial creature can open the mouth by merely relaxing the jaw adductors, the weight of jaws and associated structures acting as the motive force. So the depressor would have mainly acted in increasing the gape of the mouth once it had already opened.

The M. *add. mand. ext. prof.*, the M. *pseudotemporalis*, and the M. *pteryg. dors.* appear to have acted together as simple, powerful adductors of the jaws. The M. *add. mand. ext. sup. med.* was probably a weak adductor of the jaws and acted almost vertically in the frontal plane. The M. *add. mand. post.* was apparently also weak and may have acted chiefly to prevent disarticulation of the jaws and reduce stress in the quadrate.

The M. *add. mand. ext. prof.*, the M. *pseudotemporalis*, and the M. *pteryg. dors.*, all of which probably inserted together onto the same region of the mandible, appear to have been arranged in such a fashion as to maximize the torque exerted on the jaw. These muscles were moderate to large and, hence, presumably powerful and inserted about as far anteriorly on the jaw as was compatible with passage through the

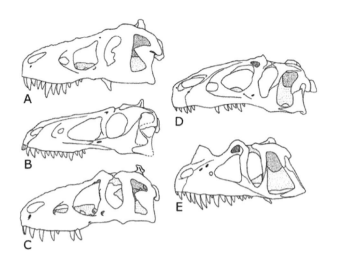

8.5. Lateral views of the skulls of the theropods treated here, drawn to the same (premaxilla to quadrate condyle) length. A) *Tyrannosaurus rex* (Osborn 1912); B) *Nanotyrannus lancensis* (Gilmore 1946, modified); C) *Daspletosaurus torosus* (Russell 1970); D) *Allosaurus fragilis* (DINO 2560, original); E) *Ceratosaurus nasicornis* (Gilmore 1920).

subtemporal fenestra. Their orientation was such as to maximize the possible lever arms. Thus these muscles presumably provided the bulk of the motivating force for biting. The *M. add. mand. ext. sup. med.* and the *M. add. mand. post.*, in contrast, seem to have been associated with maintaining the "trim" of the lower jaws while they were closing – that is, keeping them properly aligned on the skull.

The medial components of the forces exerted on the jaw by the adductors (Fig. 8.4) may well have been substantial. They would have pulled the surangular portions of the mandible medially and also rotated this region of the lower jaw about its long axis, so that the dorsal margin was displaced medially and the ventral margin laterally. When the jaws were not widely open, these medial components in living crocodilians may be resisted by the pterygoid wings, which approach closely to the medial surfaces of the mandibles. This is not so in theropods, and it would seem reasonable, therefore, to suggest that the mandible was rigid and was rigidly joined to its antimere at the symphysis. Regardless of whether or not the symphyseal joint was rigid, mandibles of at least some theropods, including *Tyrannosaurus*, probably had an intramandibular joint. This poses a problem in understanding just how the medially directed components were resisted or whether they functioned in operating the intramandibular joint.

Methods

For comparison, two-dimensional projections like those described in the previous section were constructed for the following (Fig. 8.5): *Daspletosaurus torosus*, from Russell's (1970) figure of the complete and undistorted skull of the holotype NMC 8506; *Nanotyrannus lancensis*, from the holotype skull, CMNH 7541, with modification to correct for the crushing in the quadrate region; *Allosaurus fragilis*, from the complete, articulated and undistorted skull DINO 2560; and *Ceratosaurus nasicornis*, from Gilmore's reconstruction (1920) of the holotype USNM 4735.

Comparative Interpretation

Table 8.2. Percentage differences in moment arms from those of *Daspletosaurus torosus* (accurate to ± 1%). Positive values indicate greater leverages.

Taxon	M. pterygoideus dorsalis	M. pseudotemporalis	M. add. mand. ext. sup. med.	M. add. mand. ext. prof.	M. add. mand. post.	M. dep. mand.
Tyrannosaurus rex	1–4	6–12	29–22	12–13	−9–+6	67−−14
Nanotyrannus lancensis	1–0	8–14	−19−−25	14–18	−52−−63	7−−14
Allosaurus fragilis	−37−−22	−21−−11	see text	−9−−7	−73−−88	113–25
Ceratosaurus nasicornis	−21−+2	4–32	−39−−23	17–48	0–44	47−−25

Values given are for 0° and 40° of jaw depression, respectively.

The location of the centers of the attachment areas was carried out on the actual specimens for *N. lancensis* and *A. fragilis* and was estimated from published figures and photographs and the holotype specimen for *C. nasicornis* and from published figures and photographs for *D. torosus*.

All models were constructed so that the distances between the quadrate condyle and the anterior tip of the premaxilla were equal. Hence the relative lever arms and excursions could be measured directly. The length chosen was that of the image of the skull of *Tyrannosaurus rex* in plate 1 of Osborn (1912).

Results

Although this study is concerned with *Tyrannosaurus*, in order to facilitate comparison with other large theropods, that form is not used as the standard of comparison because its expanded postorbital region (also found in *Nanotyrannus*) is not present in the great majority of other theropods. Instead, the curves for *Daspletosaurus torosus* (Fig. 8.2) are used as a standard for comparison. The results are presented in Figures 8.2 and 8.3, and those of Figure 8.2, expressed as percentage departure of lever arms from the values for *D. torosus*, are given in Table 8.2: these values were calculated from the graphs. In general, the graphs of lever arm versus amount of jaw depression are similar for all three tyrannosaurids, that of the smallest, *N. lancensis*, differing the most from the others (see Fig. 8.2). These results are considered by taxon.

TYRANNOSAURUS REX (Fig. 8.2) Most of the adductors and the depressor of *Tyrannosaurus rex* had relatively greater leverage than in *Daspletosaurus torosus*, and the M. add. mand. post. that does not, had approximately equivalent leverage.

NANOTYRANNUS LANCENSIS (Fig. 8.2) Unlike those of *Tyrannosaurus rex*, most of the jaw muscles of *Nanotyrannus lancensis* had approximately equivalent or relatively less leverage than those of *Daspletosaurus torosus*.

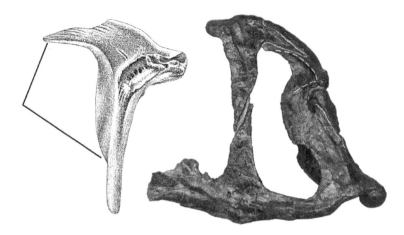

8.6. Evidence for a separate *M. adductor mandibulae externus superficialis* and *M. adductor mandibulae externus medialis* in *Allosaurus fragilis*. *Left,* the postorbital of *A. fragilis* in medial view (modified from Madsen 1976). The lines delimit a thin, smoothly surfaced shelf along the infratemporal margin of this element, here taken to be the origin scar for the *M. adductor mandibulae externus medialis.* No similar structure was seen in *Tyrannosaurus rex. Right,* the postorbital region of *A. fragilis* (AMNH 600) in lateral view. The white dashed line indicates the origin scar for the *M. adductor mandibulae externus superficialis,* in a similar position to that of *T. rex* (Molnar 2008). These two scars correspond in position to those of the two muscles in modern lizards.

ALLOSAURUS FRAGILIS (Fig. 8.2) *Allosaurus fragilis* generally exhibits substantially smaller relative leverages than the tyrannosaurids. *Allosaurus fragilis* apparently had both a *M. adductor mandibulae externus superficialis* and a *M. adductor mandibulae externus medialis,* unlike the others (Fig. 8.6). With the exception of the depressor and *M. add. mand. ext. med.,* the lever arms of the jaw muscles are about 20–35 percent less than those of *Daspletosaurus torosus,* although the *M. add. mand. post.* had much less leverage. The *M. add. mand. ext. med.* has a high curve, and for depressions less than 25°, it has the greatest lever arms, while for greater depressions it is exceeded only by the *M. pteryg. dors.* The *M. add. mand. ext. sup.,* in contrast, has the second-lowest curve over all angles of jaw depression. It is lower than that of the *M. add. mand. ext. sup. med.* of any tyrannosaurid examined, but that of the *M. add. mand. ext. med.* exceeds that of any tyrannosaurid. The average of these two curves would lie within the curves of the tyrannosaurids. The leverage of the *M. dep. mand.* is greater than in any tyrannosaurid.

CERATOSAURUS NASICORNIS (Fig. 8.2) The pattern of curves for *Ceratosaurus nasicornis* is different from those of the other taxa examined. The lever arm of the *M. pteryg. dors.* decreased more rapidly when the jaw was shut than in any other form. Likewise the curve of the *M. add. mand. ext. sup. med.* is lower than in any tyrannosaurid. The remaining adductors range from being approximately equivalent to those of *Daspletosaurus torosus* (the *M. add. mand. post.* when the jaw was shut) to having almost half again the relative leverage of those of that form (the *M. add. mand. post.* and the *M. add. mand. ext. prof.* when the jaw was widely open). The depressor had about 45 percent better leverage when the jaw was shut, but changes to falling below that of *D. torosus* when it was widely open.

Discussion

Among the tyrannosaurids, the lever arm versus jaw depression curves are more similar than they are to those of *Allosaurus* or *Ceratosaurus,* which is to be expected given the greater similarity in cranial forms. The leverages in *Tyrannosaurus rex* were similar to those of the other tyrannosaurids, and so, other things being equal, the bite might be expected to have been approximately equally powerful in all three forms. "Other things" weren't equal, of course: in *Tyrannosaurus rex* the postorbital region was laterally expanded (as in *Nanotyrannus*) and was also relatively taller than in the other tyrannosaurs (including *Nanotyrannus*), implying a relatively greater size of the adductor chamber and, presumably, of the adductor muscles. In addition, the skull of *T. rex* was larger than in the others, implying greater (absolute) sizes of the adductors. Thus *T. rex* likely would have had a substantially more powerful bite than the other tyrannosaurids. The *M. adductor mandibulae posterior*, here proposed to reduce the risk of dislocation of the jaw joint and of damage to the quadrate, also had a substantially greater mechanical advantage. This suggests that *T. rex* preyed on something that exerted powerful dislocating forces on the jaws or pulled against something tough.

Tyrannosaurus, as well as the contemporary *Nanotyrannus lancensis*, shows a widening of the postorbital region of the skull, not found in most earlier large theropods. This widening was probably correlated with an increase in the size of the adductor musculature and thus probably indicated a stronger bite in these forms than in the older taxa. A broader postorbital region, implying enlarged adductor chambers, appeared in four clades (tyrannosaurs, *Carnotaurus*, *Buitreraptor* and *Conchoraptor*) independently in the Late Cretaceous (*Buitreraptor* was Cenomanian, the others Campanian-Maastrichtian). This suggests some unrecognized parallel changes in feeding behavior or prey in these clades, but the obvious differences in cranial form between tyrannosaurs, *Conchoraptor*, and *Buitreraptor* confounds interpretation of this development.

In *Nanotyrannus lancensis* the leverage of the *M. add. mand. ext. sup. med.* is somewhat lower than for the other species because of the relatively slightly more posterior position of the infratemporal fenestra inferred in that specimen. The leverage of the *M. pteryg. dors.* approximated that of *Daspletosaurus torosus*, and the strong adductors (pseudotemporal and *M. add. mand. ext. prof.*) had slightly greater leverages than those in *D. torosus*. The *M. add. mand. post.* had a lesser mechanical advantage, suggesting that *N. lancensis* fed on some prey capable of less resistance than those fed upon by *Tyrannosaurus rex*.

No distinct differences in percentage extension are observed among the tyrannosaurs studied. It seems that none of these tyrannosaurids had a distinct advantage in opening the mouth to a greater extent than any other. Hence, if the maximum jaw depression were dependent only upon the percentage excursion of the adductors, it would have been about the same for all three forms.

This similarity of the muscle mechanics is interesting in light of the occurrences of these creatures. *Tyrannosaurus rex* and *Nanotyrannus lancensis* are both from the Hell Creek Formation and show a distinct size difference, *T. rex* having been 20–30 percent larger than *N. lancensis*, so that the two forms may have taken prey of different size. The maxillary and dentary teeth of *N. lancensis* were more laterally compressed than those of *T. rex* (Larson and Donnan 2002; Larson 2013). Whether *N. lancensis* is a valid taxon or a juvenile *T. rex*, this suggests niche partitioning in prey. Such partitioning would presumably reduce competition for food in either case.

The leverages of the tyrannosaurids are generally greater than in *Allosaurus* but generally less than in *Ceratosaurus*. The lever arm versus jaw depression curves of *A. fragilis* differ obviously from those of the tyrannosaurids. The leverages of the adductors are lower, generally by 20–30 percent, save for that of the *M. add. mand. ext. med.*, and that of the depressor is greater. This suggests that, other things being equal, *Allosaurus* would have had a weaker bite than tyrannosaurids. Again, although "other things" were almost certainly not equal, estimates of bite force for the two taxa support this implication (Erickson et al. 1996; Rayfield et al. 2001). If the tyrannosaurids replaced the allosaurids in North America, which is not clear, then a factor influencing this replacement may have been the greater mechanical efficiency of the jaw adductors in the tyrannosaurids. The depressor of *A. fragilis* has a greater lever arm than those of the tyrannosaurids because allosaurids possess a retroarticular process that was absent in the tyrannosaurids. The percentage extension of the muscles is generally lower for *A. fragilis* than for the tyrannosaurids (Fig. 8.3), suggesting that, were this the only factor in determining the maximum jaw depression, then allosaurids could have opened their mouths wider than tyrannosaurids (as independently implied by Bakker 2000).

The skull of *Ceratosaurus* is relatively deeper postorbitally and thus has a relatively shorter snout than those of the other forms. The quadrate is directed posteroventrally at an angle of 60° to the horizontal. If the jaw adductors were arranged parallel to the quadrate, this could have resulted in a relatively greater degree of jaw depression than among the other forms. However, the *M. add. mand. ext. sup. med.* scar indicates that this muscle, at least, was not aligned parallel to the quadrate, and in this orientation the adductors (save for *M. add. mand. ext. sup. med.*) appear to have been at least as mechanically efficient as those of the tyrannosaurids, and rather more so than those of *A. fragilis*. For example, at a jaw opening of 30° the leverage of the *M. pseudotemporalis* was 140 percent that in *A. fragilis*. The percentage excursions of three of the adductors of *C. nasicornis* are the greatest of the six forms examined. Thus, were this the only factor governing the degree of jaw depression, and were the adductors of all the species similar in this property, *Ceratosaurus* would have had a smaller maximum jaw depression than the others. *C. nasicornis* had a relatively deeper postorbital region, implying a relatively larger adductor chamber than in *Allosaurus fragilis*; additionally, the adductor chamber was enlarged ventroposteriorly by the posterocaudal inclination of the quadrate. This presumably implies relatively larger adductors and, for individuals of equal size, a stronger bite in *Ceratosaurus*.

Summary

Tyrannosaurus rex was compared, in terms of lever arm and extension of the jaw adductors, with the other tyrannosaurids *Nanotyrannus lancensis* and *Daspletosaurus torosus* and with the Late Jurassic *Allosaurus fragilis* and *Ceratosaurus nasicornis*. *Daspletosaurus torosus* agrees reasonably closely with *T. rex*, indicating that no great differences were apparent in the feeding mechanisms from these data. *Nanotyrannus lancensis* differs most, of these tyrannosaurids, from *T. rex*: these differences appear to have been related to the relatively lower skull of *N. lancensis*, and suggest that *N. lancensis* may have had a less powerful bite than the other tyrannosaurids examined.

There is no substantial difference in lever arms, and hence mechanical efficiency, between the tyrannosaurids examined. Thus, if *Tyrannosaurus rex* had an advantage in bite strength over the other tyrannosaurids, this was likely due to the laterally expended adductor chambers and its greater size.

The tyrannosaurids did show a marked increase of lever arm for the jaw adductors over the earlier *Allosaurus*. *Allosaurus*, in contrast, shows indications of a possibly greater maximum jaw depression. *Ceratosaurus* is a distinctly different form, exhibiting a generally greater level of mechanical efficiency of the jaw adductors than the tyrannosaurids. This, combined with an adductor chamber apparently relatively larger than in *Allosaurus* (and *Daspletosaurus*), suggests that *Ceratosaurus* had a more lethal bite than *Allosaurus* and, possibly, than some tyrannosaurids.

It should be remembered that these conclusions apply to individuals with skulls of the same size. Differences in body size would modify the conclusions.

Acknowledgments

The following people contributed materially to this study: the late E. C. Olson and P. P. Vaughn (University of California, Los Angeles), D. Whistler and K. Campbell (Los Angeles County Museum of Natural History), L. Drew (Museum of the Rockies), E. S. Gaffney (American Museum of Natural History), B. J. K. Molnar, the late S. P. Welles (University of California, Berkeley), M. J. Odano (Los Angeles County Museum of Natural History), D. A. Russell (National Museum of Canada), A. Milner (Natural History Museum, London), the late J. H. Ostrom (Yale Peabody Museum), W. Langston, Jr. (University of Texas, Austin), J. A. Madsen, Jr. (then at the University of Utah), B. Erickson (then at the Minnesota Museum of Man and Science), the late N. Hotton III and M. Brett-Surman (National Museum of Natural History), H. P. Powell and T. S. Kemp (Oxford University), J. Horner (Museum of the Rockies), P. J. Currie and J. Danis (Royal Tyrrell Museum of Palaeontology), P. Bjork (South Dakota School of Mines and Technology), and the anonymous reviewers. I very much appreciate all their assistance.

Bakker, R. T. 2000. Brontosaur killers: late Jurassic allosaurids as sabre-tooth cat analogues. *Gaia* 15:145–158.

Carr, T. D. 1999. Craniofacial ontogeny in the Tyrannosauridae (Dinosauria, Coelurosauria). *Journal of Vertebrate Paleontology* 19:497–520.

Currie, P. J. 2005. Theropods, including birds; pp. 367–397 in P. J. Currie and E. B. Koppelhus (eds.), *Dinosaur Provincial Park*. Indiana University Press, Bloomington.

Erickson, G. M., S. D. Van Kirk, J. Su, M. E. Levenston, W. E. Caler, and D. R. Carter. 1996. Bite-force estimation for *Tyrannosaurus rex* from tooth-marked bones. *Nature* 382:706–708.

Fulton, J. F. 1955. *A Textbook of Physiology*. W. B. Saunders, Philadelphia.

Gans, C., and W. J. Bock. 1965. The functional significance of muscle architecture–a theoretical analysis. *Ergebnisse der Anatomie und Entwicklungeschichte* 38:115–142.

Gilmore, C. W. 1920. Osteology of the carnivorous Dinosauria in the United States National Museum with special reference to the genera *Antrodemus* (*Allosaurus*) and *Ceratosaurus*. *United States National Museum Bulletin* 110:1–154.

Gilmore, C. W. 1946. A new carnivorous dinosaur from the Lance Formation of Montana. *Smithsonian Miscellaneous Collections* 106:1–19.

Haines, R. W. 1934. On muscles of full and of short action. *Journal of Anatomy* 69:20–24.

Iordansky, I. I. 1964. The jaw muscles of the crocodiles and some relating structures of the crocodilian skull. *Anatomische Anzeiger* 115:256–280.

Kemp, T. S. 1969. On the functional morphology of the gorgonopsid skull. *Philosophical Transactions of the Royal Society of London B* 256:1–83.

Larson, P. 2013. The case for *Nanotyrannus* In this volume.

Larson, P., and K. Donnan. 2002. *Rex Appeal*. Invisible Cities Press, Montpelier, Vermont.

Madsen, J. H., Jr. 1976. *Allosaurus fragilis*: A revised osteology. *Utah Geological Survey Bulletin* 109:1–163.

Molnar, R. E. 1991. The cranial morphology of *Tyrannosaurus rex*. *Palaeontographica A* 217:137–176.

Molnar, R. E. 2000. Mechanical factors in the design of the skull of *Tyrannosaurus rex* (Osborn, 1905). *Gaia* 15:193–218.

Molnar, R. E. 2008. Reconstruction of the jaw musculature of *Tyrannosaurus rex*; pp. 254–281 in P. Larson and C. Carpenter (eds.), Tyrannosaurus rex, *the Tyrant King*. Indiana University Press, Bloomington.

Osborn, H. F. 1912. Crania of *Tyrannosaurus* and *Allosaurus*. Memoirs of the American Museum of Natural History, n.s., 1:1–30.

Ostrom, J. H. 1969. Osteology of *Deinonychus antirrhopus*, an unusual theropod from the Lower Cretaceous of Montana. *Peabody Museum of Natural History Bulletin* 30:1–165.

Parrington, F. R. 1955. On the cranial anatomy of some gorgonopsids and the synapsid middle ear. *Proceedings of the Zoological Society of London* 125:1–40.

Pauwels, F. 1948. Die Bedeutung der Bauprinzipien des Stütz- und Bewegungsapparates für die Beanspruchung der Röhrenknochen. *Zeitschrift für Anatomie und Entwicklungsgeschichte* 114:129–166.

Rayfield, E. J., D. B. Norman, C. C. Horner, J. R. Horner, P. M. Smith, J. J. Thomason, and P. Upchurch. 2001. Cranial design and function in a large theropod dinosaur. *Nature* 409:1033–1037.

Russell, D. A. 1970. Tyrannosaurs from the Late Cretaceous of western Canada. *National Museum of Natural Science Publications in Palaeontology* 1:1–34.

Witmer, L. M. 1997. The evolution of the antorbital cavity of archosaurs: a study in soft-tissue reconstruction in the fossil record with an analysis of the function of pneumaticity. *Society of Vertebrate Paleontology Memoir* 3:1–73.

Young, J. Z. 1957. *The Life of Mammals*. Oxford University Press, Oxford.

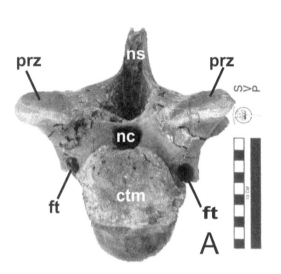 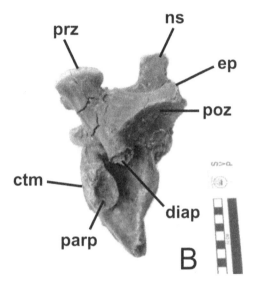

9.1. Cervical vertebra of an indeterminate tyrannosaur (TMP 2002.12.02), estimated to be C3 or C4. A) Anterior view. B) Left ventrolateral view. Abbreviations: ctm, centrum; diap, diapophysis; ep, epipophysis; ft, foramen transversarium; nc, neural canal; ns; neural spine; parp, parapophysis; poz, postzygapophysis; prz, prezygapophysis. The diapophysis articulates with the tuberculum of the cervical rib (not figured), and the parapophysis with the capitulum. Scale in cm.

Tyrannosaurid Craniocervical Mobility: A Preliminary Qualitative Assessment

Tanya Samman

Tyrannosaurs were dynamic predators, and the analysis of craniocervical mobility has implications for the biomechanics of their foraging and feeding. The cervical vertebrae of tyrannosaurids are anteroposteriorly shorter than those of many other coelurosaurs, as well as some extant birds, and the neck is correspondingly less flexible. Variation of vertebral shape along the vertebral column results in differences of mobility, with the posterior portion of the neck being much less flexible than the anterior. The software package DinoMorph™ was used, in collaboration with Dr. Kent Stevens from the Department of Computer and Information Science, University of Oregon, to digitally model tyrannosaur vertebrae as complex three-dimensional surfaces. The digital model was manipulated into various flexion poses. The "neutral pose" of theropods requires further study, and the *Tyrannosaurus rex* model needs refinement before any quantitative interpretations can be made. Soft-tissue data obtained from the study of birds helps to constrain the range-of-motion limits. Evidence from "death-pose" specimens helps to establish limits of dorsiflexion and suggests intergeneric mobility differences. Assessing the neutral pose, maximum dorsiflexion, ventriflexion, and lateral flexion gives insight into the biomechanical controls that influence the behavior and ecology of these animals.

Abstract

North American Tyrannosaurids

Introduction

The Tyrannosauridae are a lineage of large-bodied Late Cretaceous coelurosaurs (Holtz 1994, 1996, 1998 [2000]; Sereno 1997; Norell et al. 2001). Large North American genera include *Albertosaurus* Osborn, *Daspletosaurus* Russell, *Gorgosaurus* Lambe, and *Tyrannosaurus* Osborn (Russell 1970; Currie 2003). This preliminary assessment concentrates on general aspects of necks of all North American tyrannosaurids, using *Tyrannosaurus rex* and *Gorgosaurus* as representatives.

Tyrannosaur Necks

In contrast to the elongate cervicals of coelurosaurs like ornithomimids (Osmólska 1997), those of tyrannosaurids are anteroposteriorly foreshortened, as their large skulls were probably unsupportable by a long, slender neck (Carpenter 1997). Makovicky (1995) described the cervical vertebrae of tyrannosaurids in detail, noting that all tyrannosaurid cervicals are generally platycoelous, in contrast to the opisthocoelous cervical vertebrae of *Allosaurus* and other carnosaurs. Brochu (2003:63), in contrast, described tyrannosaur postaxial cervical vertebrae as "opisthocoelous, insofar as the anterior central surface is convex and the posterior surface concave." Neither classification accurately reflects the shapes of the central faces.

Figure 9.1 shows a typical cervical vertebra and some of its component parts. The first vertebra in the cervical series, the atlas (C1), is a roughly U-shaped collection of bones, with distally flared dorsal neurapophyses articulating ventrally with the anteriorly concave intercentrum, which in turn articulates with the occipital condyle of the skull. The neural spine of the second cervical, the axis (C2), is greater in size than the centrum. The axial central faces are sub-parallel, though the dorsally located convex odontoid process (atlantal pleurocentrum in Brochu 2003) and ventrally located concave intercentrum give the anterior central face a complex shape that is distinct from those of the postaxial cervicals. The prezygapophyses, which articulate with the atlantal neurapophyses, incline ventrolaterally, converge medially, and are much smaller than the postzygapophyses, which are themselves slightly smaller than the zygapophyses in the remainder of the cervical series. Differences between the anterior (C3–C5) and posterior (C6–C10) cervicals (Makovicky 1995; pers. obs.) are summarized here. In the anterior cervicals, the anterior and posterior central faces are not parallel, with the anterior face sloping more posteroventrally than the posterior face. The anterior face is wider than broad, whereas the posterior face shows the opposite condition. Anterior cervicals have horizontal zygapophyseal articulation, and the transverse process is a thin, elongate structure that is ventrally to ventrolaterally oriented. The central faces of the sixth cervical converge slightly ventrally, as it is the point of transition between the two segments of the neck. Posterior cervicals have central faces that are approximately parallel and converge dorsally in the last vertebra of the neck. The articulation of these central shapes creates in a neck with a convex anterior arc and a concave posterior arc the classic "S curve" seen in non-avian dinosaurs and birds. Postaxial tyrannosaur centra are ventrally compressed, with a ventral midline ridge. A rugose knob on the posterior end of this midline ridge, though not anteriorly located like the hypapophyses of anterior dorsal vertebrae, probably served as a soft-tissue attachment site. The prezygapophyses incline medially, with increasing degree of slope posteriorly. The transverse processes of the posterior cervicals gradually migrate dorsolaterally, with the diapophysis of the 10th cervical oriented sub-horizontally. The parapophyses are located anteroventrally in the anterior cervicals. Although the position of the parapophyses migrates slightly dorsolaterally in the last two cervicals, they remain ventral, distinguished from the dorsal position just below the diapophyses observed in dorsal vertebrae. The neural spines are thin and elongate in the anterior cervicals, becoming shorter and more laterally triangular in anterior aspect in the posterior cervicals. The second and third cervicals have a lateral expansion of the neural spine, a condition known as a "spine table" (Gauthier 1986). Epipophyses that overhang the postzygapophyses are present in the anterior cervicals, becoming less prominent posteriorly, and are reduced to small rugose prominences in the posterior cervicals. Rugosities along the anterior and posterior edges of the neural spine indicate strong soft-tissue attachments. In the vertebrae of *Tyrannosaurus rex* and *Gorgosaurus libratus*, ligaments were judged to be sufficient to maintain postural support as the neural spines are broad, with distinct scars for intraspinous ligaments (Hengst 2004). Eric Snively (pers. comm., November 2004) opined that ligaments were likely to have been sufficient for providing cervical postural support, freeing the muscles to control activity.

Function/Feeding

Tyrannosaurs, especially *Tyrannosaurus rex*, were the top North American Late Cretaceous predators, but little is actually known about the biomechanical roles of their neck and head in feeding and foraging. The functional morphology of the neck vertebrae of theropods has not been well studied, as specimens with complete cervical series are uncommon and functional studies generally focus on limb function and locomotion (e.g., Nicholls and Russell 1985; Gatesy 1990; Gatesy and Middleton 1997; Carpenter 2002; Paul 1998 [2000]). Those investigations done on feeding behavior and mechanics usually focus on the teeth and skull (e.g., Farlow and Brinkman 1994; Bakker 1998 [2000]). When the neck is considered, it is usually from the perspectives of determining phylogenetic relationships or reconstructing muscles and ligaments (Makovicky 1995; Bakker 1998 [2000]; Tsuihiji 2004a, 2004b).

Tyrannosaur teeth provide some clues about the use of the head and neck in feeding. Abler (1992) hypothesized that tyrannosaurs may have tugged at their meat more than they pushed into it, as the distal dental serrations are greater in number, more regular, and more complex. Farlow and Brinkman (1994:172) also described a feeding scenario in which, after the occlusion of the jaws trapped meat and pressed it between the teeth and against the serrations, the meat was "torn from the victim by forceful jerks of the tyrannosaur's head." Evidence for the pulling and shaking of the animal's head during feeding is taken from multidirectional surface scratches on the teeth, which suggest that head movements during feeding may have been complex (Abler 1992, 2001). Bite furrows attributed to tyrannosaurs also suggest the animals pulled away from the prey carcass during feeding (Erickson and Olson 1996). This "puncture-pull" feeding hypothesis is also supported by finite element analysis of the tyrannosaurid skull, which indicates that the cranium of *Tyrannosaurus rex* resisted the loadings caused by biting and tearing equally well (Rayfield 2004). Assessing the range of motion of the head and neck serves as a complement to these findings and provides a more complete picture of tyrannosaur feeding.

Institutional Abbreviations AMNH, American Museum of Natural History, New York; BHI, Black Hills Institute of Geological Research, Hill City, South Dakota; BMR, Burpee Museum of Natural History, Rockford, Illinois; FMNH, Field Museum of Natural History, Chicago, Illinois; MOR, Museum of the Rockies, Bozeman, Montana; SAIT, Southern Alberta Institute of Technology, Calgary; TCMI, The Children's Museum of Indianapolis, Indianapolis, Indiana; TMP, Royal Tyrrell Museum of Palaeontology, Drumheller, Alberta.

Methods

The Utility of Bird Necks

The necks of birds facilitate movements of the head for various biological roles, such as preening, foraging, exploration, and the balancing of the head during locomotion (van der Leeuw et al. 2001). Birds, as the only extant members of the clade Dinosauria, serve as the best comparative analogue for assessing the soft-tissue constraints on range of motion in their non-avian theropod relatives and are an invaluable resource for visualizing how soft tissues and bones interact at the joint articular surfaces.

Assessing Craniocervical Mobility in Tyrannosaurs

The functional morphology of theropod cervical vertebrae, including that of tyrannosaurs, has not been adequately investigated. Cervical biomechanics are difficult to study in theropod dinosaurs because small bones (e.g., those of ornithomimids) are often fragile and larger vertebrae (e.g., those of tyrannosaurids) are often unwieldy. Using a parametric computer model to digitally manipulate the bones helps to overcome these problems. This study is part of a larger study on coelurosaurian craniocervical functional morphology (Samman 2006) that builds on existing tyrannosaur neck research (Makovicky 1995; Snively 2006). This study has an emphasis on determining the range of motion of the cervical vertebrae and of the craniocervical interface, using comparative anatomical studies of theropods in both phylogenetic and functional contexts and using computer-modeling techniques to expand our knowledge of the biomechanics of feeding and foraging in North American tyrannosaurs.

Specimens

Examination of North American tyrannosaurid taxa (Table 9.1) was complemented by a comparative study of extant avian theropods (Table 9.2). When possible, original tyrannosaurid specimens were examined directly. In cases where the original material was mounted or otherwise inaccessible, high-quality research casts were examined. Bird specimens were acquired as carcasses and were obtained and used under a salvage permit held by the University of Calgary Department of Biological Sciences.

Data Collection

PHOTOGRAPHY AND MEASUREMENT (EXTINCT TAXA) Specimens of the extinct taxa were photographed digitally and measured in order to collect the data on which modeling would be based. The necks of mounted specimens were photographed in order to visualize the progression of the cervical series, but disarticulated, unmounted specimens were the main data source. Images of individual vertebrae were taken in anterior view, posterior view, right lateral view, left lateral view, dorsal view, and ventrolateral view. These images serve as a visual reference of the morphology

Specimen	Taxon
AMNH 5027	*Tyrannosaurus rex*
BHI 3033*	*Tyrannosaurus rex*
BMR P2002.4.1*	*Tyrannosaurus rex*
FMNH PR 2081*	*Tyrannosaurus rex*
MOR 555	*Tyrannosaurus rex*
TCMI 2001.90.1*	*Tyrannosaurus rex*
TCMI 2001.89.1*	*Gorgosaurus*
TMP 91.36.500	*Gorgosaurus*

Table 9.1. Extinct taxa examined in this study

*Indicates that high-quality research cast material was examined.

Common Name	Binomen
Mallard	*Anas platyrhynchos*
Ostrich	*Struthio camelus*
Trumpeter swan	*Cygnus buccinator*
Tundra swan	*Cygnus columbianus*
Snowy owl	*Bubo scandiacus (Nyctea scandiaca?)*
Bald eagle	*Haliaeetus leucocephalus*

Table 9.2. Extant taxa examined in this study

Parameter #	Parameter Description
1	Vertical height of anterior central face at midline
2	Horizontal width of anterior central face at midline
3	Vertical height of posterior central face at midline
4	Horizontal width of posterior central face at midline
5	Mediolateral distance between distal edges of prezygapophyses
6	Mediolateral distance between distal edges of postzygapophyses
7	Anteroposterior distance between distal edges of zygapophyses, right
8	Anteroposterior distance between distal edges of zygapophyses, left
9	Anteroposterior length, left prezygapophysis
10	Mediolateral width, left prezygapophysis
11	Anteroposterior length, right prezygapophysis
12	Mediolateral width, right prezygapophysis
13	Anteroposterior length, left postzygapophysis
14	Mediolateral width, left postzygapophysis
15	Anteroposterior length, right postzygapophysis
16	Mediolateral width, right postzygapophysis

Table 9.3. Parameters measured for disarticulated tyrannosaur vertebrae

of each vertebra and complement the measurement data. The measured parameters are listed in Table 9.3. Because of the large size of tyrannosaur vertebrae, measurements were taken in centimeters using a tape measure.

RADIOGRAPHY (EXTANT TAXA) X-ray images of the mallard were taken using a Hewlett-Packard 43805N Faxitron X-Ray System, employing the following settings: 70 kV, 2.75 mA, 90-second exposure. X-ray images of the ostrich, trumpeter swan, tundra swan, bald eagle, and snowy owl were taken at the SAIT Non-Destructive Testing facility with a Philips Constant Potential X-Ray Tube, using the following settings: 50–55 kV, 10 mA, 48-second exposure. Specimens, depending on degree of flexibility (from being in the freezer for various amounts of time), were posed in order to view the articulation of the centra and zygapophyses in dorsiflexion, ventriflexion, lateral flexion, and, in the case of the owl, torsion.

Modeling

In order to facilitate the assessment of range of motion in the extinct taxa, the parametric 3-D software DinoMorph™ software program (Stevens and Parrish 1999; Stevens 2002) was used, in collaboration with Dr. Kent Stevens of the Department of Computer and Information Science at the University of Oregon, USA. The articular surfaces of the joints of the neck and neck-skull interface were measured on original or high-quality research cast material, and these data were used to model the vertebrae digitally as complex three-dimensional surfaces using DinoMorph™. The digital models are created from and constrained by the results of comparative examinations of avian and non-avian taxa. The preliminary qualitative assessments of the neutral pose and range of motion of the neck were established by digitally manipulating the joints of the model, which were constrained by the results of bird examinations. When assessing range of motion, each vertebra was rotated so that the minimum overlap of the pre- and postzygapophyses was 50 percent, following Stevens and Parrish (1999).

Results and Discussion

First, an issue raised in the introduction must be addressed. Brochu (2003) described postaxial tyrannosaur cervical vertebrae as opisthocoelous, whereas they were labeled by Makovicky (1995) as platycoelous. Platycoely is defined as a condition where both articulating ends of a vertebra are slightly concave (Romer 1956). However, I observed that the anterior central face is slightly convex dorsally, concave below, ending up slightly concave ventral-most. The posterior central face, in contrast, is dorsally concave to a small degree and more convex ventrally. This condition is most pronounced in the anterior cervicals. This condition does not appear to have a technical name, and I propose that it be called "paraopisthocoely."

Specimens of *Tyrannosaurus rex* (MOR 555) and *Gorgosaurus* (TMP 91.36.500), with the neck oriented in a classic death pose (Fig. 9.2), demonstrate maximum cervical dorsiflexion, irrespective of the taphonomy of individual specimens. Weigelt (1989) proposed that the postmortem desiccation of the cervical musculature and associated shrinkage of the large ligaments resulted in this characteristic contorted position, whereas Faux and Padian (2007) attribute the position to perimortem death throes. Though flexibility was likely to have varied among dinosaur individuals, as it does in humans, this curved posture places an upper limit on cervical dorsiflexion. In addition, the dorsal curvature of the neck of an articulated death-pose specimen of *Gorgosaurus* is much more extreme than that of

Tyrannosaurus rex (Fig. 9.2), suggesting the possibility of taxonomic variation in craniocervical mobility among North American tyrannosaurid genera. While this may be an artifact of taphonomy, intergeneric mobility differences cannot be discounted, and more data on the neck position of articulated specimens are needed.

Examination of radiographic images of extant avian theropods, like the mallard examined in this study (Fig. 9.3), can elucidate intervertebral and interzygapophyseal articulations in various poses. As the birds are not alive, this can technically be considered an assessment of the limits of passive mobility. However, because the birds are generally fully fleshed and feathered, the soft tissues provide an additional constraint on mobility. This means that the qualitative limits determined by manual manipulation are theoretically possible for a live bird to achieve actively.

Boas (1929) divided the cervical column of birds into three regions of flexion, also observed by van der Leeuw et al. (2001). He discerned that the anteriormost part of the neck showed mainly ventral flexion, the middle section exhibited mainly dorsal flexion, and very limited dorsal and ventral flexion were possible in the posterior section. Joints with intermediate flexion characteristics were observed in the transitional areas between the sections. Van der Leeuw et al. (2001) noted that the number of vertebrae in each of the three regions varies in different species. In addition, van der Leeuw et al. (2001) observed extreme dorsiflexion in the joint between the head and the atlas/axis joint. They considered the atlas/axis a single unit in their range-of-motion assessments, with minimal flexion potential.

Manual manipulation of the flensed mallard (Fig. 9.4) confirmed the three regions of cervical mobility noted by Boas (1929) and van der Leeuw et al. (2001). In qualitative terms, the most posterior region of the cervical series showed very limited dorsiflexion and ventral flexion. The middle region showed good dorsal flexion and some degree of ventral flexion. The most anterior part of the neck showed poor dorsiflexion but extreme ventriflexion. Lateral flexion was not assessed, though the heterocoelous centra of birds allow for the possibility of lateral flexibility.

9.2. Specimens with neck arched in a classic "death pose," right lateral view. A) *Tyrannosaurus rex* (MOR 555). B) *Gorgosaurus* (TMP 91.36.500). The neck curves to a greater in extent in *Gorgosaurus* than *Tyrannosaurus rex*, indicating that the former taxon may have had greater mobility potential.

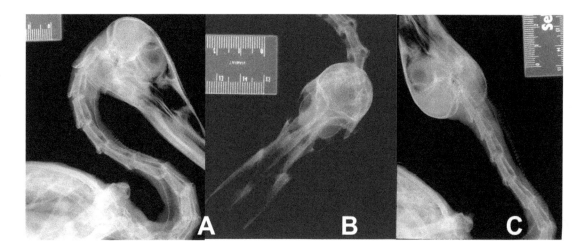

9.3. X-ray images of the mallard, *Anas platyrhynchos*. A) S typical "S curve," showing functional divisions of the neck. B) Right lateral flexion of the anterior neck and skull in dorsal view. C) Anterior neck dorsiflexive limit.

Preliminary qualitative neutral-pose and range-of-motion assessments for *Tyrannosaurus rex* (Fig. 9.5) yielded interesting results and raised uncertainty concerning the definition of a neutral pose in the theropod neck. For the purposes of this study, the neutral pose was intended to be the position of the neck in which all the centra were centered in articulation and the zygapophyses were completely overlapping (i.e., nothing was displaced, in a relative sense). The manual manipulation of pairs of vertebrae from various extinct specimens during data collection raised the question of how tyrannosaur vertebrae articulate, as it did not seem possible to orient pairs of cervicals such that both the central faces and zygapophyses were centered relative to each other. Because tyrannosaur vertebrae are large and heavy, it is difficult and awkward to manually manipulate more than a pair at a time. This is one of the factors that makes computer modeling an excellent alternative. After the preliminary model was completed, sans zygapophyses, a neutral pose was created by aligning the cervical series according to the articular shapes of the centra (Fig. 9.5A). This produced a nice S curve. However, when the zygapophyses were added to the model, it was evident that the neutral pose created by keeping the zygapophyses centered relative to each other was something entirely different; the effect resembled dorsiflexion (Fig. 5B). This is likely due to inaccuracies in capturing the complex three-dimensional curvature and orientation of the zygapophyses as well as in errors in the placement of the center of rotation for each vertebral pair. Examination of birds demonstrated that the intervertebral and interzygapophyseal soft tissues were relatively thin, acting more as a surface upon which the facets could glide rather than adding a great deal of padding to the cervical articulation points. Thus further research and refinement of the model is necessary before it can be used quantitatively, and this is why this study reports only qualitative results.

The zygapophyseal neutral pose (Fig. 9.5B) was used as the basis for preliminary qualitative inferences of mobility, as limits of flexion were constrained by a minimum overlap of the pre- and postzygapophyses of

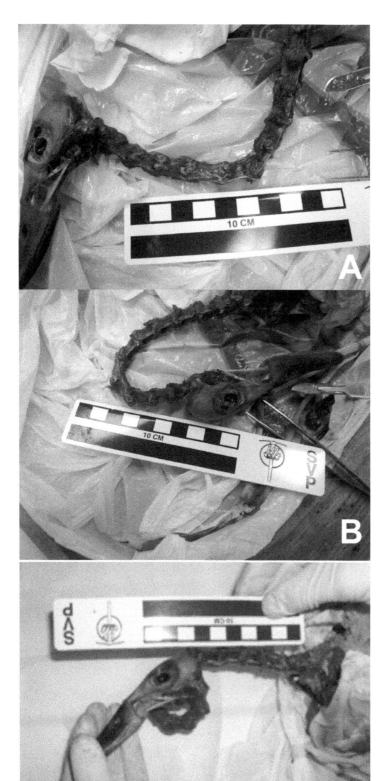

9.4. Selected results of manual manipulation of dissected/flensed mallard, *Anas platyrhynchos.* A) Dorsiflexive limits with an obvious transition point between neck sections. B) Ventriflexive limits of posterior sections of neck. C) Extreme ventriflexion in anteriormost section of neck.

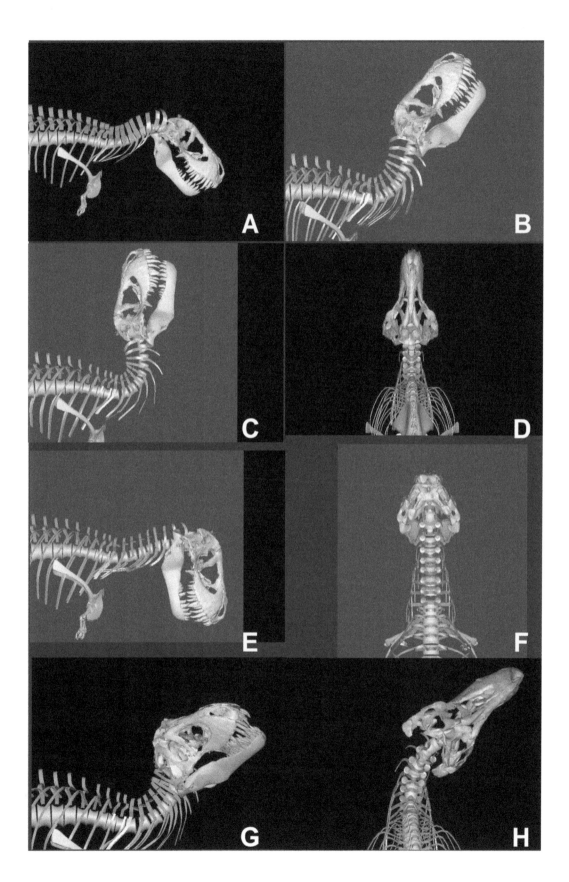

50 percent. As in birds, the *Tyrannosaurus rex* model appears to have had several functional segments in the neck, but with much less flexibility. This relative inflexibility is not surprising when the number and anteroposterior length of cervical vertebrae in the various taxa are taken into account. The mallard has 16 relatively elongate cervical vertebrae, whereas *T. rex* has only 10 foreshortened cervicals. Swans can perform amazing feats of cervical flexibility despite the torsional restriction of heterocoelous centra because they have ~24 elongate cervicals. Nonetheless, the *T. rex* model appears to be most dorsoventrally mobile at the head/atlas/axis junction and capable of greater overall flexion (dorsal, ventral, lateral) in the anterior cervicals than in the posterior ones. While trying to orient some of the bird specimens for radiography, I found it was not possible to significantly ventriflex the posteriormost part of the neck, presumably because of restriction by muscles, tendons, and ligaments. This is not an effect resulting from freezing and desiccation of the specimens as even in the fresher specimens posterior cervical manual ventriflexion was limited. The ostrich neck and the trumpeter swan had been in the freezer for several years, for example, but the tundra swan had only been collected in the spring or summer of 2005. As the tundra swan was relatively fresh and still showed the same mobility restriction, I conclude that this stiffness is due to constraints imposed by soft tissues. The *T. rex* model demonstrates constraints on the flexibility of the posterior cervicals similar to what was observed in birds.

9.5. Preliminary qualitative range-of-motion assessment results. The *Tyrannosaurus rex* model is a composite constructed using data from FMNH PR 2081 and TCMI 2001.90.1. A) Neutral pose based on central articulation. B) Neutral pose based on zygapophyseal articulation, used for range of motion assessments. C) Limits of dorsiflexion (50 percent zygapophyseal overlap), lateral view. D) Limits of dorsiflexion (50 percent overlap), dorsal view. E) Limits of ventriflexion (50 percent overlap), lateral view. F) Limits of ventriflexion (50 percent overlap), dorsal view. G) Limits of lateral flexion (50 percent overlap), lateral view. H) Limits of lateral flexion (50 percent overlap), dorsal view. The *Tyrannosaurus rex* model shows limited flexion capability in the posterior portion of the neck compared to the anterior portion.

Conclusions and Future Directions

Although the *Tyrannosaurus rex* model still requires a great deal of work before any quantitative determinations can be made, some interesting preliminary qualitative features can still be observed. The *Tyrannosaurus rex* model appears to have several functional segments of the neck, with the anterior portion showing much greater flexibility than the posterior portion, similar to the condition observed in birds in this study.

In addition, evidence from death-pose tyrannosaurid specimens indicates that intergeneric mobility differences may exist between North American tyrannosaurids. Ontogenetic variation in craniocervical mobility can be assessed by comparative studies of juvenile specimens like BMR P2002.4.1.

Finally, this study serves as the foundation upon which craniocervical mobility assessments will be performed using parametric computer-modeling techniques for other North American coelurosaurian taxa as well as other theropod dinosaurs.

Acknowledgments

I am grateful to Kent Stevens and Eric Wills (Department of Computer and Information Science, University of Oregon) for their collaboration, patience, and technical support for DinoMorph™. Access to tyrannosaurids specimens would not have been possible without the cooperation and assistance of staff from: Royal Tyrrell Museum of Palaeontology,

Alberta (especially Jim Gardner, the collections staff, and Darren Tanke, who found TMP 2002.12.02); American Museum of Natural History, New York (especially Carl Mehling); Field Museum of Natural History, Illinois (especially Bill Simpson), The Children's Museum of Indianapolis (Shane Ziemmer, Dallas Evans, Victor Porter), Indiana; Museum of the Rockies, Montana (especially Jack Horner and Ellen Lamm); Black Hills Institute of Geological Research (for the neck cast loan of BHI 3033 to Eric Snively), South Dakota; and Burpee Museum of Natural History, Illinois (for cast material of BMR P2002.4.1). The Southern Alberta Institute of Technology Non-Destructive Testing facility, Manufacturing and Automation, is thanked for a productive and innovative collaboration; special thanks to Alex Zahavich, John Moore, George Gavelis, Hielkje Klok, and Brian Hughes. Numerous colleagues from the Departments of Geology and Geophysics, University of Calgary; Biological Sciences, University of Calgary; and the Royal Tyrrell Museum of Palaeontology, Alberta, provided helpful discussions, information about specimens, and access to equipment, specimens, and resources (especially Philip Currie, Len Hills, Eva Koppelhus, Eric Snively, Tony Russell, Warren Fitch, Darla Zelenitsky, Don Henderson, Darren Tanke, Kevin Aulenback, and Michèle Asgar-Deen). The Canadian Wildlife Service, Alberta Fish and Wildlife, and the Calgary Zoo are thanked for access to bird specimens. This research was supported, in part, by funding from the University of Calgary, the Jurassic Foundation, the Dinosaur Research Institute, and contributions from the National Sciences and Engineering Research Council of Canada (NSERC) research grants awarded to L. V. Hills and A. P. Russell. Special thanks to the reviewers for their helpful comments.

Literature Cited

Abler, W. L. 1992. The serrated teeth of tyrannosaurid dinosaurs, and biting structures in other animals. *Paleobiology* 18:161–183.

Abler, W. L. 2001. A kerf-and-drill model of tyrannosaur tooth serrations; pp. 84–89 in D. H. Tanke and K. Carpenter (eds.), *Mesozoic Vertebrate Life*. Indiana University Press, Bloomington.

Bakker, R. T. 1998 [2000]. Brontosaur killers: Late Jurassic allosaurids as sabre-tooth cat analogues. *Gaia* 15:145–158.

Boas, J. E. V. 1929. Biologisch-anatomische Studien über den Hals der Vogel. *Kgl. Danske Vidensk Skrifter* 9:101–222.

Brochu, C. A. 2003. Osteology of *Tyrannosaurus rex*: insights from a nearly complete skeleton and high-resolution computed tomographic analysis of the skull. *Journal of Vertebrate Paleontology* 22:1–140.

Carpenter, K. 1997. Tyrannosauridae; pp. 766–768 in P. J. Currie and K. Padian (eds.), *Encyclopedia of Dinosaurs*. Academic Press, Orlando.

Carpenter, K. 2002. Forelimb biomechanics of nonavian theropod dinosaurs in predation. *Senckenbergiana Lethaea* 82:59–76.

Currie, P. J. 2003. Cranial anatomy of tyrannosaurid dinosaurs from the Late Cretaceous of Alberta, Canada. *Acta Palaeontologica Polonica* 48:191–226.

Erickson, G. M., and K. H. Olson. 1996. Bite marks attributable to *Tyrannosaurus rex*: preliminary description and implications. *Journal of Vertebrate Paleontology* 16:175–178.

Farlow, J. O., and D. L. Brinkman. 1994. Wear surfaces on the teeth of tyrannosaurs; pp. 165–175 in G. D. Rosenberg and D. L. Wolberg (eds.), *Dino Fest*. Paleontological Society, Indianapolis, Indiana.

Faux, C. M., and K. Padian. 2007. The opisthotonic posture of vertebrate skeletons: postmortem contraction or death throes? *Paleobiology* 33(2):201–226.

Gatesy, S. M. 1990. Caudofemoral musculature and the evolution of theropod locomotion. *Paleobiology* 16:170–186.

Gatesy, S. M., and K. M. Middleton. 1997. Bipedalism, flight, and the evolution of theropod locomotor diversity. *Journal of Vertebrate Paleontology* 17:308–329.

Gauthier, J. 1986. Saurischian monophyly and the origin of birds. *Memoirs of the California Academy of Sciences* 8:1–55.

Hengst, R. 2004. Gravity and the *T. rex* backbone. *Journal of Vertebrate Paleontology* 24(3, suppl.):69A–70A.

Holtz, T. R., Jr. 1994. The phylogenetic position of the Tyrannosauridae: implications for theropod systematics. *Journal of Paleontology* 68:1100–1117.

Holtz, T. R., Jr. 1996. Phylogenetic taxonomy of the coelurosauria (Dinosauria: Theropoda). *Journal of Paleontology* 70:536–538.

Holtz, T. R., Jr. 1998 [2000]. A new phylogeny of the carnivorous dinosaurs. *Gaia* 15:5–61.

Makovicky, P. J. 1995. Phylogenetic aspects of the vertebral morphology of Coelurosauria (Dinosauria: Theropoda). M.Sc. thesis/dissertation, Faculty of Natural Sciences, University of Copenhagen, Copenhagen.

Nicholls, E. L., and A. P. Russell. 1985. Structure and function of the pectoral girdle and forelimb of *Struthiomimus altus* (Theropoda: Ornithomimidae). *Palaeontology* 28:643–677.

Norell, M. A., J. M. Clark, and P. J. Makovicky. 2001. Phylogenetic relationships among coelurosaurian theropods; pp. 49–67 in J. Gauthier and L. F. Gall (eds.), *New Perspectives on the Origin and Early Evolution of Birds: Proceedings of the International Symposium in Honor of John H. Ostrom.* Peabody Museum of Natural History, Yale University, New Haven, Connecticut.

Osmólska, H. 1997. Ornithomimosauria; pp. 499–503 in P. J. Currie and K. Padian (eds.), *Encyclopedia of Dinosaurs.* Academic Press, Orlando.

Paul, G. S. 1998 [2000]. Limb design, function and running performance in ostrich-mimics and tyrannosaurs. *Gaia* 15:257–270.

Rayfield, E. J. 2004. Cranial mechanics and feeding in *Tyrannosaurus rex. Proceedings of the Royal Society of London B* 271:1451–1459.

Romer, A. S. 1956. *Osteology of the Reptiles.* University of Chicago Press, Chicago.

Russell, D. A. 1970. Tyrannosaurs from the Late Cretaceous of Western Canada. *National Museum of Natural Sciences Publications in Palaeontology* 1:1–34.

Samman, T. 2006. Craniocervical functional morphology of several North American coelurosaurian dinosaurs. Ph.D. Dissertation, Department of Geology and Geophysics, University of Calgary, Calgary, Alberta, 341 pp.

Sereno, P. C. 1997. The origin and evolution of dinosaurs. *Annual Review of Earth and Planetary Sciences* 25:435–489.

Snively, E. 2006. Neck musculoskeletal function in the Tyrannosauridae (Theropoda, Coelurosauria): implications for feeding dynamics. Ph.D. Dissertation, Department of Biological Sciences, University of Calgary, Calgary, Alberta, 478 pp.

Stevens, K. A. 2002. DinoMorph: parametric modeling of skeletal structures. *Senckenbergiana Lethaea* 82:23–34.

Stevens, K. A., and J. M. Parrish. 1999. Neck posture and feeding habits of two Jurassic sauropod dinosaurs. *Science* 284:798–800.

Tsuihiji, T. 2004a. The ligament system in the neck of *Rhea americana* and its implication for the bifurcated neural spines of sauropod dinosaurs. *Journal of Vertebrate Paleontology* 24:165–172.

Tsuihiji, T. 2004b. The neck of non-avian maniraptorans: how bird-like was the cervical musculature of the "bird-like" theropods? *Journal of Vertebrate Paleontology* 24(3, suppl.):122A.

van der Leeuw, A. H. J., R. G. Bout, and G. A. Zweers. 2001. Evolutionary morphology of the neck system in ratites, fowl and waterfowl. *Netherlands Journal of Zoology* 51:243–262.

Weigelt, J. 1989. *Recent Vertebrate Carcasses and Their Paleobiological Implications.* University of Chicago Press, Chicago.

Paleopathology, Paleoecology, and Taphonomy

3

10.1. Lateral view of maxillary grooves (A) in "Scotty" (RSM P2523.8). Tooth in position (B) shows apparent manner of attack.

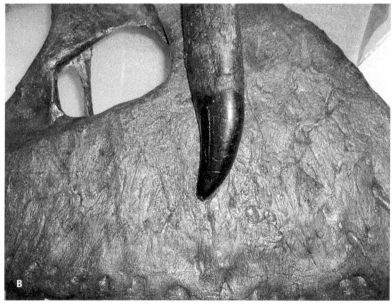

Clawing Their Way to the Top: Tyrannosaurid Pathology and Lifestyle

10

Bruce M. Rothschild

Facial scars in tyrannosaurids have been attributed to intraspecific biting behavior. Remodeled bone surrounding the lesions document survival of these attacks/interactions. While that is a reasonable hypothesis, examination of recently discovered specimens suggests an alternative explanation. The "Jane" and "Peck" *Tyrannosaurus rex* specimens have substantial evidence of trauma by sharp objects. However, the width and breadth of noted lesions match neither *Tyrannosaurus* tooth size nor those of other tyrannosaurids. The right side of the surangular of the Peck *T. rex* provides insight. It has a penetrating hole that aligns with a smaller one in the subjacent clinal. The holes are much too large for a tyrannosaurid tooth but did accommodate a *T. rex* toe claw. The potential of claws to produce bone damage has not been previously considered because of the hardness differential. Given the lack of correlation of some bone damage with tooth parameters and demonstration that claws could penetrate bone, claw damage remains the residual hypothesis.

Abstract

Brochu (2003) lists tyrannosaurids as *Albertosaurus sarcophagus*, *Alectrosaurus olseni*, *Alioramus remotos*, *Daspletosaurus torosus*, *Eotyrannus lengi*, *Gorgosaurus libratus*, *Nanotyrannus lancensis*, *Siamotyrannus isanensis* and *Tarbosaurus bataar*. *Alectrosaurus*, *Eotyrannosaurs*, *Siamotyrannus*, and *Alioramus* are not further considered due to lack of access to sufficient materials for analysis. Pathology is described herein because of the opportunity it provides for insight into behavior of the affected taxa (Avilla et al. 2004; Rothschild and Tanke 2005; Rothschild and Molnar 2009). Avilla et al. (2004) divided crocodylomorph pathology into punctures, scratches, and cracks, with scratches described as shallow and linear. Pathology in tyrannosaurids is similarly analyzed as punctures, grooves, and missing components. These terms are used here to distinguish pathology from normal topographic structures, including fenestrae, foraminae, and neurovascular grooves.

Tyrannosaurid Pathology

Facial Scars

Facial scars (Table 10.1) in tyrannosaurids (*Albertosaurus*, *Daspletosaurus*, *Gorgosaurus*, *Tarbosaurus*, and *Tyrannosaurus*) have traditionally been attributed to intraspecific biting behavior (Tanke and Currie 1998;

Table 10.1. Distribution of pathology in tyrannosaurids

	Facial	Rib	Humerus	Fibula	Vertebrae	Digit
Tyrannosaurus:						
"Jane," BMR P2002.4.1	+					+
"Peck," MOR 980	+					
« Sue, » FMNH PR 2081	+	+	b	+[a]	+[a]	+
« Stan, » BHI 3033	+	+[c]				
LACM 23844	+					
MOR 008	+					
AMNH 5027	+	+				
TMP 71.17.12						
LACM 23844	+					
"Wyrex," BHI 6230	+					
"Monty" (privately owned)	+					
"Sampson"/"Z-rex" (privately owned)	+		+[c]			
NMMNH P-3698						
Elephant Butte *T. rex,* NMMNH P-1013-1	+					
NMMNH P-27469	+					
"Scotty," RSM P2523.8	+					
Albertosaurus:						
MOR 379						+
ROM 807						+
AMNH 5432	+					
TMP 86.36.314	+					
TMP 91.10.1	+					
Gorgosaurus:						
TMP 91.36.500	+			+		+
TMP 94.12.602		+		+		
NMC 2120				+		+
Russell 1970			+			
TMP 2001.89.1	+	+	+	+[a]	+	
Daspletosaurus:						
MOR 379	+					
Russell 1970			b			
Molnar 2001						+
TMP 85.62.1	+					
Tarbosaurus:						
Blanding II-2					+	

Source: Derived from Lambe (1917); Russell (1970); Molnar (1991, 2001); Archer and Babiarz (1992); Dingus (1996); Poling (1996); Tanke (1996b); Currie (1997); Harris (1997); Kieran (1999); Williamson and Carr (1999); Brochu (2000); Rothschild et al. (2001); Tanke and Rothschild (2002); and Webster (2002).[a] Associated with infection. [b] Exostoses and tendon avulsions. [c] Inflammatory fusion.

Molnar 2001). Remodeled bone surrounding the lesions documents survival of these attacks/interactions. Dentigerous wounds were described in *Albertosaurus* by Molnar and Currie (1995). Keiran (1999) reported a subadult *Gorgosaurus libratus* TMP 91.36.500 with facial lesions that they interpreted as bites. A dorsoventrally elongate, yellowish discoloration at mid-length on the right dentary surrounds the lesion. Molnar (1991:155, 156) reported puncture injuries in *Tyrannosaurus* (both LACM 23844 surangulars and one surangular of MOR 008). Williamson and Carr (1999) described a punctured, infected *Daspletosaurus* ectopterygoid.

Tyrannosaurus rex Bite Marks

"Scotty" (RSM P2523.8) had grooves in the surface of the maxilla, which appeared to have been made by tooth-like structures, apparently representing attack from the back (Fig. 10.1). "Stan" (BHI 3033) had holes penetrating the jugal and surangular and was missing a portion of the occiput. Stan's and Scotty's lesions may have been the result of biting and dominance or mounting behavior.

Alternative Analysis in *Tyrannosaurus rex*

While biting is a reasonable hypothesis, reexamination of these and of several recently discovered specimens suggests an alternative explanation for damage in most tyrannosaurs. The Jane (BMR 2002.4.1) and Peck (MOR 980) *Tyrannosaurus rex* specimens have substantial evidence of skull trauma by sharp objects (Fig. 10.2). However, the width and breadth of the noted lesions are at variance with tyrannosaurid tooth size. There are no other carnivores with jaws large enough to have inflicted the damage, with the exception perhaps of crocodilians (Avilla et al. 2004). The latter consideration can be dismissed because of the absence of large crocodilians in Montana, the Dakotas, Saskatchewan, and Alberta during the Late Cretaceous. Further, the tyrannosaurid facial damage lacks supporting evidence of puncture by conical or ziphodont teeth (Riff and Kellner 2001; Avilla et al. 2004).

Rega and Brochu (2003) mistakenly described holes penetrating the *Tyrannosaurus rex* FMNH PR 2081 ("Sue") mandible (Fig. 10.3) as caused by a fungal infection; subsequently, Wolff et al. (2009) erroneously attributed these changes to *Trichomonas*. That is intriguing, as *Trichomonas* is a sexually transmitted disease in humans, but this was not likely the case in *T. rex*. While some species of *Trichomonas* affect birds, that parasite has not actually been documented to affect bone. Wolff et al. (2009) appropriately illustrated a hole in a bird bone but provided no evidence that it was caused by *Trichomonas*. The bird bone containing the hole also lacked the associated ingrowth of bone found in tyrannosaurids. The pathologic holes penetrating the mandible of *T. rex* present an appearance indistinguishable from that of healing trepanation in humans. These were clearly traumatic in origin, but again they lacked the elongated shape expected from tooth penetration.

Most tyrannosaurid skull pathology (the occipital crest defect in Stan, BHI 3033, excepted) is actually more compatible with a claw derivation. While manual claws can be indicted in some cases, pedal claw lesions

Claw Marks

Institutional Abbreviations AMNH, American Museum of Natural History, New York; BHI, Black Hills Institute of Geological Research, Hill City, South Dakota; BMR, Burpee Museum of Natural History, Rockford, Illinois; FMNH, Field Museum of Natural History, Chicago; LACM, Natural History Museum of Los Angeles County, Los Angeles; MOR, Museum of the Rockies, Bozeman, Montana; NMC, National Museums of Canada, Ottawa; NMMNH, New Mexico Museum of Natural History and Science, Albuquerque; ROM, Royal Ontario Museum, Toronto; RSM, Royal Saskatchewan Museum, Regina; TMP, Royal Tyrrell Museum of Palaeontology, Drumheller, Alberta.

10.2. Lateral oblique view (A) of "Peck" (MOR 980) reveals large hole in right surangular that (B) aligns with a smaller hole in the subjacent clinal. The *Tyrannosaurus* claw aligns perfectly with both holes.

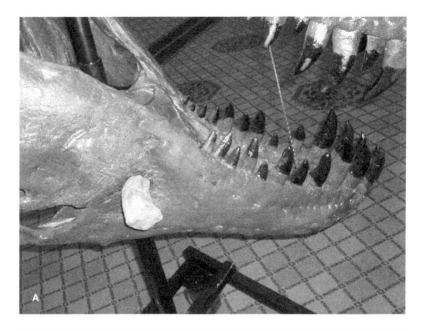

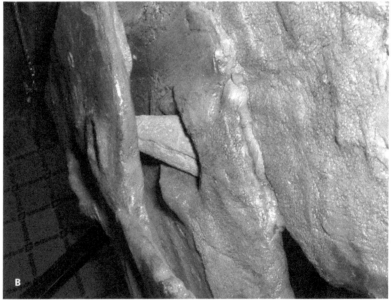

provide intriguing insights to behavior. An impediment to previous exploration of this hypothesis has been the deduction that since claws were made of keratin and keratin is softer than bone, claws would be incapable of penetrating bone. The findings in Peck (MOR 980) clearly falsify the latter perspective. The empirical observation is that a pathologic hole is present, and only the *Tyrannosaurus* toe claw fits it. The notion that soft claws could not scratch bone has recently been tested by providing tigers with an enrichment activity where bones were bolted to a log in a manner that allowed access only by paws, not by jaws or teeth: scratches and punctures were documented (B. M. Rothschild, B. Bryant, C. Hubbard,

10.3. Lateral view of mandibular holes in "Sue" (FMNH PR 2081). Ingrowth of new bone is unassociated with features of infection.

K. Tuxhorn, G. Penn Kilgore, L. Martin, and V. Naples, "The Power of the Claw," in prep.). Just as a straw, given appropriate kinetic energy, may penetrate a tree, so, too, can a claw penetrate bone.

Tyrannosaurus rex Manual Claw

Grooves on the surfaces of the nasal bones of Scotty as well as the *Tyrannosaurus* designated "Monty" have orientations suggesting they were made by the manual claw of a conspecific and that the attack was from the front. Holes with associated remodeling in Jane's maxilla suggest manual claw origin. MOR 590 has a large groove in its left maxilla and a smaller one on the lacrimal, suggesting possible manual claw damage, probably from a frontal assault. Holes penetrating the nasal of AMNH 5027 suggest manual claw damage (Fig. 10.4). A gouge in the maxilla apparently reflects a frontal attack. AMNH 2004 has a hole penetrating its preorbital. "Wyrex" (BHI 6230) has a hole penetrating its jugal. The Elephant Butte *T. rex* (NMMNH P-3698) has a hole penetrating its dentary, as does NMMNH P-27469. Stan has holes penetrating both the jugal and surangular. These isolated holes cannot be explained by tooth impingement but are compatible with a claw etiology, probably manual in origin.

The "Sir William" tyrannosaur (Stein and Triebold 2013) offers special insights. There are two linear alterations on the surface of the dentary. Both manifest healing. One is slightly excavated (as if the bone was spieled off), while the other shows a greater amount of healing. Assuming that the two reflect a single traumatic incident, it is intriguing that their midpoints are separated superiorly by 5.9 cm and inferiorly by 7.0 cm. Incompatible with tooth movement, they would be a classic observation from claw damage.

10.4. Anterior-superior view of nasal of AMNH 5027 illustrating grooves, apparently from manual claw.

Tyrannosaurus rex Pedal Claw

The Peck rex (MOR-980) is from the Hell Creek Formation of McCone County, Montana. The hole penetrating its right surangular aligns with a smaller hole in the subjacent clinal (Fig. 10.2). The hole in the surangular lacks the laterally compressed cross-sectional appearance and is much too large to have been caused by a tyrannosaurid tooth (Farlow et al. 1991), but it accommodates, and is identical in shape to a hole made by, a *Tyrannosaurus rex* toe claw. Not only is it a perfect match, but the extension of the claw through that hole matches up exactly in spatial relationship and size to the hole in the subjacent clinal.

"Samson" ("Z-rex") has a dramatic defect (Fig. 10.5), where a large portion of the left lacrimal has been ripped away, apparently exposing the underlying sinus. The geometry of the damage suggests that the attack came from the back. Additionally, a large surangular defect has associated reactive bone ingrowth. The size and shape are incompatible with teeth but are characteristic of that produced by a pedal claw.

The large hole penetrating the post-orbital of AMNH 5027 also suggests possible pedal claw origin. MOR 008 has a large hole penetrating its mandibular, compatible in size with a pedal claw. A very famous illustration of two fighting tyrannosaurs shows one on its back, with its claws aimed at the individual on top (C. R. Knight in Czerkas and Olson 1987:fig. 14). A more likely senario is the exhibit of *Tyrannosaurus rex* originally planned for the AMNH. Given the change in posture currently accepted for tyrannosaurs, from angled to more horizontal, the opportunities for direct pedal and manual contact are illustrated by deforming one of the early commercial models. This plastic deformation, to illustrate current posture perspectives (Fig. 10.6), perhaps recapitulates the

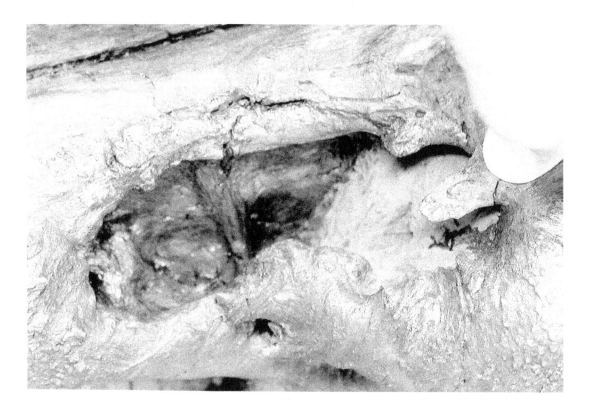

ontologic plastic deformation discussed below, related to stress fractures. Such would certainly explain the distribution of lesions, especially in the Peck *T. rex*.

10.5. Lateral-oblique view of "Samson" illustrating a large skull defect.

Claw Marks in Other Tyrannosaurids

Curiously, cranial scars appear rarer in the other tyrannosaurids, with none recognized in the controversial *Nanotyrannus*. *Albertosaurus* TMP 86.36.314 and TMP 91.10.1 (missing suborbital cheek) manifest only minimal scarring, as do *Gorgosaurus* dentary TMP 91.36.500 and TMP 2001.89.1 and *Daspletosaurus* MOR 379 and TMP 85.62.1 (with a pathologic groove in its nasal). This contrasts with transverse cuts in dentigerous elements of *Albertosaurus*, interpreted as tooth marks by Tanke and Currie (1995).

Post-cranial Scars

Evidence of trauma is also widely represented in tyrannosaurid post-cranial skeletons. This includes fractures of ribs, humeri, and even fibulae (Lambe 1917; Tanke 1996a; Currie 1997; Sotheby's Auction House 1997; Wyoming Dinosaur Center 1997; Tanke and Currie 1998; Brochu 2000; Larson 2001), sometimes complicated by infections in the form of osteomyelitis (Brochu 2000; Webster 2002).

Tanke (1996b) reported healing tyrannosaurid fibula fractures in 10–15 percent of specimens in the Royal Tyrrell collections. While rib lesions may be the result of blunt (e.g., by side of a skull) trauma and

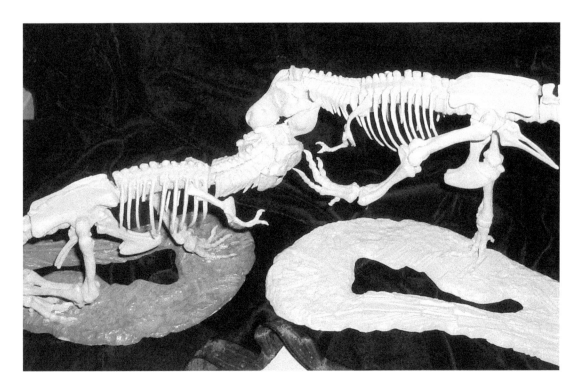

10.6. Superior-lateral view of models of *Tyrannosaurus,* illustrating relative positions in pedal claw attack.

not likely to be caused by contact with a claw, upper and lower leg and caudal vertebral evidence of trauma certainly could have been caused by conspecific bites. Rothschild and Molnar (2009) describe pedal phalangeal osteochondroma in *Gorgosaurus* TMP 91.36.500, as now noted in the *Tyrannosaurus rex* Jane. Penetrating holes in Jane's scapula and ilium (Fig. 10.7) are compatible with impact by a conspecific toe claw.

Infections are predominantly reported as isolated phenomena at puncture sites, including skull, vertebrae, scapula, ilium, ischium, humerus pedal, and manus phalanges (Wells 1984; Rothschild 1997; Williamson and Carr 1999; Tanke and Rothschild 2002). The 2.5 × 3.5 cm penetrating hole described by Molnar (2001) in the iliac blade of *Albertosaurus sarcophagus* ROM 807 is compatible with a puncture injury, and a claw is one of the few structures that could have been responsible.

Stress Fractures

Another measure of strenuous activity is occurrence of so-called stress or fatigue fractures (Rothschild and Martin 1993, 2006; Rothschild et al. 2001; Resnick 2002). Such were present in a *Tarbosaurus* metacarpal, *Gorgosaurus* TMP 2001.89.1, phalanx II-1, and a *Tyrannosaurus* phalanx (KU 1357). These are the result of repetitive activity, rather than acute trauma. These diaphyseal bumps are highly characteristic (Resnick 2002) and easily distinguished from osteomyelitis (bone infection) because of lack of bone destruction (Rothschild and Martin 1993, 2006; Resnick 2002). While this has been used as evidence for predatory rather than

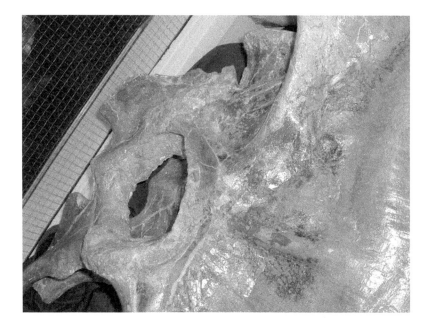

10.7. Lateral view of ilium of "Jane" (BMR P2002.4.1), with close-up of defect, apparently from pedal claw.

scavenging lifestyle, it also would be explained by frequent battles with conspecifics and the above mentioned postures.

While stress fractures are certainly pathologic, they also represent a failure of plastic deformation. Plastic deformation is the process that allows bending of bone shape. Examination of the angulation at the distal third of the *Tyrannosaurus* femur revealed bone reaction in several. Examination of Sampson's femur is most illustrative: the process partially failed, with residual evidence of a stress fracture at that level. Examination of an ontologic series will be of interest to better understand contributing factors.

Gastralia

Farlow et al. (1995), Alexander (1996), and Erickson et al. (2004) suggested that *Tyrannosaurus* probably did not run fast, speculating that a large and heavy animal such as *Tyrannosaurus* would seriously injure or accidently kill itself if it fell while running at high speed. The presence of gastralia fractures has been considered evidence that such injuries did occur and, by presumption, so did fast running.

Tanke (1996b), however, found healing gastralia fractures in an unspeciated Campanian Dinosaur Provincial Park tyrannosaurid, TMP 97.12.229. Brochu (2003) described diffuse gastralia fractures, as well as fractures of presacral 15–22; right rib 15, 18–22; the right coracoid; and the right fibula in Sue (FMNH PR 2081). Lambe (1917) reported healed gastralia fractures in *Gorgosaurus*. Molnar (Molnar 2001; Rothschild and Molnar 2009) noted that *Gorgosaurus* (TMP 94.12.602) had a fractured gastralium with pseudoarthrosis. The latter is the result of an injury in which the fracture components do not fuse, instead forming a false joint. While such injury could have been from a "belly landing," it also is compatible with trauma from a conspecific.

Implications

In examining tooth marks and claw marks, is there any correlation with size or morphotype? While I have examined only a limited number of individuals (in which sex/morphotype is suggested), I found that lesions attributed to manual claw penetration were common and present equally in both morphotypes, independent of animal size/maturity. Lesions attributable to pedal claw injury were limited to the male or gracile morphotype among the examined tyrannosaurs. Does this represent dominance behavior or bellicose attitude of males or at least of male victims? It would seem unlikely for species this bellicose to have been simply scavengers. The frequencies of facial, mandibular, and post-cranial pathologies in tyrannosaurids contrasts with the low frequency noted in scavengers (e.g., *Oviraptor*).

Acknowledgments

Thanks to Nate Murphy, Judith River Dinosaur Institute, Duane and Linda Sibley, Matt Lamanna, Phil Fraley, J. D. Stewart, Eugene Gaffney, Darren Tanke, Jack Horner, Phil Currie, and Pete and Neal Larson.

Literature Cited

Alexander, R. 1996. *Tyrannosaurus* on the run. *Nature* 379:121.

Archer, B., and J. P. Babiarz. 1992. Another tyrannosaurid dinosaur from the Cretaceous of northwestern New Mexico. *Journal of Paleontology* 66:690–691.

Avilla, L., R. Fernandes, and D. F. Ramos. 2004. Bite marks on a crocodylomorph from the Upper Cretaceous of Brazil: evidence of social behavior? *Journal of Vertebrate Paleontology* 24:971–973.

Brochu, C. A. 2000. Postcranial axial morphology of a large *Tyrannosaurus rex* skeleton. *Journal of Vertebrate Paleontology* 20:32A.

Brochu, C. A. 2003. Osteology of *Tyrannosaurus rex:* insights from a nearly complete skeleton and high-resolution computed tomographic analysis of the skull. *Journal of Vertebrate Paleontology* 22 (4, suppl.):1–138.

Czerkas, S. J., and E. C. Olson. 1987. *Dinosaurs Past and Present.* Vol. 1. University of Washington Press, Seattle.

Currie, P. J. 1997. *Gorgosaurus*? hip and tail; p. 3 in Field Experience–Summer 1996. *Royal Tyrrell Museum of Palaeontology Field Experience* 96 update, 4 pp.

Dingus, L. 1996. *Great Fossils at the American Museum of Natural History–Next of Kin.* Rizzoli, New York.

Erickson, G. M., P. J. Makovicky, P. J. Currie, M. A. Norell, S. A. Yerby, and C. A. Brochu. 2004. Gigantism and comparative life-history parameters of tyrannosaurid dinosaurs. *Nature* 430:772–775.

Farlow, J. O., W. L. Abler, and P. J. Currie. 1991. Size, shape and serration density of theropod dinosaur lateral teeth. *Modern Geology* 16:161–198.

Farlow, J. O., M. B. Smith, and J. M. Robinson. 1995. Bone mass, bone "strength indicator," and cursorial potential of *Tyrannosaurus rex. Journal of Vertebrate Paleontology* 15:713–725.

Harris, J. D. 1997. A Reanalysis of Acrocanthosaurus atokensis, Its Phylogenetic Status, and Paleobiogeographic Implication, Based on a New Specimen from Texas. Master's thesis, Southern Methodist University, Dallas, Texas.

Keiran, M. 1999. *Discoveries in Palaeontology: Albertosaurus–Death of a Predator.* Raincoast Books, Vancouver, British Columbia.

Lambe, L. 1917. The Cretaceous theropodous dinosaur *Gorgosaurus. Canada Department of Mines Memoir, Geological Series,* no. 100.

Larson, P. L. 2001. Pathologies in *Tyrannosaurus rex:* snapshots of a killer's life. *Journal of Vertebrate Paleontology* 21:71A–72A.

Molnar, R. E. 1991. The Cranial Morphology of *Tyrannosaurus rex. Paleontographica A* 217:137–176.

Molnar, R. E. 2001. Theropod paleopathology: a literature survey; pp. 337–363 in D. H. Tanke and K. E. Carpenter (eds.), *Mesozoic Vertebrate Life.* Indiana University Press, Bloomington.

Molnar, R. E., and P. J. Currie. 1995. Intraspecific fighting behavior inferred from toothmark trauma on skulls and teeth of large carnosaurs (dinosaurs). *Journal of Vertebrate Paleontology* 15:55A.

Poling, J. 1996. The Dinosauria Homepage. www.dinosauria.com/gallery/darren/htm~/fibula/jpg (website discontinued).

Rega, E. A., and C. A. Brochu. 2003. Paleopathology of a mature *Tyrannosaurus rex* skeleton. *Journal of Vertebrate Paleontology* 21:92A.

Resnick, D. 2002. *Diagnosis of Bone and Joint Disorders.* Saunders, Philadelphia.

Riff, D., and A. W. Kellner. 2001. On the dentition of *Baurusuchus pachecoi* Price (Crocodyliformes, Metasuchia) from Upper Cretaceous of Brazil. *Boletim do Museu Nacional,* n.s.,*Geologica* 59:1–15.

Rothschild, B. M. 1997. Dinosaurian paleopathology; pp. 426–448 in J. O. Farlow and M. K. Brett-Surman (eds.), *The Complete Dinosaur.* Indiana University Press, Bloomington.

Rothschild, B. M., and L. D. Martin. 1993. *Paleopathology: Disease in the Fossil Record.* CRC Press, New York.

Rothschild, B. M., and L. D. Martin. 2006. *Skeletal Impact of Disease.* New Mexico Museum of Natural History, Albuquerque.

Rothschild, B. M., and R. E. Molnar. 2009. Tyrannosaurid pathologies as clues to nature and nuture in the Cretaceous; pp. 287–306 in P. Larson and K. Carpenter (eds.), Tyrannosaurus rex, *The Tyrant King.* Indiana University Press, Bloomington.

Rothschild, B. M., and D. H. Tanke. 2005. Theropod paleopathology: state of the art review; pp. 351–365 in K. Carpenter (ed.), *Carnivorous Dinosaurs.* Indiana University Press, Bloomington.

Rothschild, B. M., D. Tanke, and T. Ford. 2001. Theropod stress fractures and tendon avulsions as a clue to activity; pp. 331–336 in D. Tanke and K. Carpenter (eds.), *Mesozoic Vertebrate Life.* Indiana University Press, Bloomington.

Russell, D. A. 1970. Tyrannosaurs from the late Cretaceous of Western Canada. *National Museum of Natural Sciences Publications in Paleontology* 1:1–34.

Sotheby's Auction House. 1997. *Tyrannosaurus rex:* a highly important and virtually complete fossil skeleton. Sale no. 7045. Sotheby's Auction House catalog, New York.

Stein, W. W., and M. Triebold. 2013. Preliminary analysis of a sub-adult tyrannosaurid skeleton from the Judith River Formation of Petroleum County, Montana; pp. 54–77 in J. Michael Parrish, Ralph E. Molnar, Philip J. Currie, and Eva B. Koppelhus (eds.) *Tyrannosaurid Paleobiology.* Indiana University Press, Bloomington.

Tanke, D. 1996a. Leg injuries in large theropods, etc. Posted to Dinosaur Mailing List, March 9, 1996. Available at http://www.cmnh.org/fun/dinosaurarchive/1996Mar/0150.html. Accessed December 1, 2000.

Tanke, D. 1996b. Tyrannosaur paleopathologies, updates. Posted to Dinosaur Mailing List, November 8, 1996. Available at http://www.cmnh.org/fun/dinosaur,archive/1996Nov/0208.html. Accessed December 1, 2000.

Tanke, D. H., and P. J. Currie. 1995. Intraspecific fighting behavior inferred from toothmark trauma on skulls and teeth of large carnosaurs (Dinosauria). *Journal of Vertebrate Paleontology* 15:55A.

Tanke, D. H., and P. J. Currie. 1998. Head-biting behavior in theropod dinosaurs: paleopathological evidence. *Gaia* 15:167–184.

Tanke, D. H., and B. M. Rothschild. 2002. An annotated bibliography of dinosaur paleopathology and related topics, 1838–1999. *New Mexico Museum of Natural History and Science Bulletin* no. 20.

Webster, D. 2002. Debut Sue. *National Geographic* 197(6):24–37.

Wells, S. P. 1984. *Dilophosaurus wetherilli* (Dinosauria, Theropoda) osteology and comparisons. *Palaeontolographica A* 185:85–180.

Williamson, T. E., and T. D. Carr. 1999. A new tyrannosaurid (Dinosauria: Theropoda) partial skeleton from the Upper Cretaceous Kirtland Formation, San Juan Basin, NM. *New Mexico Geology* 21(2):42–43.

Wolff, E. D., S. W. Salisbury, J. R. Horner, and D. J. Varricchio. 2009. Common avian infection plagued the tyrant dinosaurs. *PLoS One* 4:e7288. doi:10.1371/journal.pone.0007288.

Wyoming Dinosaur Center. 1997, February. The case of the hole in the head. *Bones* (newsletter of the Big Horn Basin Foundation), p. 5.

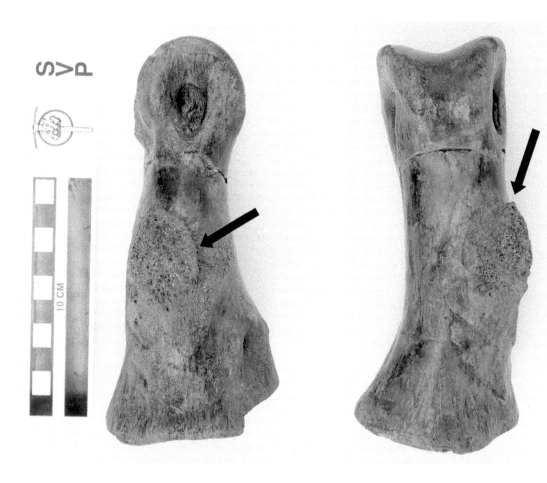

11.1. Left pedal digit II phalanx I. Side (*left*) and dorsal (*right*) views show an eccentric osseous protuberance (arrows).

Brodie Abscess Involving a Tyrannosaur Phalanx: Imaging and Implications

11

Christopher P. Vittore, MD, and Michael D. Henderson

Osteomyelitis is rarely found in dinosaur fossils. When it is identified, the bone lesion shows characteristic, chronic changes of disorganized osseous overgrowth resulting in distortion of the bone. We present a focal bone lesion compatible with osteomyelitis involving a tyrannosaur phalanx. Computed tomographic (CT) imaging disclosed typical findings of a type of subacute osteomyelitis known as a Brodie abscess. This has not been reported in a dinosaur previously. CT scanning of recovered bone fossils is advised for further assessment, particularly if there is any visible surface anomaly. Osteomyelitis, Brodie abscess, and the potential impact on an animal with this lesion are discussed.

Abstract

Introduction

During the summer of 2002, field crews from the Burpee Museum of Natural History collected the partially articulated skeleton of a juvenile tyrannosaurid, nicknamed "Jane" (BMR P2002.4.1), approximately 60 m above the base of the Hell Creek Formation (latest Maastrichtian) of Carter County, Montana. The collection site exposes a fining-upward sequence of clastic sediments that record an active channel and the subsequent formation of an oxbow lake. One hundred forty-five skeletal elements representing about 52 percent of the skeleton (by bone count) of an approximately 7 m long juvenile tyrannosaurid were recovered near the base of this sequence. Recovered skeletal elements show excellent preservation. The dinosaur lay on its right side on top of a point of bar sand. Sixteen proximal caudal vertebrae arced over the back, while the neck was pulled back with the skull positioned over the hips. This posture, the common avian/dinosaur "death pose," is most likely a result of perimortem muscle contractions (Faux and Padian 2007). The skull, pectoral girdle, ribs, presacral vertebrae and distal caudal vertebrae were disarticulated. However, these elements were generally found near their life positions. Although the cause of death is unknown, it was apparently not a result of predation or agonistic interaction given the completeness of the remains and their relatively undisturbed nature. Burial took place after decomposition was advanced, and the animal largely skeletonized, as evidenced by the extensive disarticulation. Sediments surrounding the skeleton (a clay ball conglomerate) are consistent with burial during a flood event, which probably occurred several weeks to months after the death of the animal.

During preparation, it was noticed that one of the pedal phalanges had a focal abnormality along the surface (Fig. 11.1). This was further investigated with standard radiography and CT imaging. We subsequently identified characteristics consistent with a Brodie abscess, a subacute type of pus-forming osteomyelitis. Though this type of bone infection has been identified in humans as far back as the Neolithic period (Lagier and Baud 1983), there is no report in the literature of such a lesion in a dinosaur bone. Given the unique characteristics and implications of a Brodie abscess, the lesion is described and discussed here.

Methods

Imaging

Standard film-screen radiography of the phalanx was performed, obtaining frontal, lateral, and oblique views. The bone was then scanned with a Light Speed multi-detector row CT scanner (GE Medical Systems, Milwaukee, WI). Scanning was done helically with the following acquisition parameters: tube voltage, 120 kV; tube current, 300 mA; scan time, 0.8 s; slice thickness, 1.25 mm; reconstruction interval, 0.6 mm; field of view, 10.5 cm. Image reconstruction was performed with a high-resolution algorithm (E3 bone; GE Medical Systems). Multiplanar reconstructions and three-dimensional image models were created on a workstation (Vitrea 2, Vital Images, Plymouth, Minnesota).

Diagnosis

The initial axial CT images and the subsequent reformatted images in various planes were reviewed. The external and internal characteristics of the lesion were compared with those of known pathologic entities documented in the medical literature

Description of the Material

Material

Gross examination of the left pedal digit II phalanx I shows an eccentric lesion at the mid to proximal diaphysis on the dorsomedial aspect of the bone (Fig. 11.1). The lesion measures $63 \times 43 \times 11$ mm (length × width × depth). The external surface of the lesion is predominantly an irregular, convex bone proliferation. The surface becomes smoother immediately adjacent to the bone. There is no associated alteration in the longitudinal bone axis or in the bone length. The ends of the phalanx are normal. The other recovered bones were without pathologies except for partially healed puncture lesions along the left maxilla and nasal bone (Peterson et al. 2009). There is no evidence of associated bone infection at these sites.

Imaging

Radiographs show that the phalanx lesion is composed of cancellous bone similar to that seen elsewhere within the confines of the anatomic

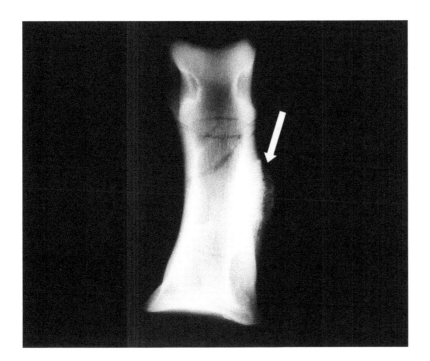

11.2. Radiograph demonstrates that the lesion along the medial shaft of the phalanx (arrow) is composed of disorganized osseous proliferation. A subtle darker region within the lesion near the bone cortex is difficult to see with certainty.

cortical bone, but with a more poorly organized trabecular pattern (Fig. 11.2). The cortex underlying the lesion appears intact. The radiographs suggest a small dark region centrally in the bone proliferation. This is best seen on the frontal radiographic view. Internal characteristics of the dark region cannot be discerned because of the extensive bone overlap. The radiographs also show sharply marginated decreased density filling the central diaphysis. This is compatible with the absence of cancellous bone in the region, typical for a tyrannosaur phalanx. Numerous linear radiolucencies consistent with post-burial fracturing are incidentally noted.

CT scanning allowed further characterization of the lesion (Fig. 11.3), confirming its composition of disorganized, cancellous bone trabeculae. However, the underlying cortical bone is not intact; rather, it has a 7 mm discontinuity located beneath the osseous protrusion. This area is filled with the proliferative bone reaction. Within the proliferative bone and superficial to the cortical discontinuity is an elongated channel-like cavity measuring 21 × 6 × 4 mm, oriented parallel to the longitudinal bone axis (Fig. 11.4). The cavity has thin, sharply defined margins and contains two tiny bone fragments. Reformatted CT images demonstrate the relationship of the cavity to the overlying bone surface (Fig. 11.5).

Discussion

The few reports of osteomyelitis affecting dinosaur bones in the published literature describe proliferative bone changes and draining sinus tracts typical of chronic osteomyelitis (Lindblad 1954; Gross et al. 1993; Laws 1995; Carpenter 1998; Marshall et al. 1998; McWhinney et al. 1998; Rega and Brochu 2001; Tanke and Rothschild 2002). Molnar performed a

11.3. Noncontiguous axial CT scans through the lesion. A) The osseous trabeculae composing the lesion are disorganized and fill a focal discontinuity of cortical bone. The cortical bone does not merge smoothly with the outer surface of the lesion, as occurs with osteochondroma. Instead, the lesion overlies the cortical bone. B) A sharply marginated cavity is present within the lesion. C) One of the tiny bone fragments, consistent with a sequestrum, is visible inside the cavity (arrow).

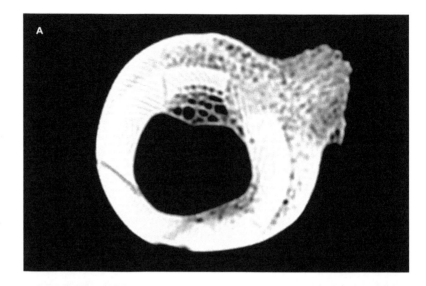

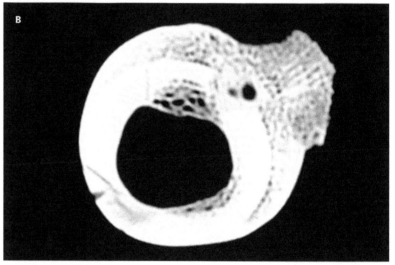

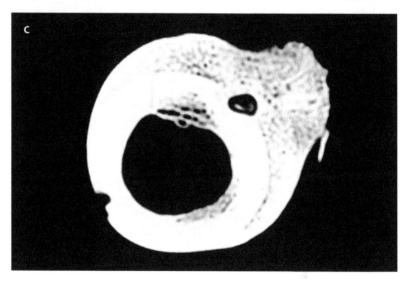

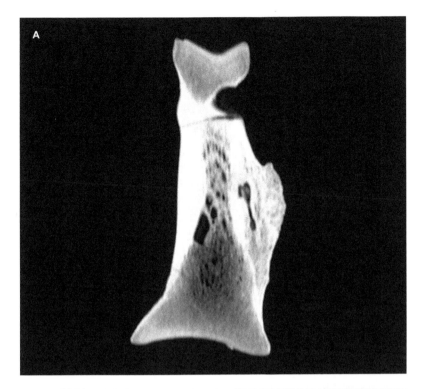

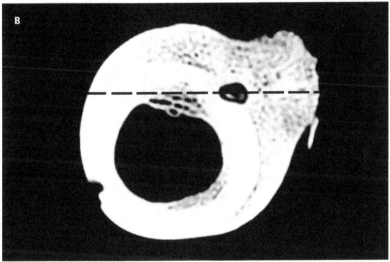

11.4. A) A coronal reformatted CT image shows the internal structure of the lesion as well as the elongated cavity with internal bone fragments (sequestra). B) Cross-referenced with an axial section through the lesion, the dashed line indicates the level of the coronal reformation.

literature survey of theropod paleopathology that disclosed 119 pathologies (Molnar 2001). Most were listed as injuries or due to an undetermined cause. Only seven were classified as infection. Reported pathologies were less common in smaller theropods. The skeletal distribution of the pathologies showed that hind-limb involvement was second only to axial skeleton involvement, and the most common hind-limb bone involved was the phalanx. We present an additional case of a theropod bone infection involving the phalanx of a small tyrannosaur. However, this lesion is more specific. The internal characteristics, best shown by CT scan, are

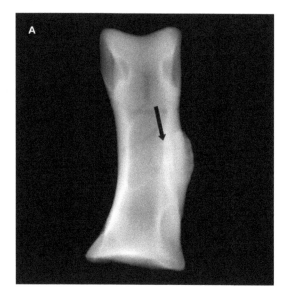 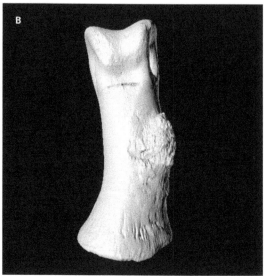

11.5. The axial CT image dataset can be reformatted on a workstation to produce additional image models: A) A three-dimensional see-through display gives a radiograph appearance but does not suffer from overpenetration burnout in regions of lesser density and underpenetration in regions of greater density, as the conventional radiograph does. Improved visualization of the cavity in the lesion results (arrow; cf. figure 11.2). B) Shaded surface display shows the external surface overlying the cavity.

typical for a pus-forming, subacute bone infection. In modern medical literature, this is referred to as a "Brodie abscess." Previously, the oldest reported case of a Brodie abscess was in a human tibia from the Neolithic period estimated to be 5000 years old (Lagier and Baud 1983).

On first examination of the phalangeal abnormality, the external characteristics were suspicious for an osteochondroma, the most common bone tumor in humans. An osteochondroma is a developmental, benign lesion resulting from the separation of a fragment of growth plate cartilage. These lesions have (1) cortical bone margins that merge smoothly with the adjacent normal bone cortex and (2) internal cancellous bone in continuity with the underlying medullary cavity (Fig. 11.6). Subsequent imaging of the subject lesion showed the internal characteristics are clearly not compatible with an osteochondroma. Cortical bone is present underneath the lesion, and there is a central, sharply marginated, elongated, channel-like cavity containing two tiny bone fragments (Figs. 11.3, 11.4). These findings are consistent with the type of subacute bone infection known as a Brodie abscess.

The imaging appearance of a Brodie abscess is a sclerotic bone lesion with an internal cavity. A cloak of proliferative new bone formation, called an involucrum, surrounds the infection and is formed by periosteal reaction. Involucrum is more commonly found in immature bone, where the periosteum is stronger and can provide a barrier to the spread of infection. The configuration of the cavity typically is channel-like or serpentine. Three weeks after the onset of disease, one or more pieces of avascular bone, called sequestra, may become isolated within the cavity (Kharbanda and Dhir 1991). Sequestrum formation is more likely when the abscess is located at the diaphysis. In an institutional review of 25 pathologically proved Brodie abscesses, 4 of the 5 lesions exhibiting sequestra involved the diaphysis (Miller et al. 1979).

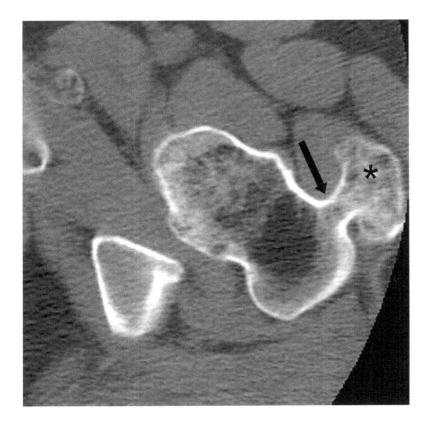

11.6. CT scan of an osteo-chondroma arising from the greater trochanter of a human left femur. The typical findings of an osteochondroma (asterisk) are demonstrated. The cortical margin of the bone merges smoothly with the edge of the lesion (arrow), and the internal cancellous bone shows uninterrupted extension into the medullary cavity.

Courtesy of Terrence Demos, MD, Loyola University Medical Center, Maywood, Illinois.

Differential diagnostic considerations for a sclerotic, diaphyseal bone lesion with an associated region of decreased density include osteoid osteoma and stress fracture. An osteoid osteoma is a benign bone-forming neoplasm. It has a small, round region of decreased density rather than the elongated configuration of the cavity in this case. An osteoid osteoma can also have a sequestrum. However, the bone fragment is typically smooth and centrally located in the less dense portion of the lesion, rather than the irregular, eccentric sequestrum seen in the cavity of a Brodie abscess.

The decreased density in a stress fracture is elongated, and such a lesion has been reported in a ceratopsian phalanx (Rothschild 1988). However, a stress-fracture lucency is hairline thin, rather than having the thicker, channel-like appearance in the subject bone. Also, the lucency in a stress fracture is perpendicular, rather than parallel, to the bone's length. The characteristics of the lesion on the subject bone are thus consistent, not with either of these alternatives, but rather with an indolent infection.

While plain film radiography was helpful in further characterizing the lesion by giving a glimpse into the interior, it was still suboptimal in its demonstration of the elongated cavity deep inside the lesion. Also, it did not show the breach in the underlying cortex, and the sequestra could not be reliably visualized. CT scanning was instrumental in arriving at the correct diagnosis, as it clearly showed these internal details. Studies on humans have come to the same conclusions regarding the superior

11.7. Osteomyelitis resulting from spread of infection from adjacent soft tissues. 1) Infective material in soft tissues adjacent to the bone. Periostitis can result without cortical invasion. 2) Cortical invasion has developed with compromise of the outer cortical surface. There is spread of infection through haversian and Volkmann canals as well as periosteal new bone reaction. 3) Inner cortical margin transgressed with involvement of medullary bone.

Adapted from Resnick and Niwayama (1981).

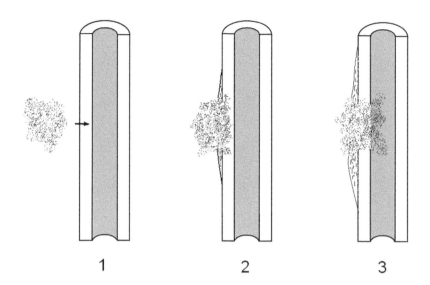

1   2   3

performance of CT scanning over plain film radiographic evaluation for osteomyelitis assessment (Wing and Jeffrey 1985).

Histologically, a bone infection incites an acute inflammatory reaction. Fluid and pus accumulate within the rigid confines of the bone and increase pressure, compromising regional blood flow and causing tissue death. Bone trabeculae break down, and infection spreads within the bone via tiny anatomic channels, the haversian and Volkmann canals. Portions of bone can become isolated from blood flow and separate, resulting in sequestra (Waldvogel et al. 1970).

When osteomyelitis arises from adjacent soft-tissue infection, fascial planes are disrupted early with development of soft-tissue abscesses. These extend into the periosteum by initially invading its outer and more fibrous portion (Fig. 11.7). Further accumulation of pus, subperiosteal bone resorption and cortical disruption follow (Resnick and Niwayama 1981). The bone responds with disorganized, reparative osseous proliferation, resulting in gross sclerotic change.

Osteomyelitis can be classified based on the pathogenesis: (1) osteomyelitis secondary to a contiguous focus of infection or from a direct puncture injury; (2) osteomyelitis of blood-borne (hematogenous) origin; and (3) osteomyelitis associated with peripheral vascular disease. In a published series of 247 patients with osteomyelitis, 47 percent were in the first category associated with adjacent infection or puncture injury. The second and third categories made up 19 percent and 34 percent, respectively (Waldvogel et al. 1970).

For the tyrannosaur discussed here, the contiguous focus or direct penetration etiology seems a likely mechanism for development of osteomyelitis, particularly given the activities of a predatory lifestyle. There is an increased risk of developing osteomyelitis with compound fractures, which involve an associated open wound and thus contamination with skin microorganisms, but there is no evidence for an underlying fracture on this subject bone. An injury resulting from stepping onto something sharp

11.8. A model of fighting dryptosaurs by Charles Knight shows one potential mechanism for injury to the pedal phalanx. The positioning of the animal on its back puts it at risk for a bite to the foot. *Permission from Linda Hall Library of Science, Engineering & Technology, Kansas City, Missouri.*

could produce soft-tissue infection that could spread to the adjacent bone. But the lesion is located on the side of the bone. In fact, the position is more dorsal than ventral along the side of the bone. Thus, the lesion location makes stepping onto a sharp object a less likely etiology for osteomyelitis.

Alternatively, the toe could have been scratched or bitten during an aggressive activity such as intra- or interspecies fighting, mating, or killing prey. An injury from a claw or spike could go down to the bone or produce subsequent soft-tissue infection with potential abscess development, resulting in contiguous spread of infection to the bone. If a claw inflicted the injury, the foot may have been stepped on with the animal upright versus an injury related to an activity with varying positions such as a fight. A tooth-inflicted injury is another consideration, and soft-tissue infections that arise from animal bites are known to be especially troublesome for development of osteomyelitis since the bite is often deep, allowing direct bone inoculation (Lavine et al. 1974). This mechanism exposes the wound to oral pathogens as well as those on the injured skin surface. If the foot was bitten, this animal could have been on its back defending itself with clawed foot raised (Fig. 11.8).

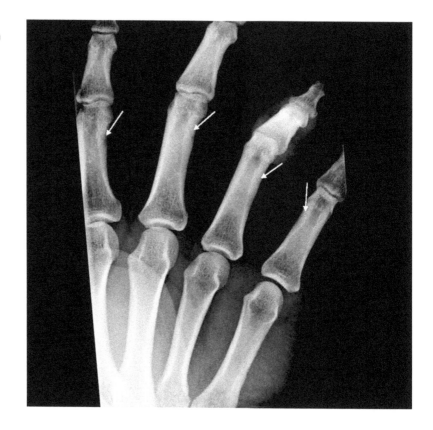

11.9. Radiograph of human hand shows nutrient foramina (arrows) as faint darker lines extending diagonally through the cortical bone of several phalanges. These narrow vascular tunnels could conceivably play a part in the development of a diaphyseal Brodie abscess if a clump of blood borne bacteria lodged at this location.

However, a hematogenous route of infection is the typical pathogenesis for a Brodie abscess. Miller et al. (1979) report a series of 25 Brodie abscesses, and none of these had a history of adjacent soft-tissue infection or penetrating trauma at the site of lesion development. A nutrient foramen can be identified extending through the cortex of some bone shafts (Fig. 11.9). This vascular tunnel may be related to the diaphyseal location of some Brodie abscesses, as a clump of bacteria could conceivably lodge at this small-vessel location. Blood-flow changes due to the regional vascular configuration may also be important. As the end arteries in the metaphysis of an immature long bone are felt to favor deposition of hematogenously delivered infective bacteria by way of sluggish blood flow, perhaps blood flow is also altered sufficiently along an artery in a nutrient foramen as a result of vessel branch pattern. In addition, the tyrannosaur phalanx has a typical "empty" space devoid of bone trabeculae centrally in the diaphysis. It is possible that whatever tissue was in this location also contributed to regional blood-flow alteration and resulted in subsequent infection in the adjacent cortex

One report of a series of 62 patients with hematogenous osteomyelitis disclosed 8 patients with multiple bones involved (Waldvogel et al. 1970). However, these cases were not described as Brodie abscesses, which are typically solitary. No other recovered bones of this tyrannosaur showed evidence for infection. Hematogenous osteomyelitis cases have been reported concomitant with other tissue involvement, including subcutaneous abscess, endocarditis, pericarditis, empyema, meningitis, septic arthritis, and uveitis (Waldvogel et al. 1970). Conversely, a Brodie abscess is usually an isolated lesion with only minimal clinical findings.

Potential sources for hematogenous dissemination of infective organisms can be as innocuous as skin pustules or furuncles (Dickson 1944). Infections involving genitourinary, gastrointestinal, biliary, and respiratory systems can also be a source of bacterial contamination of the blood stream. Even the transient introduction of bacteria into the bloodstream during tooth brushing has been suggested as an inciting event for hematogenous osteomyelitis (Schiller 1988). There is extensive evidence for substantial dental trauma in theropods (Tanke and Currie 1998), and negligible oral trauma in this predator must have occurred as well. Therefore, transient bacteremia in this tyrannosaur remains a strong consideration as an event leading to a Brodie abscess.

The Brodie abscess is named for Sir Benjamin Brodie, who first documented the lesion in the surgical literature in 1832 when he described a series of three cases involving the tibia (Brodie 1832). The lesion is more often seen in children, where it is of hematogenous origin and located centrally in the metaphysis of a long bone. This is believed to result from the unique terminal artery configuration in this area prior to growth-plate closure and associated stagnant blood flow. Diminished phagocytic activity in the regional vasculature may also make this portion of a bone more susceptible to infection (Schenk et al. 1968). Besides long-bone metaphyses, Brodie abscesses can occur in flat or irregular bones and can be located in the diaphysis (Gledhill 1973), as occurred in the phalanx reported here. In Miller's series of 25 Brodie abscesses, one third were located in the diaphysis (Miller et al. 1979). The intraosseous abscess may arise as a result of reduced bacterial virulence or increased host resistance to infection (Wiles 1951). In humans, the infecting organism is usually the common skin bacteria, *Staphylococcus aureus* and, occasionally, *S. epidermidis*. However, *Escherichia coli*, *Salmonella typhi*, *Treponema pallidum*, *T. pertenue*, *Actinomyces*, *Histoplasma*, and hemolytic *Streptococci* have also been reported (Harris and Kirkaldy-Willis 1965). Fungal or tuberculous bone infections will not typically evoke a reactive bone formation. As human Brodie abscesses usually arise from bacterial infection, this also could easily have been the type of pathogen responsible for the infection of the tyrannosaur phalanx. Bacteria are the oldest known fossils, dating back 3.5 billion years, and there is fossil evidence readily comparable to modern forms of bacteria (Moodie 1918). Dinosaurs are known to have lived intimately with bacteria since silicified bacteria were found in pore canal systems of dinosaur eggs (Hirsch 1994). Evidence for bacteria in dinosaur bones has also been previously reported when fossilized bacterial mucilage films were found in a vertebra of an *Iguanodon* (Clark and Barker 1993). The finding of a Brodie abscess in a tyrannosaur indicates the similarity of the inflammatory reaction mounted by these animals' immune systems to that of modern vertebrates.

Patients with a Brodie abscess typically complain of mild-to-moderate pain over a period of several months, often described as a persistent ache. Spontaneous remissions occur commonly and can be prolonged. Fever is often absent. There may be swelling with focal tenderness. Other times, these findings are also absent (Harris and Kirkaldy-Willis 1965). A limp and moderate muscle atrophy can result when there is lower limb involvement (Abril and Castillo 2000). Most cases of Brodie abscess come to the attention of a physician within a few months of symptom onset and are subsequently treated. Therefore it is worthwhile to note the information provided by Brodie in his initial description of a patient who had been suffering from such a lesion for more than 12 years:

> There was a considerable enlargement of the lower extremity of the right tibia. . . . The integuments at this part were tense, and they adhered closely to the surfaces of the bone. The patient complained of a constant pain referred to the enlarged bone and neighboring parts. The pain was always sufficiently distressing; but he was also liable to more severe paroxysms in which his sufferings were described as most excruciating. These paroxysms recurred at irregular intervals, confining him to his room for many successive days and being attended with a considerable degree of constitutional disturbance. [The patient indicated the disease] rendered his life miserable. (Brodie 1832:239)

This man came to Dr. Brodie requesting, and receiving, amputation because of his ongoing suffering despite various other treatments. An examination of the amputated limb resulted in the discovery of an intraosseous abscess cavity the size of a walnut. Another of his patients described a spontaneous, acute pain making him unable to even put his foot to the ground. Regional soft-tissue abscesses later formed and broke; some remained open for a considerable time. For 8–10 years subsequently the patient had "occasional attacks of pain, lasting one or two days at a time; the intervals between them being of various duration, and in one instance, not less than nine months" (Brodie 1832:247). Remitting and relapsing symptoms such as those described above would have had a significant effect on a tyrannosaur's daily activities and overall health. There is a report in the literature of a horse exhibiting lameness for 3 months prior to discovery of a Brodie abscess involving the lateral epicondyle of the humerus (Huber and Grisel 1997). For a tyrannosaur, lameness would have obvious detrimental implications given the impact on chasing prey, and diet could be expected to suffer. If the animal was immature when the lesion first developed, the recurring symptoms could become manifest by a deviation from the expected growth curve. If the lesion was involved in the demise of this dinosaur, it would be an indirect factor, as sepsis does not typically occur with a Brodie abscess. Given time, the lesion could progress, however, and it is noteworthy that osteomyelitis carried an overall mortality rate of 15–25 percent in humans prior to the advent of antibiotics (Dickson 1945).

Osteomyelitis is rarely found in dinosaur fossils, and this is the first report of a Brodie abscess in a dinosaur. A Brodie abscess is a subacute, pus-forming type of osteomyelitis. The discovery of such a lesion emphasizes the similarity between the inflammatory reaction mounted by the immune system of a tyrannosaur and that of modern vertebrates. While this lesion should not have been a direct cause of death, it likely made a significant impact on the health of the animal and could account for a disturbance in growth. CT scanning was instrumental in making the diagnosis by showing proliferative cancellous bone filling a cortical bone discontinuity, an elongated central cavity and bone sequestra. Any potential pathology detected in fossilized bone should be further investigated with CT scanning for optimum assessment.

Summary

We thank Cindy Blitch, RT, for providing her time and technical expertise regarding CT scanning, and Rockford Health System for allowing the use of its medical imaging equipment for this work.

Acknowledgments

Literature Cited

Abril, J. C., and F. Castillo. 2000. Brodie's abscess of the hip simulating osteoid osteoma: case report. *Orthopedics* 23(3):285–287.

Brodie, B. C. 1832. An account of some cases of chronic abscess of the tibia. *Medico-Chirurgical Transactions* 17:239–249.

Carpenter, K. 1998. Evidence for predatory behavior by carnivorous dinosaurs. *Gaia* 15:135–144.

Clark, J. B., and M. J. Barker. 1993. Diagenesis in *Iguanodon* bones from the Wealden Group, Isle of Wight, Southern England. *Kaupia: Darmstädter Beiträge zur Naturgeschichte* 2:57–65.

Dickson, F. D. 1944. Hematogenous osteomyelitis. Beaumont Foundation Lecture series, no. 23:7–47. Wayne County Medical Society, Detroit.

Dickson, F. D. 1945. The clinical diagnosis, prognosis and treatment of acute hematogenous osteomyelitis. *Journal of the American Medical Association* 127:212–217.

Faux, C. M., and K. Padian. 2007. The opisthotonic posture of vertebrate skeletons: postmortem contraction or death throes? *Paleobiology* 33:201–226.

Gledhill, R. B. 1973. Subacute osteomyelitis in children. *Clinical Orthopedics and Related Research* 96:57–69.

Gross, J. D., T. H. Rich, and P. Vickers-Rich. 1993. Bone Infection: chronic osteomyelitis in a hypsilophodontid dinosaur in Early Cretaceous, Polar Australia. *Research and Exploration: A Scholarly Publication of the National Geographic Society* 9(3):286–293.

Harris, N. H., and W. H. Kirkaldy-Willis. 1965. Primary subacute pyogenic osteomyelitis. *Journal of Bone and Joint Surgery* 47:526–532.

Hirsch, K. F. 1994. The fossil record of vertebrate eggs; pp. 269–94 in S. K. Donovan (ed.), *The Palaeobiology of Trace Fossils*. Johns Hopkins University Press, Baltimore, Maryland, 308 pp.

Huber, M. J., and R. Grisel. 1997. Abscess on the lateral epicondyle of the humerus as a cause of lameness in a horse. *Journal of the American Veterinary Medical Association* 211:1558–1561.

Kharbanda, Y., and R. Dhir. 1991. Natural course of hematogenous pyogenic osteomyelitis (a retrospective study of 110 cases). *Journal of Postgraduate Medicine.* 37:69–75.

Lagier, R., and C. Baud. 1983. Brodie's abscess in a tibia dating from the Neolithic period. *Virchows Archiv A: Pathology. Pathologische Anatomie* 401:153–157.

Lavine, L. S., H. D. Isenberg, W. Rubins, and J. I. Berkman. 1974. Unusual osteomyelitis following superficial dog bite. *Clinical Orthopaedics and Related Research* 98:251–253.

Laws, R. R. 1995. Description and analysis of the multiple pathological bones of a sub-adult *Allosaurus fragilis* (MOR 693). *Geological Society of America Abstracts with Programs, Rocky Mountain Section, 47th Annual Meeting* (May 18–19, 1995, Bozeman, Montana) 27(4):43.

Lindblad, G. E. 1954. Field notes, Steveville, Alberta. (Field notes on file at the Royal Tyrrell Museum of Paleontology Library, Drumheller, Alberta).

Marshall, C., D. Brinkman, R. Lau, and K. Bowman. 1998. Fracture and osteomyelitis in PII of the second pedal digit of *Deinonychus antirrhopus* (Ostrom) an Early Cretaceous "raptor" dinosaur (abstract, Palaentological Association, 42nd Annual Meeting, Abstracts and Posters, December 16–19, 1998, University of Portsmouth). *Palaeontological Newsletter* 39:16.

McWhinney, L. A., B. M. Rothschild, and K. Carpenter. 1998. Post-traumatic chronic osteomyelitis in *Stegosaurus* dermal spikes (abstract). *Journal of Vertebrate Paleontology* 18(3):62A.

Miller, W. B., W. A. Murphy, and L. A. Gilula. 1979. Brodie abscess: reappraisal. *Radiology* 132:15–23.

Molnar, R. E. 2001 Theropod paleopathology: a literature survey; pp. 337–363 in D. Tanke and K. Carpenter (eds.), *Mesozoic Vertebrate Life: New Research Inspired by the Paleontology of Philip J. Currie*. Indiana University Press, Bloomington.

Moodie, R. L. 1918. Pathological lesions among extinct animals: a study of the evidences of disease millions of years ago. *Surgical Clinics of Chicago* 2:219–231.

Peterson, J. E., M. D. Henderson, R. P. Scherer, and C. P. Vittore. 2009. Face biting on a juvenile tyrannosaurid and behavioral implications. *Palaios* 24:780–784.

Rega, E. A., and C. A. Brochu. 2001. Paleopathology of a mature *Tyrannosaurus rex* skeleton (abstract). *Journal of Vertebrate Paleontology* 21(3):92A.

Resnick, D., and G. Niwayama. 1981. Section XIII. Infectious diseases: 62. Osteomyelitis, septic arthritis, and soft tissue infection: the mechanisms and situations; pp. 2058–2079 in *Diagnosis of Bone and Joint Disorders*, vol. 3. W. B. Saunders, Philadelphia.

Rothschild, B. M. 1988. Stress fracture in a ceratopsian phalanx. *Journal of Paleontology* 62(2):302–303.

Schenk, R. K., J. Wiener, and D. Spiro. 1968. Fine structural aspects of vascular invasion of the tibial epiphyseal plate of growing rats. *Acta Anatomica (Basel)* 69:1–17.

Schiller, A. L. 1988. Bones and joints; pp. 1304–1393 in E. Rubin and J. L. Farber (eds.), *Pathology*, 1st ed. Lippincott, Philadelphia.

Tanke, D. H., and P. J. Currie. 1998. Head-biting behavior in theropod dinosaurs: paleopathological evidence. *Gaia* 15:167–84.

Tanke, D. H., and B. M. Rothschild. 2002. Dinosores: an annotated bibliography of dinosaur paleopathology and related topics–1838–2001. *New Mexico Museum of Natural History and Science Bulletin*, no. 20:1–96.

Waldvogel, F. A., G. Medoff, and M. N. Swartz. 1970. Osteomyelitis: a review of clinical features, therapeutic considerations and unusual aspects. *New England Journal of Medicine* 282:198–206.

Wiles, P. 1951. *Essentials of Orthopaedics*. J & A Churchill Ltd, London.

Wing, V. W., and R. B. Jeffrey, Jr. 1985. Chronic osteomyelitis examined by CT. *Radiology* 154:171–174.

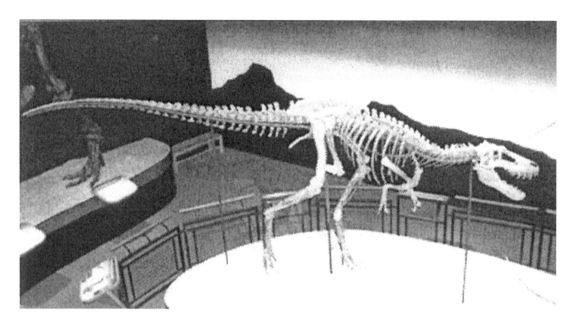

12.1. Restored skeleton of juvenile tyrannosaurid Jane in Burpee Museum of Natural History, Rockford, Illinois.

Using Pollen, Leaves, and Paleomagnetism to Date a Juvenile Tyrannosaurid in Upper Cretaceous Rock

12

William F. Harrison, †Douglas J. Nichols,
Michael D. Henderson, and Reed P. Scherer

The juvenile tryrannosaurid from the Hell Creek Formation (Upper Cretaceous: Maastrichtian) in southeastern Montana, informally named "Jane" (BMR P2002.4.1), is determined to be from a zone in the formation that dates to about 66 Ma. The stratigraphic position of the Jane site is established on the basis of palynology and paleobotany by comparison with correlative sections in southwestern North Dakota and is supported by paleomagnetic data. The palynological and paleobotanical data tightly constrain the age and stratigraphic position of this unique fossil.

Abstract

In June 2001, an expedition from the Burpee Museum of Natural History in Rockford, Illinois, discovered the skeleton of a juvenile tryrannosaurid (BMR P2002.4.1; see Fig. 12.1), approximately 7 m in length, in the Hell Creek Formation (Upper Cretaceous: Maastrichtian) in northwestern Carter County, southeastern Montana (45°46'N, 104°56'W; see Fig. 12.2). The specimen, nicknamed "Jane," was initially identified as either a young *Tyrannosaurus rex* (Carr 2005; Henderson 2005; Parrish et al. 2005) or a *Nanotyrannus lancensis* (Larson 2005), which was known from a single skull found earlier in the same Montana county.

A diverse assemblage of microvertebrates, invertebrates, and fossil plants was associated with the tyrannosaur (Henderson and Harrison 2008). This assemblage, coupled with the extreme rarity of juvenile tyrannosaurid specimens, make this specimen, and the site from which it was recovered, especially significant. However, stratigraphic problems made it difficult to identify the precise level of the Jane site within the Hell Creek Formation.

Introduction

The Jane site exposes a fining-upward sequence in a partial section of the Hell Creek Formation in an isolated, 20 m high butte (Fig. 12.3). The top of the Hell Creek Formation that contains the Cretaceous-Tertiary (K/T) boundary with its firm date of 65.51 Ma (Hicks et al. 2002) is missing from the butte. A series of buttes within view of the Jane site may contain the K/T boundary, but we were not given access to the area.

Stratigraphic Setting

12.2. Location of the Jane site in southeastern Montana.

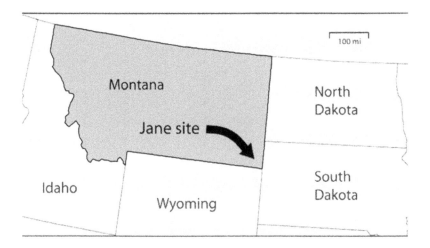

12.3. Isolated outcrop of part of Upper Cretaceous Hell Creek Formation that contains the Jane site.

Additionally, the base of the formation is deeply buried beneath the site. A borehole log from the general area recorded a thickness of 150 m for the formation (Belt et al. 1997). A magnetic reversal (the polarity subchron C30n–C29r boundary) was located in the nearby Blacktail Creek area of Montana, with an average depth of 30 m below the top of the formation (Belt et al. 1997). Just 75 km northeast of the Jane site in southwestern North Dakota the entire formation, including the K/T boundary and the subchron C30n–C29r boundary, is exposed (Nichols and Johnson 2002). In spite of discontinuous outcrops and rapid lateral thinning of strata in and around the Jane site, an 8 m thick measured section at the site shows lithologies and various significant horizons that could be used to interpret stratigraphic and relative age relations (Fig. 12.4).

Stratigraphic and taphonomic evidence indicates that shortly after death, Jane's carcass was deposited on a sandy point bar and was buried within a 40 cm thick lens of poorly sorted silt, sand, and clay balls (Fig. 12.4). This clay ball conglomerate contains abundant plant and animal fossils and diagenetically produced siderite (Fig. 12.5).

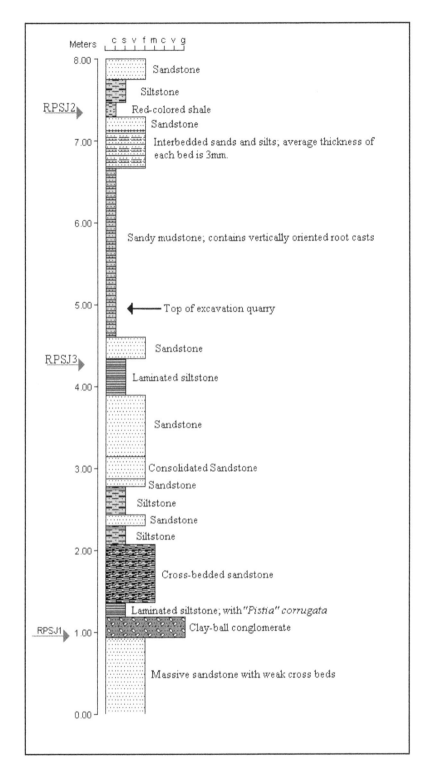

12.4. Stratigraphic section of Hell Creek Formation at the Jane site. Symbol RPSJ1 indicates levels at which paleomagnetic samples were collected within this part of the section. The Jane skeleton was found on the massive sandstone unit with weak cross beds at the base of the section.

12.5. Polished section of clay ball conglomerate that yielded fossil pollen and spores.

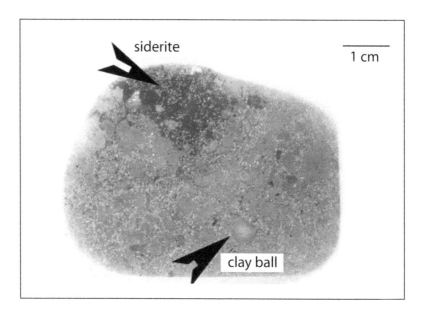

Palynology

The clay balls and siderite within the clay ball conglomerate were also found to contain fossil pollen and spores (Fig. 12.6) that provide the basis for reconstructing the paleoenvironment and determining the stratigraphic position of the Jane site within the Hell Creek Formation. Global Geolab of Medicine Hat, Alberta, prepared slides of the vegetational residue from the clay balls and siderite for palynological analysis. The pollen and spores recovered are typical of the *Aquilapollenites* palynofloral province, which is a characteristic of the uppermost Cretaceous of western North America to northeast China. This province is identified by the presence of pollen of the genera *Aquilapollenites* and *Wodehouseia* (Fig. 12.6), both of which are now extinct.

After the pollen and spores from the Jane site were identified, they were compared with those collected by Nichols (2002) from the entire Hell Creek strata in southwestern North Dakota, where the formation is 100 m thick, in contrast to the 150 m thickness of the formation in southeastern Montana. Although the Hell Creek Formation contains only one palynostratigraphic zone (the *Wodehouseia spinata* Assemblage Zone), Nichols (2002) recognized that during the time of its deposition, some individual pollen species disappeared and others appeared. This observation made possible the subdivision of the *W. spinata* Assemblage Zone in southwestern North Dakota (Nichols 2002) and suggested that the stratigraphic position of the Jane site in southeastern Montana could be determined from its pollen assemblage, which is essentially identical to that recognized in an interval from 28 m to 35 m below the K/T boundary in North Dakota. The primary basis for this conclusion is the presence at the Jane site of the key species *Aquilapollenites collaris*, *A. conatus*, and *A. marmarthensis* and the absence of the key species *Liliacidites altimurus* and *Porosipollis porosis* (Fig. 12.7).

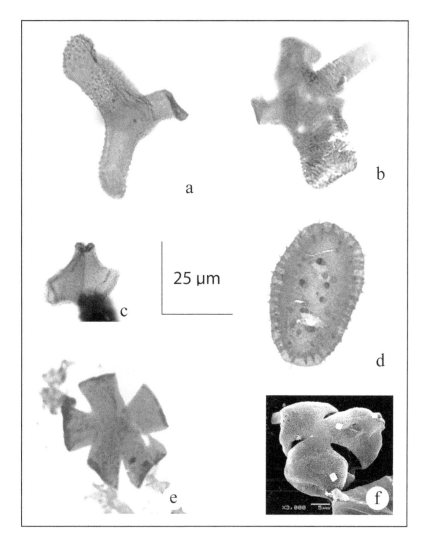

12.6. Fossil pollen from the Jane site. A) *Aquilapollenites collaris.* B) *A. conatus.* C) *A. marmarthensis.* D) *Wodehouseia spinata.* E) *Leptopecopites pocockii.* F) *Tricolpites microreticulatus.*

25 µm

Paleobotany

Fossil leaves collected from the Jane site corroborate the stratigraphic position indicated by the pollen and support the determination of its geologic age. These biostratigraphically important leaves came from the surface of the point bar on which Jane's skeleton lay and from the sandy clay ball stratum that buried the bones (Fig. 12.4). Once the leaves were identified to genus and species, they were compared to those collected by Johnson (2002) from the entire Hell Creek Formation at the same North Dakota site where Nichols (2002) had mapped the distribution of pollen. The leaves recovered from the Jane site were found to correspond to a very narrow megafloral zone, which Johnson (2002) identified as Zone HCIIb. He characterized the HCIIb zone as having a paucity of species and a great abundance of *"Dryophyllum" subfalcatum* and *"Vitis" stantoni* leaves. This characterization precisely describes the megaflora recovered from the point bar and clay ball conglomerate at the Jane site. To date, we have recovered 152 specimens, almost all of which are *"D." subfalcatum* and *"V." stantoni.*

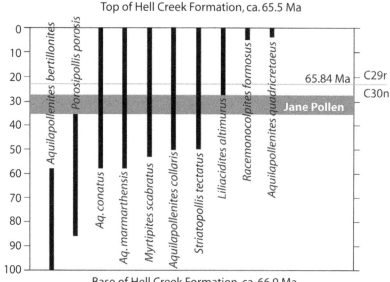

12.7. Stratigraphic ranges of key species of pollen in the Hell Creek Formation identified by Nichols (2002) in North Dakota served to subdivide the *Wodehouseia spinata* Assemblage Zone and permitted the correlation of pollen from the Jane site in Montana with a narrow band just below the boundary of the subchron reversal C30n–C29r with its date of 65.84 Ma. Thickness measurements are in meters based on North Dakota section.

Of the 152 leaves recovered, 75 are *"Dryophyllum" subfalcatum* (Fig. 12.8A), 68 are *"Vitis" stantoni* (Fig. 12.8B), 3 are *"Rhamnus" cleburni* (Fig. 12.8C), and 1 is *Marmarthia trivialis* (Fig. 12.8D). In addition, a cluster of conifer needles and a single cone of *Parataxodium* (Fig. 12.9) were found. *"Rhamnus" cleburni* is found in both HCIIa and HCIIb (Johnson 2002), whereas *M. trivialis* is typical of zone HCIII, which overlies HCIIa. *Parataxodium* is found throughout the Hell Creek Formation (Johnson 2002). Johnson's HCIIb paleobotanical zone falls within an interval from 28 m to 35 m below the K/T boundary in North Dakota, the same interval to which the pollen assemblage from the Jane site corresponds (Fig. 12.10).

Further paleobotanical confirmation of the stratigraphic position of Jane site within the Hell Creek Formation came from a 33 cm thick stratum just above the clay ball conglomerate that contains Jane's bones. This stratum is literally packed with leaves of *"Pistia" corrugata*, an aquatic monocot. According to Johnson (2002), *"P." corrugata* has been found only at sites within the lower three-fourths of the Hell Creek Formation in North Dakota, in strata that record polarity subchron C30n, never at sites in the upper one fourth that record polarity subchron C29r. Thus, in accordance with the data from North Dakota, the presence of *"Pistia"* leaves at the Jane site indicates that the site lies somewhere within the lower three fourths of the formation in Montana.

The absence of leaves of *"Celastrus" taurenensis* at the Jane site greatly delimits the site to near the top of the lower three-fourths of the formation. According to Johnson (2002), *"Celastrus"* leaves are found from the base of the formation to within 36.9 m of the top in North Dakota. If we continue to follow the North Dakota model, the presence of *"Pistia"* and the absence of *"Celastrus"* leaves at the Jane site place the site in a narrow stratigraphic interval in Montana equivalent to that

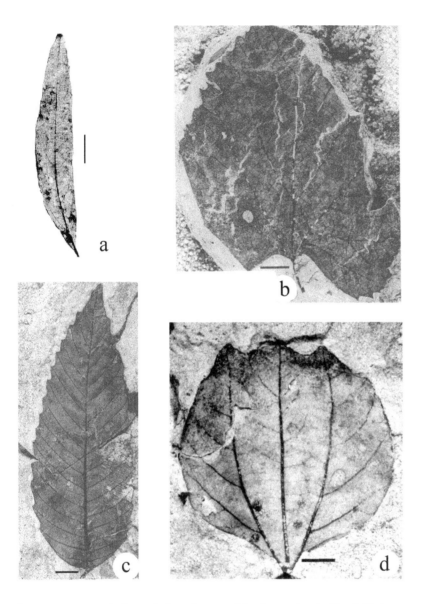

12.8. Fossil leaves from the Hell Creek Formation in North Dakota that were also identified from the Jane site in Montana: a) *"Dryophyllum" subfalcatum,* b) *"Vitis" stantoni,* c) *"Rhamnus" cleburni,* d) *Marmarthia trivialis.*

Specimens © and d) from Johnson (2002).

between 24 m and 36.9 m below the top of the formation in North Dakota (Fig. 12.11).

The restricted stratigraphic interval of the Jane site based on all the paleobotanical data is in agreement with the palynostratigraphic data, which place the Jane site in Montana at the equivalent interval between 24 m and 36.9 m below the top of the Hell Creek Formation in North Dakota (see Fig. 12.11). It is necessary to qualify these stratigraphic positions with the word "equivalent" because of the apparent differences in the total thickness of the formation in southwestern North Dakota (100 m) and southeastern Montana (150 m).

12.9. *Parataxodium* cone from the clay ball conglomerate at the Jane site.

3 cm

12.10. Leaf assemblage at the Jane site correlates to Johnson's (2002) HCIIb leaf zone in North Dakota, which is the same narrow zone just below the subcron reversal C30n–C29r boundary with which the pollen from the Jane site was found to correlate. Thickness measurements are in meters based on North Dakota section.

Top of Hell Creek Formation, ca. 65.5 Ma

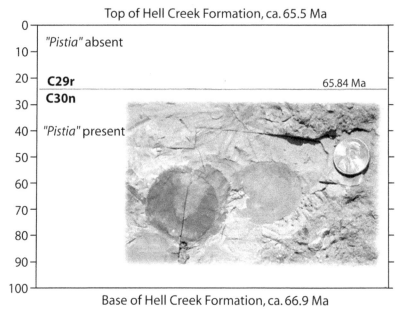

"Pistia" absent

C29r — 65.84 Ma

C30n

"Pistia" present

Base of Hell Creek Formation, ca. 66.9 Ma

Magneto-stratigraphy

Placing the Jane site within a magnetostratigraphic framework also corroborates its stratigraphic relations and supports geologic age interpretation. Twelve paleomagnetic samples were collected from fine-grained sedimentary units at the Jane site from 1.5 m beneath the fossil-bearing unit to the top of the butte (Fig. 12.4). The Paleomagnetics Laboratory of the University of California, Davis, analyzed the samples and found all to be of normal polarity, with the exception of samples from the clay ball conglomerate stratum, which demonstrated chaotic orientation of magnetic grains, as would be expected from a mass-flow deposit. These

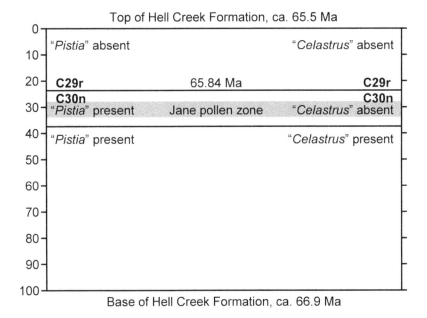

Top of Hell Creek Formation, ca. 65.5 Ma

"*Pistia*" absent	"*Celastrus*" absent
C29r 65.84 Ma	**C29r**
C30n	**C30n**
"*Pistia*" present Jane pollen zone	"*Celastrus*" absent
"*Pistia*" present	"*Celastrus*" present

0
10
20
30
40
50
60
70
80
90
100

Base of Hell Creek Formation, ca. 66.9 Ma

12.11. Distribution of "*Pistia*" corrugata and "*Celastrus*" taurenensis leaves in the Hell Creek Formation in south-western North Dakota and equivalent stratigraphic position of pollen from the Jane site. Thickness measurements are in meters based on North Dakota section.

results corroborate the palynological and paleobotanical data, which predicted that the Jane site would be within an interval of normal polarity (polarity subchron C30n).

The polarity subchron reversal C30n–C29r has been identified at an average depth of 24 m below the top of the Hell Creek Formation in North Dakota (Hicks et al. 2002), which is 4–11 m above the palynostratigraphic subzone containing pollen present at the Jane site (Fig. 12.7). The C30n–C29r reversal is dated at 65.84 Ma (Hicks et al. 2002). Given the differences in the thickness of the formation in North Dakota and Montana, the stratigraphic distance of the Jane site below the C30n–C29r reversal can be assumed to be proportionally greater, but it is evident that the strata at the Jane site were deposited only a short time before the 65.84 Ma reversal.

Based on palynostratigraphic and thickness relations, and in accordance with the magnetostratigraphic age model of Hicks et al. (2002), an age range of 65.9–66.0 Ma can be inferred for the stratigraphic interval of the Hell Creek Formation represented at the Jane site. Fig. 12.12 is a reconstruction of Jane at that time.

Conclusion

By comparing both pollen and leaves from the Jane site in southeastern Montana with established palynofloral and megafloral zones in North Dakota, strata at the Jane site can be correlated with that part of the Hell Creek Formation that occupies the interval 28–35 m below the K/T boundary in western North Dakota. The combined palynological and paleobotanical data indicate that the time frame for the Jane site is slightly older than the C30n–C29r paleomagnetic reversal, which has a firm date

12.12. Jane at ca. 66 Ma.

Artistic reconstruction by M. W. Skrepnick.

of 65.84 Ma, so an age range of 65.9–66.0 Ma can be inferred for when the juvenile tyrannosaurid Jane lived and died.

Acknowledgments

We thank K. Johnson, Denver Museum of Nature and Science, for his assistance in identification and interpretation of the megaflora and for a critical review of the manuscript of this paper. We also thank C. Turner, U.S. Geological Survey, for her critical review of the manuscript. Thanks to K. Verosub for performing paleomagnetic measurements in his laboratory, to J. Warnock for preparing the stratigraphic column, and to M. Skrepnick for the reconstruction of Jane in life. Many thanks also to J. Cooke-Plagwitz, B. Ball, G. Jacky, J. S. Miller, and M. P. Olney for technical assistance.

The authors (WFH, MDH, and RPS) mourn the unexpected loss of their colleague and fellow author Douglas Nichols on February 14, 2010, and dedicate this chapter to him.

Literature Cited

Belt, E. S., J. F. Hicks, and D. A. Murphy. 1997. A pre-Lancian regional unconformity and its relationship to Hell Creek paleogeography in southeastern Montana. *University of Wyoming, Contributions to Geology* 31:1–26.

Carr, T. D. 2005. Independent assessment of the identity of juvenile tyrannosaurs. *The Origin, Systematics, and Paleobiology of Tyrannosauridae: a symposium hosted jointly by Burpee Museum of Natural History and Northern Illinois University, Abstracts*, p. 23.

Henderson, M.D. 2005. Nano no more: the death of the pygmy tyrant. *The Origin, Systematics, and Paleobiology of Tyrannosauridae: a symposium hosted jointly by Burpee Museum of Natural History and Northern Illinois University, Abstracts*, p. 17.

Henderson, M. D., and W. F. Harrison. 2008. Taphonomy and environment of deposition of a juvenile tyrannosaurid skeleton from the Hell Creek Formation (latest Maastrichtian) of southeastern Montana; pp. 82–90 in P. Larson and K. Carpenter (eds.), Tyrannosaurus rex, *the Tyrant King*. Indiana University Press, Bloomington.

Hicks, J. F., K. R. Johnson, J. D. Obradovich, L. Tauxe, and D. Clark. 2002. Magnetostratigraphy and geochronology of the Hell Creek and basal Fort Union Formations of southwestern North Dakota and a recalibration of the age to the Cretaceous–Tertiary boundary; pp. 25–55 in J. H. Hartman, K. R. Johnson, and D. J.

Nichols (eds.), The Hell Creek Formation and the Cretaceous–Tertiary boundary in the northern Great Plains: an integrated continental record of the end of the Cretaceous. *Geological Society of America Special Paper* no. 361.

Johnson, K. R. 2002. Megaflora of the Hell Creek and lower Fort Union Formation in the western Dakotas: vegetational response to climate change, the Cretaceous–Tertiary boundary event and rapid marine transgression; pp. 329–391 in J. H. Hartman, K. R. Johnson, and D. J. Nichols (eds.), The Hell Creek Formation and the Cretaceous–Tertiary boundary in the northern Great Plains: an integrated continental record of the end of the Cretaceous. *Geological Society of America Special Paper* no. 361.

Larson, P. 2005. The case for *Nanotyrannus. The Origin, Systematics, and Paleobiology of Tyrannosauridae; a symposium hosted jointly by Burpee Museum of Natural History and Northern Illinois University, Abstracts,* p. 18.

Nichols, D. J. 2002. Palynology and palynostratigraphy of the Hell Creek Formation in North Dakota: a microfossil record of plants at the end of the Cretaceous time; pp. 393–456 in J. H. Hartman, K. R. Johnson, and D. J. Nichols (eds.), The Hell Creek Formation and the Cretaceous–Tertiary boundary in the northern Great Plains: an integrated continental record of the end of the Cretaceous. *Geological Society of America Special Paper* no. 361.

Nichols, D. J., and K. R. Johnson. 2002. Palynology and microstratigraphy of Cretaceous-Tertiary boundary sections in southwestern North Dakota; pp. 95–143 in J. H. Hartman, K. R. Johnson, and D. J. Nichols (eds.), The Hell Creek Formation and the Cretaceous–Tertiary boundary in the northern Great Plains: an integrated continental record of the end of the Cretaceous. *Geological Society of America Special Paper* no. 361.

Parrish, M. J., M. D. Henderson, and K. A. Stevens. 2005. Functional implications of ontogenetic change in the glenohumeral joint within the Tyrannosauridae. *The Origin, Systematics, and Paleobiology of Tyrannosauridae; a symposium hosted jointly by Burpee Museum of Natural History and Northern Illinois University, Abstracts,* p. 21.

13.1. Proposed model of a *Tyrannosaurus–Triceratops* collision. The *Triceratops* is represented as a rectangular mass, while the *Tyrannosaurus* applies rotational momentum in order to pivot it about the rotational axis established with the feet on the opposite side and tip it over. In this diagram, F_T is the force applied by the *Tyrannosaurus* to point A (or point A1 as an alternate impact point with a larger *Tyrannosaurus* and/or a smaller *Triceratops*), and F_G is the force of gravity; B and C represent two points at which the rear feet touch the ground, establishing a reference plane for the calculations described in the text.

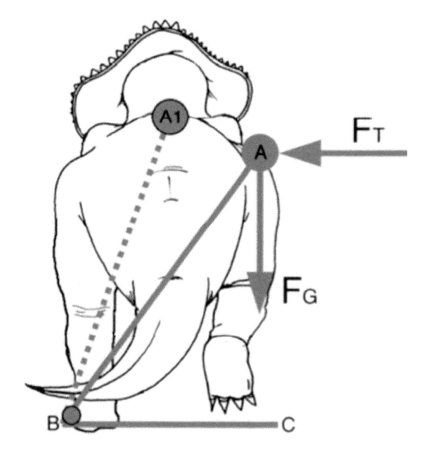

The Biomechanics of a Plausible Hunting Strategy for *Tyrannosaurus rex*

13

David A. Krauss and †John M. Robinson

We present here a biomechanical analysis of a hunting strategy that *Tyrannosaurus rex* could have employed effectively. The modern analogy for this hunting strategy is "cow tipping," in which reckless people ambush and tip cows over. Although this analogy seems odd, it is apt. Anatomical analysis of *Triceratops* indicates that, like a cow, if it were knocked over on its side it would have experienced difficulty in getting up. It seems likely that tyrannosaurs could have exploited this weakness in a hunting strategy.

A series of physical analyses were conducted to test the hypothesis that this hunting method was possible for *Tyrannosaurus rex*. The results indicate that an adult tyrannosaur moving at moderate speed would have produced more than enough force to knock a large *Triceratops* off its feet. Further, it may only have needed to maintain pressure against the side of the *Triceratops* for as little as 2–3 seconds to tip it over.

The most interesting aspect of this theory is that it explains most of the unique features of tyrannosaur anatomy. Specifically, their small arms seem to be an adaptation allowing them to grasp their prey's back while pushing it with the pectoral region of their torso. Their large heads would have helped in tipping their prey over, and their large mouths and bone-piercing teeth would have made bites to the side more effective and lethal. According to this "ceratopsian-tipping" hypothesis, *Tyrannosaurus* would have ambushed its prey from cover, knocking it over and rendering it vulnerable, then killing it with a swift bite to the rib cage.

Abstract

Introduction

Although generally described as a predator (Osborn 1905; Paul 1987, 1988; Molnar and Farlow 1990; Carpenter 1997; Carpenter and Smith 2001), *Tyrannosaurus* has also been described as a scavenger (Lambe 1917; Horner and Lessem 1993), and debate over its primary mode of feeding has persisted since the early 1900s. Most recently, the view of *Tyrannosaurus* as a scavenger has predominated based on structural (Horner and Lessem 1993; Farlow 1994) and bioenergetic (Farlow 1976, 1994; Ruxton and Houston 2003) arguments. Like many large carnivores, *Tyrannosaurus* may well have scavenged when the opportunity arose, but it is possible that it was primarily a hunter. What is missing from the argument in favor of tyrannosaur hunting is a viable strategy for an adult animal. Various hunting strategies for *Tyrannosaurus* have been proposed, including the quick strike of a land shark–like predator (Paul 1987) and a group hunting strategy, like that used by lions (Currie 1998), but neither of these hypotheses is entirely convincing. Here we present a modified version of the "land-shark" type of ambush hunting strategy that is not only physically possible for an adult *T. rex* but also explains some of the unusual aspects of its anatomy.

Tyrannosaurs are unusual among theropods for several reasons. Most notable are their proportionally small arms and large head. Explaining the reduced arm size has long been a problem in paleontology. The hunting model put forth here presents one possible explanation. In this case, tyrannosaurs would use their arms to grapple with prey, not in the traditional sense of holding or slashing, but rather to grip the smooth back of a ceratopsian and prevent it from dislodging the attacking tyrannosaur. Short arms could be extended fully, thus reducing the risk of injury to the attacker. This hunting strategy would also place directional selection on large head size, which would provide a mechanical advantage in attacking ceratopsian prey, as will be seen below.

A modern analog for the hunting strategy being proposed is the practice known as "cow tipping," in which two or more people (usually intoxicated teenagers) sneak up on a sleeping cow and, by applying force to one side of the cow's back, knock it over on its side. Once felled, the cow is incapable of getting back up because the structure of its legs does not allow it to place its feet under its center of gravity. A ceratopsian dinosaur would have been similarly incapacitated if knocked over on its side, and it seems plausible that tyrannosaurs evolved to take advantage of this difficulty. A ceratopsian should have been able to recover from a position of lying on its side by rolling back and forth as modern rhinoceroses do, but it would have taken several seconds during which time it would have been highly vulnerable. At this point we wish to emphasize that "cow tipping" is a cruel, dangerous, and illegal activity, and we do not condone it in any way. It does, however provide an appropriate model for tyrannosaur hunting.

One of the most serious criticisms of a predatory lifestyle for *Tyrannosaurus rex* is that it was probably not a high-speed runner (Farlow et al. 1995, 2000; Hutchinson and Garcia 2002; Brusatte et al. 2010). It seems unlikely that it could have chased down and caught speedier prey species like hypsilophodonts or even hadrosaurs. However, even the relatively low speeds calculated for tyrannosaurs would be sufficient to attack a ceratopsian that would probably be restricted to similar speeds from ambush. A tyrannosaur could have lunged from cover, generating enough force to knock over a ceratopsian at even a very moderate speed, provided it took its prey by surprise. It would take only a matter of seconds for a tyrannosaur to make contact with a ceratopsian and push it over, after which the ceratopsian would be significantly incapacitated, and the tyrannosaur could dispatch it with relative ease.

There is ample evidence that *Tyrannosaurus* ate *Triceratops* and/or other ceratopsians (Erickson and Olson 1996; Erickson et al. 1996; Chin

et al. 1998), so *Triceratops* was selected as the sample prey animal for this study. Also, as the largest and heaviest of the ceratopsians, it would have been the most difficult prey species for *Tyrannosaurus* to tip over. The cow-tipping strategy should have been effective against any large ceratopsian. Once knocked over on its side, a ceratopsian would have had great difficulty in getting back up. Large modern mammals keep their legs underneath their bodies when at rest on the ground. If they find themselves lying on their sides, they must roll onto their backs, fold their legs under their ventral surfaces, and then roll back onto their legs in order to get back up. We made numerous observations of rhinoceroses, camels, and large bovids in zoos, videotaping this behavior as part of this study. This process usually takes from 3 to 10 seconds. *Triceratops* would have had to perform this action pattern (which may have been hindered by its frill) to get back up, during which time the *Tyrannosaurus* could deliver a killing bite to its vulnerable flank.

Methods and Results

In order for a tyrannosaur to tip over a ceratopsian, it would have needed to exert enough force to overcome its prey's inertia and maintain this pressure for enough time to knock it over. In order to test the hypothesis that such an action was possible, we performed a series of calculations based on measurements of mounted museum specimens and mass estimates derived from the literature. These calculations were designed to be highly conservative—that is, to make it as difficult as possible for a tyrannosaur to tip over a triceratops.

In order to calculate the force a tyrannosaur could generate in a charge, it is necessary to know its mass and speed. Estimates for both of these factors vary widely. An average of values for the mass of an adult *Tyrannosaurus* found in the literature surveyed (Colbert 1962; Alexander 1985; Anderson et al. 1985; Peczkis 1994; Henderson 1999; Seebacher 2001) was 6300 kg. Although the more recent best estimates are above 7000 kg (Henderson 1999; Seebacher 2001), the average value of 6300 kg was used for most calculations as it was more conservative and made this hunting method less feasible.

Estimates of tyrannosaur speed also vary widely, with 11 m/s being the lowest estimate of a tyrannosaur's top speed found in the literature (Hutchinson and Garcia 2002). As a conservative estimate, a speed of 7.5 m/s was used in these analyses. This speed was also chosen as a safe speed at which a *Tyrannosaurus* could impact a *Triceratops* without causing injury to itself. Upon impact with a *Triceratops*, a *Tyrannosaurus* would experience a significant reduction in velocity. Assuming that the initial velocity of the Triceratops is 0 m/s, then the deceleration would be enough to cause potentially serious injury to the tyrannosaur. At speeds below 7.5 m/s, the deceleration would be less than 3.4 g, which has been calculated as the greatest impact force a *Tyrannosaurus* could withstand without injury (Alexander 1996).

The average mass of *Triceratops* taken from literature values is 5846 kg, but this number is skewed by exceptionally heavy early estimates (Colbert 1962), and more recent calculations place *Triceratops* at 3938 kg (Henderson 1999) and 4963 kg (Seebacher 2001). We assumed a mass of 5000 kg for *Triceratops*, as this heavy estimate would make it less likely that a *Tyrannosaurus* could have tipped it over. We assumed that the velocity of the *Triceratops* would be zero. This assumption is fair because, even if the *Triceratops* were engaged in forward motion at the moment of impact, it would have zero velocity in the lateral direction, which is where the *Tyrannosaurus* would most likely approach to attack.

Upon impact, the tyrannosaur would have essentially attempted to pivot the triceratops around a line formed by its two feet on the opposite side from the tyrannosaur (Fig. 13.1). The effective lever arm is the line drawn between the pivot point and the point of impact on the back of the *Triceratops*. It is likely that the tyrannosaur would intentionally hit its prey at or near its pelvis, as this is the highest point on the animal and is a low flexibility area where the force will be transferred most directly into rotational motion (Fig. 13.2A). Estimates of the length of this lever arm were therefore based on measurements from the outer edge of the hind foot to the opposite topside of the pelvis of mounted *Triceratops* skeletons on display at the American Museum of Natural History and the Smithsonian Museum of Natural History. A fair bit of variation is inherent in such calculations based on the style of mounting, estimates of muscle mass, and so forth.

When colliding with a stationary *Triceratops*, the *Tyrannosaurus* would experience a significant deceleration, running the risk of injury caused by the collision. Alexander (1996) calculated that a *Tyrannosaurus* could withstand a deceleration equivalent to approximately 3.4 g without sustaining injury. Using the model described below, a tyrannosaur could have successfully tipped a ceratopsian at an attacking speed of only 3.6 m/s, well within the literature values for *Tyrannosaurus* speed estimates. At this speed, assuming reasonable values for the plastic deformation of the bodies of both dinosaurs involved in the collision, the *Tyrannosaurus* would experience a deceleration of approximately 2.3 g well within the safe zone for avoiding serious injury. Increase in speed would have increased deceleration, but there is still plenty of room for variation within the calculated limits of tyrannosaur anatomy.

Model for a *TyrannosaurusTriceratops* Collision

Assuming an inelastic collision in which a *Triceratops* rotates around an axis connecting its two feet on the opposite side from the impact, the *Triceratops* can be represented simplistically as a rectangular mass M_2, base length $2L$, and height H. The mass of *Tyrannosaurus* is represented as mass M_1. The center of mass is assumed to be at height D above the axis of rotation and located sagittally on the centerline. There will be a large, brief impulsive force exerted on the feet of the *Triceratops* by the

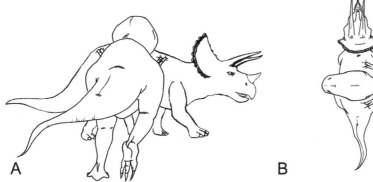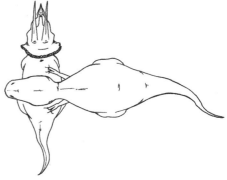

A

B

ground during the brief collision time. However this force has no lever arm around the axis of rotation and therefore no torque, thus resulting in a conservation of angular momentum during the collision leading to the equation

$$M_1 v_0 H = I\omega_0 + M_1 v_f H. \tag{1}$$

The left side of the equation represents the angular momentum of the tyrannosaur of mass M_1 and initial velocity v_0, assuming for simplicity a height H for the tyrannosaur's center of gravity. The first term on the right is the rotational angular momentum of the *Triceratops*, its moment of inertia around the axis being designated by I and its initial angular velocity just after the collision being designated by ω_0. The last term represents the residual angular momentum of the tyrannosaur with v_f designating its residual velocity.

The assumption of an inelastic collision implies that the horizontal component of the tangent velocity, $\omega_0 R$ of the contact point on the rotating *Triceratops*, equals the horizontal residual velocity v_f of the tyrannosaur. That is, the animals remain in contact for at least a short time after the collision. This condition is expressed by

$$v_f = \omega_0 R \sin\theta. \tag{2}$$

Substituting Equation (2) into Equation (1) allows us to solve for ω_0 as follows:

$$\omega_0 = \frac{M_1 v_0 H}{I + M_1 RH \sin\theta}. \tag{3}$$

The *Triceratops* will tip over if the rotational kinetic energy associated with ω_0 exceeds the increase in its gravitational potential energy required to raise its center of gravity to and slightly beyond a position directly over its axis of rotation. This condition is expressed in Equation (4):

$$\frac{1}{2} I \omega_0^2 \geq M_2 g (r - D), \tag{4}$$

where $r = \sqrt{D^2 + l^2}$.

13.2. Proposed strategy of a *Tyrannosaurus* attacking ceratopsian (*Triceratops*) prey. Tyrannosaurs may well have employed a technique analogous to modern "cow tipping" in ambushing ceratopsian prey species. By rushing from cover, a tyrannosaur could push against the pelvis of a ceratopsian thus knocking it over, rendering it vulnerable to further attack. (A)Ceratopsian-tipping, side view and (B) overhead view.

Another critical concern is the deceleration a_d experienced by the tyrannosaur during the collision:

$$a_d = \frac{\Delta v}{\Delta t}, \tag{5}$$

where

$$\Delta v = v_f - v_0 \tag{6}$$

and

$$\Delta t = \frac{\Delta x}{\left[\frac{1}{2}\left(v_f + v_0\right)\right]}, \tag{7}$$

where Δx is the sum of the plastic deformations of the two animals during the collision, and v_f is given by Equation (2). Using the parameters for such a collision derived above and substituting these values into the equations produces a simulation of a possible *Tyrannosaurus–Triceratops* interaction:

$H = 2.36$ m,
$l = .93$ m,
$M_1 = 6300$ kg,
$M_2 = 5000$ kg,
$R = 3$ m,
$\theta = 52°$,

and

$v_0 = 7.5$ m/s.

In addition, D is assumed to be 2 m, and Δx is assumed to be .2 m. The sum of deformations Δx is consistent with a deformation of .1 m for each animal following the estimations of Farlow et al. (1995). There remains the problem of estimating the moment of inertia I of the *Triceratops*. By the parallel axis theorem,

$$I = Ic.g. + M_2 r^2, \tag{8}$$

where $Ic.g.$ is the moment of inertia around the center of gravity, which we can estimate by using the formula for the $Ic.g.$ of a rectangular prism of mass M_2:

$$Ic.g. = \frac{1}{l^2} M_2 \left(4l^2 + H^2\right). \tag{9}$$

This is a crude approximation for $Ic.g.$, however; the model predictions are rather insensitive to $Ic.g.$ owing to the dominance of the second term in Equation (8).

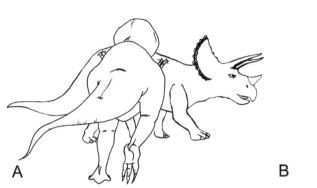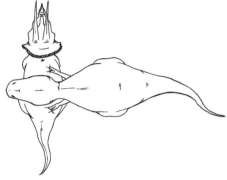

A   B

ground during the brief collision time. However this force has no lever arm around the axis of rotation and therefore no torque, thus resulting in a conservation of angular momentum during the collision leading to the equation

$$M_1 v_0 H = I\omega_0 + M_1 v_f H. \tag{1}$$

The left side of the equation represents the angular momentum of the tyrannosaur of mass M_1 and initial velocity v_0, assuming for simplicity a height H for the tyrannosaur's center of gravity. The first term on the right is the rotational angular momentum of the *Triceratops*, its moment of inertia around the axis being designated by I and its initial angular velocity just after the collision being designated by ω_0. The last term represents the residual angular momentum of the tyrannosaur with v_f designating its residual velocity.

The assumption of an inelastic collision implies that the horizontal component of the tangent velocity, $\omega_0 R$ of the contact point on the rotating *Triceratops*, equals the horizontal residual velocity v_f of the tyrannosaur. That is, the animals remain in contact for at least a short time after the collision. This condition is expressed by

$$v_f = \omega_0 R \sin\theta. \tag{2}$$

Substituting Equation (2) into Equation (1) allows us to solve for ωo as follows:

$$\omega_0 = \frac{M_1 v_0 H}{I + M_1 R H \sin\theta}. \tag{3}$$

The *Triceratops* will tip over if the rotational kinetic energy associated with ω_0 exceeds the increase in its gravitational potential energy required to raise its center of gravity to and slightly beyond a position directly over its axis of rotation. This condition is expressed in Equation (4):

$$\frac{1}{2}I\omega_0^2 \geq M_2 g(r - D), \tag{4}$$

where $r = \sqrt{D^2 + l^2}$.

13.2. Proposed strategy of a *Tyrannosaurus* attacking ceratopsian (*Triceratops*) prey. Tyrannosaurs may well have employed a technique analogous to modern "cow tipping" in ambushing ceratopsian prey species. By rushing from cover, a tyrannosaur could push against the pelvis of a ceratopsian thus knocking it over, rendering it vulnerable to further attack. (A)Ceratopsian-tipping, side view and (B) overhead view.

Another critical concern is the deceleration a_d experienced by the tyrannosaur during the collision:

$$a_d = \frac{\Delta v}{\Delta t}, \qquad (5)$$

where

$$\Delta v = v_f - v_0 \qquad (6)$$

and

$$\Delta t = \frac{\Delta x}{\left[\frac{1}{2}\left(v_f + v_0\right)\right]}, \qquad (7)$$

where Δx is the sum of the plastic deformations of the two animals during the collision, and v_f is given by Equation (2). Using the parameters for such a collision derived above and substituting these values into the equations produces a simulation of a possible *Tyrannosaurus–Triceratops* interaction:

$H = 2.36$ m,
$l = .93$ m,
$M_1 = 6300$ kg,
$M_2 = 5000$ kg,
$R = 3$ m,
$\theta = 52°$,

and

$v_0 = 7.5$ m/s.

In addition, D is assumed to be 2 m, and Δx is assumed to be .2 m. The sum of deformations Δx is consistent with a deformation of .1 m for each animal following the estimations of Farlow et al. (1995). There remains the problem of estimating the moment of inertia I of the *Triceratops*. By the parallel axis theorem,

$$I = Ic.g. + M_2 r^2, \qquad (8)$$

where $Ic.g.$ is the moment of inertia around the center of gravity, which we can estimate by using the formula for the $Ic.g.$ of a rectangular prism of mass M_2:

$$Ic.g. = \frac{1}{l^2} M_2\left(4l^2 + H^2\right). \qquad (9)$$

This is a crude approximation for $Ic.g.$, however; the model predictions are rather insensitive to $Ic.g.$ owing to the dominance of the second term in Equation (8).

Results

For the above choice of model parameters, we obtain the following results from the preceding equations:

$$\omega_0 = 1.77 \text{ rad/s},$$
$$v_f = 4.18 \text{ m/s},$$

and

$$a_d = 97.1 \text{ ms}^{-2} = 9.9 \text{ g}.$$

Furthermore, the left side of Equation (4) greatly exceeds the right side. These results indicate that using these parameters a *Tyrannosaurus* could tip over a *Triceratops* but that the deceleration from a v_0 of 7.5 m/s would probably have been lethal to the *Tyrannosaurus* as well. These results lead to the question, What is the minimum v_0^1 required to tip a *Triceratops*, and would it be safe enough for a *Tyrannosaurus* to survive the impact? This question can be answered by replacing the inequality in Equation (4) with an equality and solving Equations (3) and (4) for v_0^1 :

$$v_0^1 = \frac{I + M_1 R H \sin\theta}{M_1 H} \sqrt{\frac{2M_2 g(r - D)}{I}}. \tag{10}$$

These equations yield the following predictions:

$$v_0^1 = 3.6 \text{ m/s},$$
$$\omega_0 = 0.855 \text{ rad/s},$$
$$v_f = 2.02 \text{ m/s},$$

and

$$a_d = 22.8 \text{ ms}^2 = 2.33 \text{ g}.$$

We see that the tyrannosaur can attack at a much lower speed than the assumed maximum and reduce the deceleration to safe levels. The collision force F and torque τ for such a minimal speed collision can be found as follows:

$$F = M_1 a_d = 1.44 \times 10^5 N,$$

and

$$\tau = FH = 3.4 \times 10^5 N \cdot m.$$

The latter torque is much greater than the resting gravitational torque τ_g of the *Triceratops*, which is calculated by

$$\tau_g = M_2 g l = .46 \times 10^5 N \cdot m$$

in even the minimal-speed scenario. Increases in speed would certainly increase the tyrannosaur's chances of a successful attack and can well be accommodated within the limits of these equations. A great deal of variability is possible with larger tyrannosaurs attacking smaller ceratopsians at higher speed as long as the deceleration remains less than 3.4 g. The mathematical conclusion is that it was possible for *Tyrannosaurus* to successfully attack and tip over a *Triceratops*.

Discussion

It is clear that, at least on paper, an adult *Tyrannosaurus rex* would have had the ability to tip over a *Triceratops*, rendering it vulnerable. An adult, and even a sub-adult *Tyrannosaurus*, could have generated more than enough force to tip over an adult *Triceratops*. However, these numbers should not be taken at face value. Depending on how the skeleton is reconstructed, the high point on the back of a *Triceratops* may have been as much as 2.8 m high, whereas modern reconstructions of *Tyrannosaurus rex* place its pectoral girdle at only 2–2.2 m above the ground. If tyrannosaurs pushed from their chests, as we argue below, it seems less likely that they would have been able to knock over a large adult *Triceratops*. That does not mean that this "cow-tipping" hunting method would have been impossible. Smaller individuals and other species of ceratopsian would have been highly vulnerable to such an attack, and we suggest that these animals may have been an important enough prey source to have significantly influenced the evolution of the tyrannosaurids.

Once on its side, a *Triceratops* (or any other ceratopsian) would have been nearly helpless, much like a cow. With most of its mass resting on the proximal portions of two of its legs, those limbs would have been effectively pinned underneath it and would have hindered its efforts to roll over onto its feet. Observations of several large modern animals (rhinoceroses, camels, horses, bison, and cows) were made in order to determine how they deal with this potential vulnerability. In all cases, the modern animals observed folded their legs under their bodies when lying down, thus keeping them under their centers of gravity. This posture enables them to rise quickly and easily. Rhinoceroses (arguably the best model for ceratopsians) cross one of their front legs under their body in order to attain this posture. Many of these animals do sometimes lie on their sides in order to bathe in dust or mud, and all have some difficulty in getting back up. To rise from a position lying on their sides, rhinoceroses rock back and forth, taking the weight off of their legs and allowing them to fold the legs back under their bodies. They must also achieve enough momentum in their rocking to roll back over on top of their legs. Observations of this procedure in animals at the Bronx, San Diego, and Denver zoos produced times ranging from 3 to 10 seconds under ideal conditions. It is this amount of time when the flank of the ceratopsian prey would be exposed that a *Tyrannosaurus* could deliver a killing bite.

As yet, there is no direct paleontological evidence to support this theory, nor is this behavior likely to be preserved in the fossil record. Ample evidence supporting this theory can be found in the unique anatomy of the tyrannosaurs. Compared to other theropods, tyrannosaurs have unusual tooth structure. Tyrannosaur teeth have an unusually wide cross section (Carpenter 1997) that seems to be an adaptation for puncturing bones. Biomechanical analyses by Snively and Russell (2007a, 2007b) have demonstrated aspects of the skull and neck that further characterize *Tyrannosaurus* bite mechanics as typical of a puncture and pull/shake feeding style. Such findings can be used to argue in favor of a scavenging habit as scavenging animals have to deal with bones and require reinforced rostral structures (Snively et al. 2006) and shaking ability to rip pieces from a large carcass. These analyses also fit the hunting model presented here. If tyrannosaurs were "tipping" ceratopsians, the most vulnerable and exposed portion of their prey would be the flank. A bite to the flank would be most effective if it were delivered to the rib-cage area, thus increasing the potential to damage the lungs with the broken bones. Selection for stouter teeth and jaw structures capable of puncturing bone would have been placed on tyrannosaurs in this circumstance. Subsequent enhancement of such a feature, enabling them to puncture larger bones and take bigger bites, is a logical corollary.

The mouth itself is unusual in *Tyrannosaurus*. The muzzle is also reinforced vertically, creating exceptional biting power (Molnar and Farlow 1990; Snively et al. 2006; Snively and Russell 2010), a fact reflected by its limited ability to expand (Horner and Lessem 1993). Therrien et al. (2005) further demonstrated that the mechanics of the tyrannosaur jaw have their closest modern analog in molluscivorous monitor lizards that must crunch shells as part of their diet. It has been argued (Horner and Lessem 1993) that bone-crunching ability implies a scavenging habit, but this is not necessarily the case. Modern wolves and hyenas both scavenge but are primarily predators and have teeth specialized to crunch bones. If anything, a "cow-tipping" hunting strategy involving a bite to the flank through the ribs would place directional selection pressure on tyrannosaurs for larger, stronger mouths in order to effectively kill their prey. The large flat surface of the prey's side would necessitate a wider gape and stronger jaw to generate the force needed to bite it effectively, crushing the ribs in the process.

This hunting strategy would also place directional selective pressure on the evolution of the unusually large head of *Tyrannosaurus*. *Tyrannosaurus* has long been noted for its extremely large head. A brief survey done as part of this study found that the height of the head of *Tyrannosaurus* as a proportion of its body length was more than 4 standard deviations larger than the mean for all other theropods, although the length of its head as a proportion of body length was closer to the norm. If *Tyrannosaurus* was a "cow tipper," then this hunting strategy could have produced directional selection for a larger head. As noted above, a larger mouth would increase the effectiveness of a bite to the side of a ceratopsian, and a larger head produces a larger mouth. Additionally, a larger head would actually create a mechanical advantage for tipping a ceratopsian. Tyrannosaurs would not have been able to effectively use their small arms in the process of pushing over a ceratopsian. Rather, such an effort would have been most effective by pushing from the chest and probably from the area of the pectoral girdle that is the most heavily reinforced area of the body. If this were the case, then the head of the tyrannosaur would have extended to the far side of the ceratopsian's center of gravity (Fig. 13.2B). As more weight would have been placed on the far side of its prey's center of gravity, it would have gained a mechanical advantage for tipping it over. Thus the head of tyrannosaurs could, again, be placed under directional selection for proportionally large size.

By far the most perplexing aspect of tyrannosaur anatomy has been the proportionally small size of their forelimbs. Although the arms of tyrannosaurs are proportionally small, they were very strong, each probably being capable of lifting 180 kg (Carpenter and Smith 2001). However, this strength is concentrated in the flexor, rather than the extensor, direction. In the "cow-tipping" hunting strategy it is assumed that the tyrannosaur is pushing its prey with its chest, not its arms. However, while pushing it ran the risk of slipping or being shaken off by its struggling prey. It is logical that a tyrannosaur would have grasped its prey's back to prevent slipping while it pushed its prey over. As the ceratopsian shifted, throwing the *Tyrannosaurus*, for example, to the left, the predator could have flexed its right arm to hold its place while it pushed and would correspondingly flex its left arm if the ceratopsian tried to thrust it to the right. If the arms of the tyrannosaur were fully extended, or nearly so, while this action was taking place, the tyrannosaur would run the least risk of having its elbows wrenched and sustaining serious joint injury. The actual orientation and length of *T. rex* arms are ideal for this strategy. There is some variation dependent on how one reconstructs the orientation of the scapulocoracoid in *Tyrannosaurus*. However, in nearly all reconstructions at full extension with the chest of the tyrannosaur pressed against the side of a ceratopsian, the fingers would have reached just about to the spine of its prey, allowing for a potentially useful grip. The "cow-tipping" hunting strategy would therefore select for tyrannosaurs with very strong but relatively short arms, with an ideal length very close to that observed in preserved specimens. A reduction to two forward-facing fingers might have also reduced the risk of injury to a third, more laterally oriented digit, although this inference may be a bit of a stretch.

Finally, *Tyrannosaurus rex* is also noted for having proportionally large feet for a theropod. Again, this pattern would be consistent with the "cow-tipping" model. Once a tyrannosaur made contact with a ceratopsian, it would need to push it over while not being displaced by it. Large feet would provide better traction for the tyrannosaur, improving its ability to exert pressure against its prey. Thus tyrannosaurs may have been subject to directional selection for large feet as well.

It is not our intent here to argue that *Tyrannosaurus rex* was not a scavenger. We doubt that a tyrannosaur would have passed up a free meal. There are too many precedents for opportunistic scavenging among large modern predatory reptiles, mammals, and birds. Rather, we hope to have presented a convincing argument that *T. rex* could have been primarily a predator and to show that a predatory habit rather than a scavenging lifestyle more readily explains its unusual anatomy. *Tyrannosaurus rex* almost certainly could not have chased down prey and would most likely have lost a head-to-head battle with *Triceratops*. However, this analysis has shown that it could have been a successful ambush hunter. Analysis of a juvenile *Tarbosaurus* skull (Tsuihiji et al. 2011) has shown significant ontological changes in the feeding mechanisms of tyrannosaurids, and it seems likely that ceratopsian tipping would have been a hunting strategy employed by adult animals.

So, to paint a final picture of the hunter as set forth by the "ceratopsian tipping" model: Imagine a meadow in the late Cretaceous with woods along its margins. A group of ceratopsians ambles through the meadow feeding placidly. As they near the woods, their scent carries to a lurking *Tyrannosaurus*. It waits patiently for them to draw near, perhaps within 20–30 m (the distance typical of modern lions). When they get in range, the tyrannosaur locks onto a sub-adult of about the right size and lunges from cover. In only a few seconds, the tyrannosaur has covered the distance between them, calculating innately for the movement of the ceratopsian as it begins to run. With a crash it knocks into the ceratopsian's hip, immediately grabbing onto its prey's spine with its forelimbs. As the ceratopsian struggles, it is already off balance, and, with one more step, the tyrannosaur has it tilting, and with a second step it is on its side on the ground. As its prey struggles and rolls on the ground, the tyrannosaur lunges in delivering a massive bite to the rib cage. At this point, other members of the family group are rallying to the aid of the fallen member and the tyrannosaur runs rapidly back to the woods, but its job is done. The ceratopsian is mortally wounded, and, when the rest of the herd eventually moves on, the tyrannosaur returns to feed on its kill.

Acknowledgments

We thank Carl Mehling for his help in allowing us access to the specimens on display at the American Museum of Natural History and Dr. Michael Brett-Surman for his help in gaining access to the specimens at the Smithsonian Museum of Natural History. W also thank Dr. Kevin Harrison for his helpful comments on this manuscript. We thank Michael Parrish for his editing of this manuscript and are especially grateful to James Farlow for significant comments and improvements on the model presented here. Finally, we thank Cynthia Frelund for her insights into cow tipping.

Literature Cited

Alexander, R. McN. 1985. Mechanics of posture and gait of some large dinosaurs. *Zoological Journal of the Linnean Society* 83:1–25.

Alexander, R. McN. 1996. *Tyrannosaurus* on the run. *Nature* 379:121.

Anderson, J. F., A. Hall-Martin, and D. A. Russell. 1985. Long-bone circumference and weight in mammals, birds and dinosaurs. *Journal of Zoology London* 207:53–61.

Brusatte, S. L., M. A. Norrell, T. D. Carr, G. M. Erickson, J. R. Hutchinson, A. M. Balanoff, G. S. Bever, J. N. Choiniere, P. J. Makovicky, and X. Xu. 2010. Tyrannosaur paleobiology: new research on ancient exemplar organisms *Science* 329:1481–1485.

Carpenter, K. 1997. Tyrannosauridae; pp. 766–768 in P. J. Currie and K. Padian (eds.), *Encyclopedia of Dinosaurs.* Academic Press, Orlando.

Carpenter, K., and M. Smith. 2001. Forelimb Osteology and Biomechanics of *Tyrannosaurs rex;* pp. 90–116 in D. H. Tanke and K. Carpenter (eds.), *Mesozoic Vertebrate Life.* Indiana University Press, Bloomington.

Chin, K., T. T. Tokaryk, G. M. Erickson, and L. C. Calk. 1998. A king-sized theropod coprolite. *Nature* 393:680–682.

Colbert, E. H. 1962. The weights of dinosaurs. *American Museum Novitates* 2076:1–16.

Currie, P. J. 1998. Possible evidence of gregarious behavior in tyrannosaurids. *Gaia* 15:271–277.

Erickson, G. M., and K. H. Olson. 1996. Bite marks attributable to *Tyrannosaurus rex:* preliminary description and implications. *Journal of Vertebrate Paleontology* 16:175–178.

Erickson, G. M., S. D. Van Kirk, J. Su, M. E. Levenston, W. E. Caler, and D. R. Carter. 1996. Bite-force estimation for *Tyrannosaurus rex* from tooth-marked Bones. *Nature* 382:706–708.

Farlow, J. O. 1976. A consideration of the trophic dynamics of a Late Cretaceous large-dinosaur community (Oldman Formation). *Ecology* 57:841–857.

Farlow, J. O. 1994. Speculations about the carrion-locating ability of tyrannosaurs. *Historical Biology* 7:159–165.

Farlow, J. O., M. B. Smith, and J. M. Robinson. 1995. Body mass, bone "strength indicator," and cursorial potential of *Tyrannosaurus rex. Journal of Vertebrate Paleontology* 15:713–725.

Farlow, J. O., S. M. Gatesy, T. R. Holtz, Jr., J. R. Hutchinson, and J. M. Robinson. 2000. Theropod locomotion. *American Zoologist* 40:640–663.

Henderson, D. M. 1999. Estimating the masses and centers of mass of extinct animals by 3-D mathematical slicing. *Paleobiology* 25:88–106.

Horner, J. R., and D. Lessem. 1993. *The complete* T. rex. Simon & Schuster, New York.

Hutchinson, J. R., and M. Garcia. 2002. *Tyrannosaurus* was not a fast runner. *Nature* 415:1018–1021.

Lambe, L. M. 1917. The cretaceous theropodous dinosaur *Gorgosaurus. Memoirs of the Geological Survey of Canada* 100:1–84.

Molnar, R. E., and J. O. Farlow. 1990. Carnosaur paleobiology; pp. 210–224 in D. B. Weishampel, P. Dodson, and H. Osmólska (eds.), *The Dinosauria.* University of California Press, Berkeley.

Osborn, H. F. 1905. *Tyrannosaurus* and other cretaceous carnivorous dinosaurs. *Bulletin of the American Museum of Natural History* 21:259–265.

Paul, G. S. 1987. Predation in the meat eating dinosaurs; pp. 171–176 in P. J. Currie and E. H. Koster (eds.), *4th Symposium on Mesozoic Terrestrial Ecosytems Short Papers.* Tyrell Museum of Palaeontology, Drumheller, Alberta.

Paul, G. S. 1988. *Predatory Dinosaurs of the World.* Simon & Schuster, New York.

Peczkis, J. 1994. Implications of body-mass estimates for dinosaurs. *Journal of Vertebrate Paleontology* 14:520–533.

Ruxton, G. D., and D. C. Houston. 2003. Could *Tyrannosaurus rex* have been a scavenger rather than a predator? An energetics approach. *Proceedings of the Royal Society of London B* 270:731–733.

Seebacher, F. 2001. A new method to calculate allometric length-mass relationships of dinosaurs. *Journal of Vertebrate Paleontology* 21:51–60.

Snively, E., and A. P. Russell. 2007a. Craniocervical feeding mechanics of *Tyrannosaurus rex. Paleobiology* 33:610–638.

Snively, E., and A. P. Russell. 2007b. Functional variation of neck muscles and their relation to feeding style in Tyrannosauridae and other large theropod dinosaurs. *Anatomical Record* 290:934–957.

Snively, E., D. M. Henderson, and D. S. Phillips. 2006. Fused and vaulted nasals of tyrannosaurid dinosaurs: implications for cranial strength and feeding mechanics. *Acta Palaeontologica Polonica* 51:435–454.

Therrien, F., D. M. Henderson, and C. B. Ruff. 2005. Bite me: biomechanical models of theropod mandibles and implications for feeding behavior; pp. 179–238 in K. Carpenter (ed.), *The Carnivorous Dinosaurs.* Indiana University Press, Bloomington.

Tsuihiji, T., M. Watabe, K. Tsogtbaatar, T. Tsubamoto, R. Barsbold, S. Suzuki, A. Lee, R. Ridgely, Y. Kawahara, and L. Witmer. 2011. Cranial osteology of a juvenile specimen of *Tarbosaurus baatar,* (Theropoda, Tyrannosauridae) from the Nemegt Formation (Upper Cretaceous) of Bugin Tsav, Mongolia. *Journal of Vertebrate Paleontology* 31:497–517.

Wolfson, R., and J. M. Pasachoff. 1999. *Physics for Scientists and Engineers,* 3rd ed. Addison-Wesley, Reading, Massachusetts.

14.1. Skeleton of *Tyrannosaurus rex,* represented by "Stan."

Photograph courtesy of the Black Hills Institute of Geological Research.

A Closer Look at the Hypothesis of Scavenging versus Predation by *Tyrannosaurus rex*

14

Kenneth Carpenter

Controversy surrounds the feeding behavior of the large theropod *Tyrannosaurus:* Was it an obligate scavenger, a predator, or an opportunist that scavenged as well as hunted? Evidence for an obligate scavenging lifestyle is examined: Enlarged olfactory lobes, allegedly for carrion detection, are shown to also occur in extant non-scavenging predators and in other dinosaurs. Eyesight may have been poor in low light but otherwise acute. Hindlimb segment lengths are significantly greater than potential prey, thus *Tyrannosaurus* could outrun prey. Despite the proportionally short arms, various stress fractures of the furcula show that the forelimbs generated great stresses, as would be expected with holding struggling prey. Finally, evidence of failed attacks on prey demonstrates conclusively that *Tyrannosaurus* was a predator.

Abstract

Introduction

The large theropod *Tyrannosaurus rex* was named in 1905 by Henry Fairfield Osborn for a partial skeleton excavated in 1902 and 1905 by Barnum Brown. The skeleton was found in the Hell Creek Formation exposed along Hell Creek, Garfield County, Montana. The skeleton was found on Sheba Hill on the Max Sieber Ranch located about 209 km northwest of Miles City and 19 km south of the Missouri River. In selecting the name, Osborn sought to convey his impression of this dinosaur as a "tyrant lizard king" and implied that it was a predator by stating that it was "undoubtedly the chief enemy of the Ceratopsia and Iguanodontia" (Osborn 1906:281). Later, he wrote that "*Tyrannosaurus* is the most superb carnivorous mechanism among the terrestrial Vertebrata, in which raptorial destructive power and speed are combined" (Osborn 1916:762). Osborn (1910:8) wrote that "*Tyrannosaurus* of the Cretaceous [was] fitted by nature to attack and prey upon the largest of their herbivorous contemporaries." Brown (1915:271) elaborated on this belief in his fanciful reconstruction of *Tyrannosaurus* behavior: "A huge herbivorous dinosaur *Trachodon*, coming on shore for some favorite food has been seized and partly eaten by a giant *Tyrannosaurus.*" This idea that *Tyrannosaurus* was a predator has long persisted in the popular culture, even up to the present, as seen in the 1993 blockbuster movie *Jurassic Park*, where *Tyrannosaurus* is seen eating a lawyer and chasing people.

Insitutional Abbreviations DMNH, Denver Museum of Natural History (now Denver Museum of Nature and Science), Denver, Colorado; MOR, Museum of the Rockies, Bozeman, Montana; TCM, The Children's Museum, Indianapolis, Indiana; UMNH, Utah Museum of Natural History, Salt Lake City.

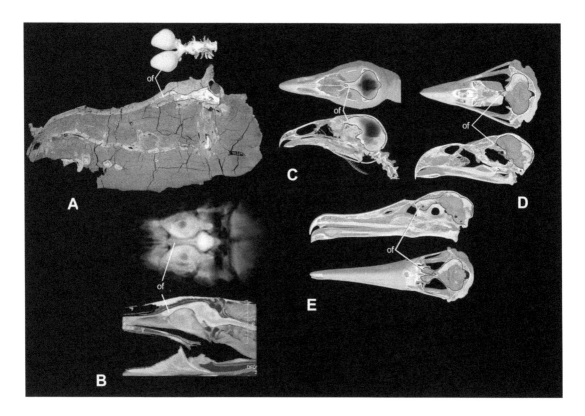

14.2. Comparisons of the olfactory region in (A) *Tyrannosaurus,* (B) *Caiman* sp., (C) *Coragyps atratus* (black vulture), (D) *Haliaeetus leucocephalus* (bald eagle,), and (E) *Diomedea immutabilis* (Laysan albatross). The anterior and ventral margins (shown as dashed lines) of the olfactory lobe in *Tyrannosaurus* are unknown; although clearly wide, they may not be as long as shown in top view.

Adapted from Brochu (2000). (A) Computed tomography data from Brochu (2003). (C–E) Courtesy of Tim Rowe, Digimorph.

Osborn's reasons for considering *Tyrannosaurus* a predator are numerous: its large size as a carnivore (Osborn 1905, 1906); its large, robust, massive skull (Osborn 1906, 1912, 1916); its large, powerful, "raptorial" teeth Osborn 1916); its manus with powerful recurved claws (Osborn 1916); and its powerful hindlimb, also with recurved claws (Osborn 1916). Many of these characters are still cited as evidence for *Tyrannosaurus* being a predator (e.g., Haines 2000) and can be seen in the skeleton (Fig. 14.1).

A challenge to the predator hypothesis was made by Horner and Lessem (1993) and Horner (1994), who contended that *Tyrannosaurus* was ill equipped to be a predator. Therefore, it must have been a scavenger. They claim that *Tyrannosaurus* had a keen sense of smell, poor eyesight, was unable to run fast, and had arms too short to reach its mouth, concluding that it was "hard to image how it caught its dinner" (Horner and Lessem 1993:208). Some additional support for the carrion-locating abilities of *Tyrannosaurus* was presented by Farlow (1994). He noted that carrying the head high would allow *Tyrannosaurus* to see carrion at greater distances than smaller theropods and that the elevated position of the nostril also enabled it to smell carrion more quickly than a small theropod would. Ruxton and Houston (2002) speculated that there was sufficient carrion in the Cretaceous to sustain *Tyrannosaurus* as a scavenger. These various statements are examined more closely below.

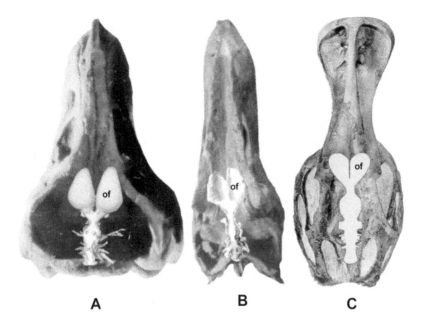

A   B   C

14.3. Large olfactory lobes (of) may be a common feature of dinosaurs, as demonstrated by comparing that of *Tyrannosaurus* (A) with another predator, *Allosaurus,* (B) and prey *Edmontosaurus* (C). All olfactory lobes are approximations. Skulls are shown to same length to show the large size of the olfactory lobes relative to skull size.

The endocast of Tyrannosaurus *is adapted from Brochu (2000). The endocast of* Allosaurus *is based on UMNH VP7435 and the olfactory lobe from the underside of the skull roof of DMNH 1943. The* Edmontosaurus *endocast is a composite between Lull and Wright (1942) and Ostrom (1961).*

Olfactory Region

Brochu (2000; 2003) substantiated the claim by Horner and Lessem (1993) that the olfactory region of *Tyrannosasurus* skull is large (Fig. 14.2A). Horner and Lessem (1993:216) actually refer to the olfactory lobes as "abnormally large," despite the fact that the actual size is unknown because the anterior and ventral margins of the capsule housing the lobe are unossified. Nevertheless, they note that the vulture also has large olfactory lobes, which indeed they are (Fig. 14.2B). The lobes are considerably more prominent than in the eagle (Fig. 14.2C), which uses vision to hunt. Horner and Lessem (1993) state that the enlarged olfactory lobes of *Tyrannosaurus* imply that it had a heightened sense of smell in order to locate carrion, as does the vulture.

To support such a hypothesis, the enlarged olfactory lobe would have to be restricted to scavengers, which can be tested by examining the size of the lobes in extant vertebrates. As can be seen in Figure 14.2, crocodilians have large lobes (Fig. 14.2B), as does the albatross (Fig. 14.2E), both which are predators. It would therefore seem that olfactory lobe size alone is not an adequate test for a scavenging diet. As for the statement that the olfactory lobe is "abnormal" in size, this can only be substantiated in comparison to other dinosaurs. As can be seen in Figure 14.3, large olfactory lobes are common in dinosaurs and may be characteristic of them. The large olfactory lobe of predators may be just as likely to be used to smell prey as to locate carrion. In contrast, large olfactory lobes in prey may be used for smelling the approach of predators.

Eyesight

Based on orbital size, Horner (1994:161) stated that *Tyrannosaurus* had "beady, little eyes." The large (~13 cm anteroposterior length) orbit indicates a large eyeball (Farlow and Holtz 2002), the size of which could

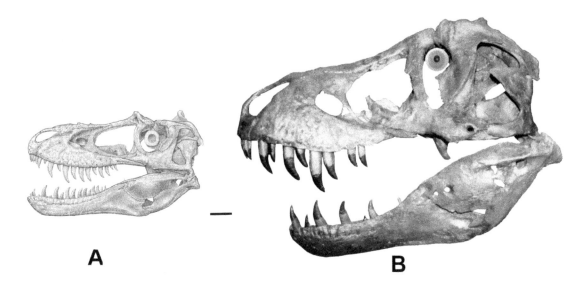

A　　　　　　　　　　**B**

14.4. Sclerotic rings for *Tyrannosaurus* are unknown but can be approximated from that of *Gorgosaurus* (A) by scaling the orbital lengths (B). These data can be used to estimate eyeball diameter (11–12 cm), pupil diameter (2.5 cm), and focal length (7.7–9.6 cm). These results indicate that *Tyrannosaurus* had poor low-light vision (see text for details). If the length of the skull of *Gorgosaurus* were simply scaled to that of *Tyrannosaurus,* the orbit would be 21.5 cm in diameter and the sclerotic ring, 12.5 cm, which are values greater than the actual orbital length. Scale 10 cm.

(A) Adapted in part from Currie (2003).

be approximated from the sclerotic ring if it were known. It is known for the tyrannosaurid *Gorgosaurus* (Fig. 14.4A; Currie 2003), and, assuming that the size of the sclerotic ring scales with the size of the orbit, then the sclerotic ring can be approximated for *Tyrannosaurus*. Surprisingly, the orbit size of the much smaller, juvenile *Gorgosaurus* is about the same size as that of the *Tyrannosaurus* "Stan," suggesting that eyeball size did not change significantly with increased skull size. The inferred sclerotic ring of *Tyrannosaurus* has a diameter of ~7 cm and an internal aperture diameter of ~3.5 cm. The pupil diameter is about 90 percent of the lens diameter, which in extant reptiles is 55–80 percent of the diameter of the sclerotic ring aperture (Montani et al. 1999). For Stan (skull length = 142 cm), the pupil diameter may have been about 2.5 cm. The iris diameter is close to the diameter of the sclerotic ring, and the eyeball diameter is about 1.4–1.5 times larger. For Stan, that would be an eyeball between 11–12 cm in diameter. All of these numbers are approximations and cannot be verified because no eye tissue is known. Nevertheless, they do seem reasonable based on what is known about the eyes of extant reptiles.

The focal length for *Tyrannosaurus* is more difficult to determine. Eyeball depth/width data are available for the Caribbean flamingo (*Phoenicopterus ruber*), which is about 70 percent. Montani et al. (1999) used 80 percent in their calculations for extinct ichthyosaurs; therefore, it seems safe to assume that eyeball depth for *Tyrannosaurus* is ~7.7–9.6 cm. The f-number as a reflection of vision for Stan's eye is $f/D = 3$–3.8, where f = focal length and D = pupil diameter. Because diurnal animals have f-numbers of 2.1 or higher (Montani et al. 1999), the results would indicate that *Tyrannosaurus* had poor low-light vision and therefore hunted during the day.

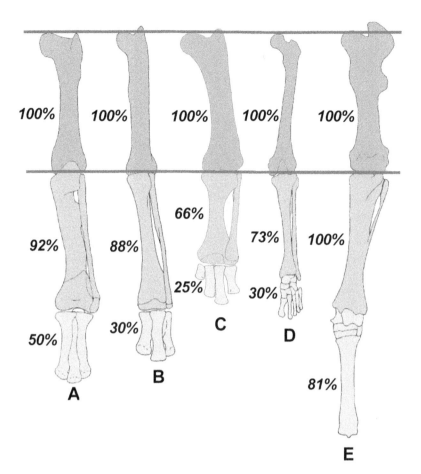

100% 100% 100% 100% 100%

92% 88% 66% 73% 100%

50% 30% 25% 30%

C D

A B

81%

E

14.5. Comparison of the relative proportions of hindlimb segments scaled to femur length in (A) *Tyrannosaurus*, (B) *Edmontosaurus*, (C) *Triceratops*, (D) *Homo*, and (E) *Equus*. The lower leg proportions for *Tyrannosaurus* are far greater than the potential prey, *Edmontosaurus* and *Triceratops*, indicating that it could outrun them.

Speed

The maximum speed of *Tyrannosaurus* is highly controversial, with estimates ranging from as low as 18 km/h (Hutchinson and Garcia 2002) to a maximum of 50 km/h (Paul 2000, 2008). Horner and Lessem (1993) advocate a lower speed based on the femur-tibia proportion and argue that *Tyrannosaurus* must be a scavenger because it was too slow to catch its prey. However, *Tyrannosaurus* is digitigrade, which requires inclusion of the metatarsal in speed calculations, as it is for horses and other cursorial mammals (Pike and Alexander 2002).

The importance of the metatarsal is seen in Figure 14.5, where various limb proportions are scaled to femur length. As may be seen, the tibia and metatarsals of *Tyrannosaurus* are significantly larger than those of potential prey *Edmontosaurus* and *Triceratops*. This proportional addition is especially important in increasing step length (Fig. 14.6). Regardless of the maximum speed of *Tyrannosaurus*, it merely had to be just a little faster than its prey, which it certainly could be with a longer step.

One aspect of locomotion that has not been addressed is the elastic strain energy stored in the tendons of the feet and tail. The importance of elastic strain energy stored in the tendons of the feet has been discussed for various mammals (Alexander 1984; Dimery and Alexander 1985; Farley et al. 1993). It is equally likely that it was significant in *Tyrannosaurus*

14.6. Illustration showing the distance the pelvis (and, hence, the body) of *Tyrannosaurus* moved per step is approximately 3.75 m, compared with a human, which is about 0.8 m.

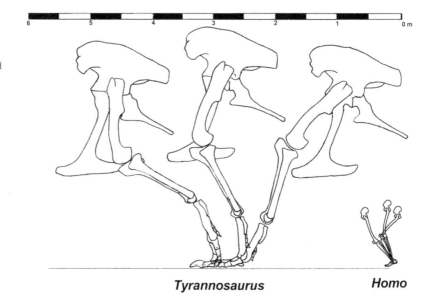

Tyrannosaurus　　　　*Homo*

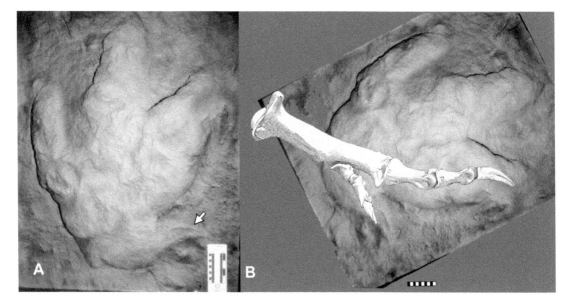

14.7. Cast of the footprint of *Tyrannosaurus* (A) showing the hallux (digit I) impression (arrow). Considering its position relative to metatarsal II (see Fig. 14.1), the foot must have assumed a nearly horizontal position (B). A similar position for a theropod footprint was reported by Gatsey et al. (1999). Scale in cm.

as well, with possible evidence seen in a footprint that shows the impression of digit I (Fig. 14.7A). The position of the impression relative to the other digits and the position of digit I relative to metatarsal II indicate that the metatarsals had to have been sloped, not erect, relative to the ground surface (Fig. 14.7B) at some phase during stride. Gatesy et al. (1999) concluded a similar footfall for a Triassic theropod. With the foot in such a position, the tendons from the *M. gastrocnemius*, which originated on the posterior-distal end of the femur and inserted on the posterior (= plantar) surface of the metatarsals, were stretched, thereby storing elastic energy (Fig. 14.8). This energy was released as the foot pushed forward and upward in a manner analogous to what is seen during ostrich locomotion

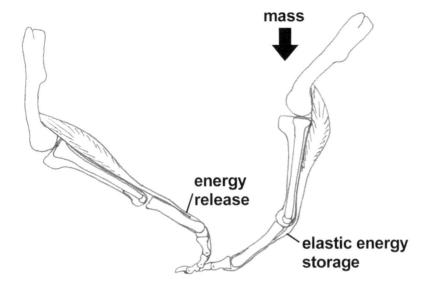

14.8. Storage and release of elastic energy by the gastroc-nemial tendon can increase the power output by the hind legs.

(Alexander et al. 1979). Elastic energy is also stored and released in the interspinal ligaments and tendons of the tail (and, to a much lesser extent, in the dorsals). The tail may be viewed as a third-class lever, with the fulcrum at the base of the tail, the motive force provided by the interspinal ligaments, and the resistance force being gravity, which causes the distal end of the tail to sag. During locomotion, the tail would have oscillated up and down, with energy storage in the interspinal ligaments when the tail moves ventrally and energy release when the tail moves dorsally. The force vectors during release are dorsal and anterior, thus pushing the body mass forwards. (A second component that adds to the lateral pelvic sway is the bilateral movement of the tail relative to the sagittal axis. The result is that the tip of the tail circumscribes a horizontal figure 8 during locomotion).

Short Arms

The short arms of *Tyrannosaurus* have long been controversial, with proponents advocating no use or some use (summarized by Carpenter and Smith 2001; Carpenter 2002; Lipkin and Carpenter 2008). Although short relative to body size (Fig. 14.1), the forelimbs are actually as long as a human arm (Fig. 14.9). Horner and Lessem (1993) argued that the shortness of the arms precluded their use in bringing prey to the mouth. Elsewhere I have shown that no theropod is able to extend its forelimbs very far forward (Fig. 14.10; see Carpenter 2002), therefore the arms were not used in any theropod to bring food to the mouth (*Deinonychus* would have had to tuck its head posteroventrally to reach its hands, but no studies of the cervicals have yet been made to determine this feasibility).

There must have been a functional reason for short arms developing during tyrannosauroid evolution. The primitive condition, seen in *Dilong* and *Guanlong*, is a humerus being more than half the femur

14.9. The forelimb of *Tyrannosaurus* is as long as a human arm, yet the bones are significantly more massive and have more prominent muscle-insertion scars. Scale 10 cm.

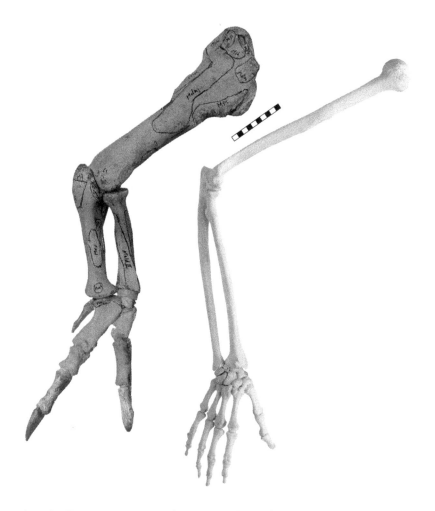

length (Fig. 14.11; see Xu et al. 2004, 2006). By the time tyrannosaurids appeared in the middle Campanian, arm reduction was stabilized, with no further reduction in their segments throughout the rest of tyrannosaurid evolution.

Analysis of reconstructed forelimb musculature indicates a powerful arm, thereby indicating that it was not a useless appendage (Lipkin and Carpenter 2008). Various pathologies on the forelimb also support active use (Lipkin and Carpenter 2008), especially stress fractures in the furcula (Fig. 14.12), as well as musculature avulsions on the humerus (Lipkin and Carpenter 2008). Stress fractures occur as a result of repetitive loading on bone, which leads to mechanical failure and microfracturing (Resnick 2002). This repetitive loading in the furculum is best explained as being caused by intermittent compression of the furculum between the scapulae as the arms grappled with struggling prey. Severe avulsion of muscles inserting on the deltopectoral crest of one specimen, MOR 980, may be due to violent stress from holding a struggling prey or to repetitive overuse (El-Khoury et al. 1997; Stevens et al. 1999). Either way, these and other forelimb pathologies indicate active use of the arms (Lipkin and Carpenter 2008).

Deinonychus

Allosaurus

Tyrannosaurus

Catching Dinner

Although it may be hard to imagine how *Tyrannosaurus* caught its prey (Horner and Lessem 1993), evidence of failed predation on potential prey proves predation. This evidence includes partially healed bite marks on the tail of an adult *Edmontosaurus* (Fig. 14.13; Fig. 15.3 this volume; Carpenter 2000) and on the face of an adult *Triceratops* (Happ 2008). That such predation is typical of large tyrannosaurids is evidenced by a new specimen of hadrosaur with evidence of a failed attack by *Daspletosaurus* (see Murphy et al. 2013). Such specimens disprove the pure scavenging hypothesis of Ruxton and Houston (2002). The specimens also show that *Tyrannosaurus* had the capacity of biting through bone, as previously hypothesized from the teeth and skull architecture (Farlow et al. 1991; Meers 2002; Rayfield 2004, 2005; Therrien et al. 2005).

Conclusion

There is no doubt that *Tyrannosaurus* was a carnivore, as attested by coprolite content (Chin et al., 1998), tooth marks on bone (Erickson and Olson 1996; Erickson et al. 1996), tooth analysis (Farlow et al. 1991), and cranial mechanics (Meers 2002; Rayfield 2004, 2005; Therrien et al. 2005). What has been controversial is whether the prey was scavenged or hunted, or, as Horner and Lessem (1993) admitted, perhaps *Tyrannosaurus* was a lazy opportunist that hunted but would not pass up a free meal. Such a position is supported by Farlow and Holtz (2002), as well as by

14.11. Chart showing the relative proportion of forelimb and hindlimb bones. Note the relatively greater proportions of the humerus and radius in the primitive tyrannosauroid *Guanlong,* compared to later tyrannosaurids. Once the short forelimb was attained, no further limb reduction occurred during tyrannosaurid evolution.

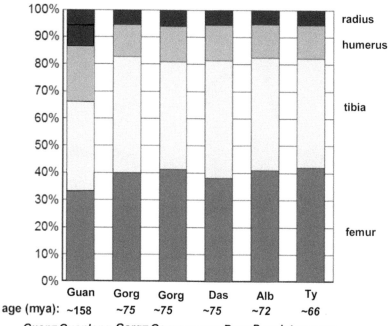

Guan=*Guanlong* Gorg=*Gorgosaurus* Das=*Daspletosaurus*
Alb= *Albertosaurus* Ty= *Tyrannosaurs*

Lipkin and Carpenter (2008), and is a position supported here. However, the abundance of pathologies indicates that tyrannosaurids did live an extreme lifestyle, as advocated by Paul (2008).

Acknowledgments

Thanks to Peter Larson and the Black Hills Institute of Geological Research for measurements and photographs; Tim Rowe and Digimorph for use of computed tomographic images of birds; Lorrie McWhinney and the Radiology Department of Kaiser Permanente, Denver, for CAT scans and MRIs of the caiman; Jeff Stephenson, Denver Museum of Nature and Science, for access to bird skulls; Michael Henderson and Scott Williams, Burpee Museum of Natural History, for the invitation to the tyrannosaurid conference; and Michael J. Parrish for his editorial work. Thanks to Philip Manning for an enjoyable discussion of energy storage and release during locomotion while standing between the hindlimbs of Stan. We had independently concluded that there was elastic energy storage in the tail and foot. And finally, thanks to Jim Farlow for review comments.

Literature Cited

Alexander, R. M. 1984. Elastic energy storage in running vertebrates. *American Zoologist* 24:85–94.
Alexander, R. M., G. M. O. Maloiy, R. Njau, and A. S. Jayes. 1979. Mechanics of running of the ostrich (*Struthio camelus*). *Journal of Zoology, London A* 187:169–178.
Brochu, C. A. 2000. A digitally-rendered endocast for *Tyrannosaurus rex. Journal of Vertebrate Paleontology* 20:1–6.
Brochu, C. A. 2003. Osteology of *Tyrannosaurus rex:* Insights from a nearly

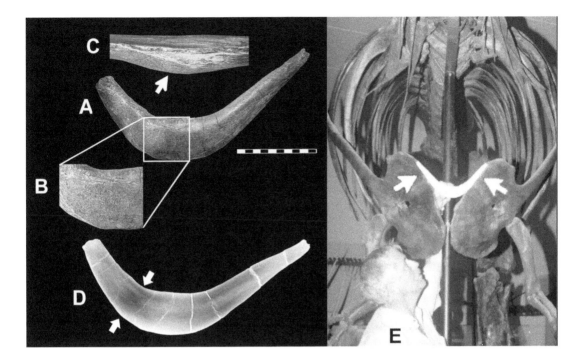

complete skeleton and high-resolution computed tomographic analysis of the skull. *Society of Vertebrate Paleontology Memoir* 7:1–138.

Brown, B. 1915. *Tyrannosaurus,* the largest flesh-eating animal that ever lived. *Museum Journal* 15:271–279.

Carpenter, K. 2000. Evidence of predatory behavior by carnivorous dinosaurs; pp. 135–144 in B. P. Pérez-Moreno, T. Holtz, Jr., J. L. Sanz, and J. Moratalla (eds.), Aspects of Theropod Paleobiology. *Gaia: Revista de Geociencias, Museu Nacional de Historia Natural, Lisbon,* vol. 15.

Carpenter, K. 2002. Forelimb biomechanics of nonavian theropod dinosaurs in predation. *Senckenbergiana Lethaea* 82:59–76.

Carpenter, K., and M. Smith. 2001. Forelimb osteology and biomechanics of *Tyrannosaurus rex;* pp. 90–116 in D. H. Tanke and K. Carpenter (eds.), *Mesozoic Vertebrate Life.* Indiana University Press, Bloomington.

Chin, K., T. T. Tokaryk, G. M. Erickson, and L. C. Calk. 1998. A king-sized theropod coprolite. *Nature* 393:680–682.

Currie, P. J. 2003. Cranial anatomy of tyrannosaurid dinosaurs from the Late Cretaceous of Alberta, Canada. *Acta Palaeontologica Polonica* 48:191–226.

Dimery, N. J., and R. M. Alexander. 1985. Elastic properties of the hind foot of the donkey, *Equus asinus. Journal of Zoology, London A* 207:9–20.

El-Khoury, G. Y., W. W. Daniel, and M. H. Kathol. 1997. Acute and chronic avulsive injuries. *Radiological Clincs of North America* 35:747–766.

Erickson, G. M., and K. H. Olson. 1996. Bite marks attributable to *Tyrannosaurus rex:* preliminary description and implications. *Journal of Vertebrate Paleontology* 16:175–178.

Erickson, G. M., S. D. van Kirk, J. Su, M. E. Levenston, W. E. Caler, and D. R. Carter. 1996. Bite force estimation for *Tyrannosaurus rex* from tooth-marked bone. *Nature* 382:706–708.

Farley, C. T., J. Glasheen, and T. A. McMahon. 1993. Running springs: Speed and animal size. *Journal of Experimental Biology* 185:71–86.

Farlow, J. O.1994. Speculations about the carrion-locating ability of tyrannosaurs. *Historical Biology* 7:159–165.

Farlow, J. O., and T. R. Holtz, Jr. 2002. The fossil record of predation in dinosaurs. *Paleontological Society Papers* 8:251–265.

Farlow, J. O., D. L. Brinkman, W. L. Abler, and P. J. Currie. 1991. Size, shape, and serration density of theropod dinosaur lateral teeth. *Modern Geology* 16:161–198.

Gatesy, S. M., K. M. Middleton, F. A. Jenkins, and N. H. Shubin. 1999. Three-dimensional preservation of foot movements in Triassic theropod dinosaurs. *Nature* 399:141–144.

14.12. A) Furcula of *Tyrannosaurus* (TCM 2001.90.1) showing evidence of stress fracture on the posterior side (note also the missing proximal end of the left ramus). B) Close-up showing filagree pattern characteristic of a mild osteomyelitis infection. C) Extosis of the stress fracture in dorsal view. D) Radiolucent thickening as seen in X-ray. E) Position of the furcula bridging the scapula (arrows) means that a great deal of bending stress would result from tension developed from struggling prey.

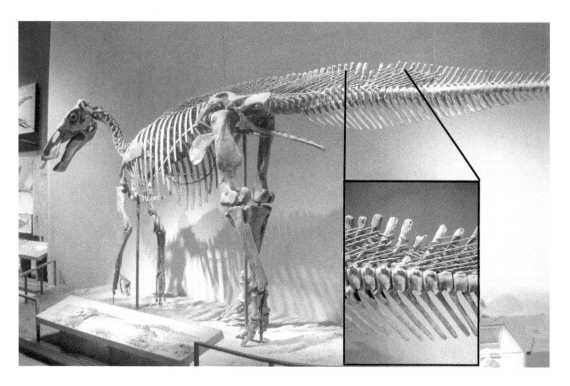

14.13. Evidence of failed predation by *Tyrannosaurus* is seen in the tail section of *Edmontosaurus*.

See Carpenter (2000).

Haines, T. 2000. *Walking with Dinosaurs: A Natural History.* Dorling Kindersley, New York.

Happ, J. 2008. An analysis of predator–prey behavior in a head-to-head encounter between *Tyrannosaurus rex* and *Triceratops;* pp. 354–368 in P. Larson and K. Carpenter (eds.), Tyrannosaurus rex, *the Tyrant King.* Indiana University Press, Bloomington.

Horner, J. R. 1994. Steak knives, beady eyes, and tiny little arms (a portrait of *T. rex* as a scavenger); pp. 157–164 in G. D. Rosenberg and D. L. Wolberg (eds.), Dino Fest. *Paleontological Society Special Publication,* no. 7.

Horner, J. R., and D. Lessem. 1993. *The Complete* T. Rex. Simon & Schuster, New York.

Hutchinson, J. R., and M. Garcia. (2002). Tyrannosaurus was not a fast runner. *Nature* 415:1018–1021.

Lipkin, C., and K. Carpenter. 2008. Looking again at the forelimb of *Tyrannosaurus rex;* pp. 166–90 in P. Larson and K. Carpenter (eds.), Tyannosaurus rex, *the Tyrant King.* Indiana University Press, Bloomington.

Lull, R. S., and N. E. Wright. 1942. Hadrosaurian dinosaurs of North America. *Geological Society of America Special Paper* 40:1–242.

Motani, R., B. M. Rothschild, and W. Wahl. 1999. Large eyeballs in diving ichthyosaurs. *Nature* 402:747.

Meers, M. R. 2002. Maximum bite force and prey size of *Tyrannosaurus rex* and their relationships to the inference of feeding behavior. *Historical Biology* 16:1–12.

Murphy, N. L., K. Carpenter, and D. Trexler. 2013. New evidence for predation by a large tyrannosaurid; pp. 278–285 in J. Michael Parrish, Ralph E. Molnar, Philip J. Currie, and Eva B. Koppelhus (eds.) *Tyrannosaurid Paleobiology.* Indiana University Press, Bloomington.

Osborn, H. F. 1905. *Tyrannosaurus* and other Cretaceous carnivorous dinosaurs. *Bulletin of the American Museum of Natural History* 21:259–265.

Osborn, H. F. 1906. *Tyrannosaurus,* Upper Cretaceous carnivorous dinosaur (second communication.). *Bulletin of the American Museum of Natural History* 22:150–165.

Osborn, H.F. 1910. The *Tyrannosaurus. American Museum Journal* 10:3–8.

Osborn, H. F. 1912. Crania of *Tyrannosaurus* and *Allosaurus. Memoirs of the American Museum of Natural History,* n.s., 1:1–30.

Osborn, H. F. 1916. Skeletal adaptations of *Ornitholestes, Struthiomimus, Tyrannosaurus. Bulletin of the American Museum of Natural History* 35:761–771.

Ostrom, J. H. 1961. Cranial morphology of the hadrosaurian dinosaurs of North America. *Bulletin of the American Museum of Natural History* 122:33–186.

Paul, G. S. 2000. Limb design, function and running performance in ostrich-mimics and tyrannosaurs; pp. 257–70 in B. P. Pérez-Moreno, T. Holtz, Jr., J. L. Sanz, and J. Moratalla (eds.), Aspects of Theropod Paleobiology. *Gaia: Revista de Geociencias, Museu Nacional de Historia Natural, Lisbon,* vol. 15.

Paul, G. S. 2008. The extreme lifestyles and habits of the gigantic tyrannosaurid super-predators of the Late Cretaceous of North America and Asia; pp. 306–52 in P. Larson and K. Carpenter (eds.), Tyannosaurus rex, *the Tyrant King.* Indiana University Press, Bloomington.

Pike, A. V. L., and R. M. Alexander. 2002. The relationship between limb-segment proportions and joint kinematics for the hind limbs of quadrupedal mammals. *Journal of Zoology, London* 258:427–433.

Rayfield, E. J. 2004. Cranial mechanics and feeding in *Tyrannosaurus rex. Proceedings of the Royal Society of London B* 271:1451–1459.

Rayfield, E. J. 2005. Aspects of comparative cranial mechanics in the theropod dinosaurs *Coelophysis, Allosaurus* and *Tyrannosaurus. Zoological Journal of the Linnean Society* 144:309–316.

Resnick, D. 2002. *Diagnosis of Bone and Joint Disorders.* Saunders, Philadelphia.

Ruxton, G. D., and Houston, D. C. 2002. Could *Tyrannosaurus rex* have been a scavenger rather than a predator? An energetics approach. *Proceedings of the Royal Society of London B* 270:731–733.

Stevens, M. A., G. Y. El-Khoury, M. H. Kathol, E. A. Brandser, and S. Chow.1999. Imaging features of avulsion injuries. *RadioGraphics* 19:655–672.

Therrien, F., D. M. Henderson, and C. B. Ruff. 2005. Bite me: biomechanical models of theropod mandibles and implications for feeding behavior; pp. 179–237 in K. Carpenter (ed.), *The Carnivorous Dinosaurs.* Indiana University Press, Bloomington.

Xu, X., M. A. Norell, X. Kuang, X. Wang, Q. Zhao, and C. Jia. 2004. Basal tyrannosauroids from China and evidence for protofeathers in tyrannosauroids. *Nature* 431:680–684.

Xu, X., J. M. Clark, C. A. Forster, M. A. Norell, G. M. Erickson, D. A. Eberth, C. Jia, and Q. Zhao. 2006. A basal tyrannosauroid dinosaur from the Late Jurassic of China. *Nature* 439:715–718.

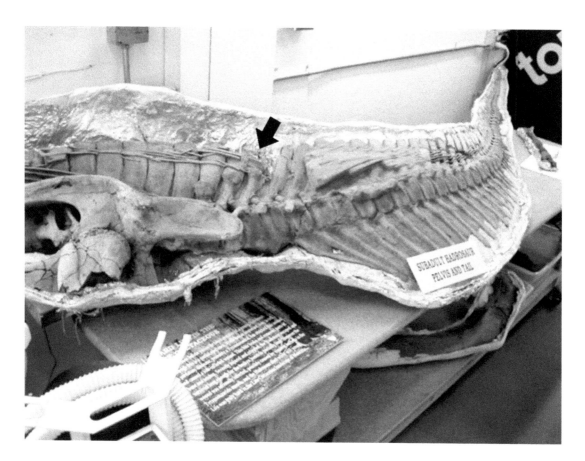

15.1. Partial pelvic and caudal regions of *Brachylophosaurus* cf. *canadensis,* JRF 1002, showing evidence of a failed attack by a predator, possibly *Daspletosaurus.* Arrow denotes the neural spine that was bitten off.

New Evidence for Predation by a Large Tyrannosaurid

15

Nate L. Murphy, Kenneth Carpenter, and David Trexler

A partial skeleton of a hadrosaur believed to be *Brachylophosaurus canadensis* shows evidence of a failed attack by a large theropod, possibly *Daspletosaurus* sp. The injury consists of damage to the neural spines of the last sacral vertebra and first two caudal vertebrae. Remodeled bone, even at the site of the traumatic amputation of the caudal neural spine, demonstrates that the individual survived the attack. In addition, the attack came from the rear, suggesting that the *Brachylophosaurus* was fleeing the attacker. This specimen adds to the growing body of knowledge that tyrannosaurids were capable of active predation.

Abstract

Introduction

The relative importance of predation and scavenging by tyrannosaurid theropods has been controversial. Horner and Lessem (1993) have argued that *Tyrannosaurus* was an obligate scavenger because of the large body size, short arms, large eyes, and large olfactory lobes. However, Carpenter (2000) described a failed attack on an adult hadrosaur, *Edmontosaurus.* Carpenter argued from morphological features that the bite was most parsimoniously ascribed to *Tyrannosaurus.* The features of *Tyrannosaurus* that Horner and Lessem (1993) cited as evidence for scavenging apply equally to other tyrannosaurids as well. By inference then, the other tyrannosaurids (e.g., *Gorgosaurus, Daspletosaurus* and *Albertosaurus*) were scavengers in the absence of evidence to the contrary. That contrary evidence is now slowly emerging. Wegweiser et al. (2004) report evidence of an attack on a lambeosaurine hadrosaur by an unidentified tyrannosaurid. That specimen consists of a partially healed rib bearing the impression of a tooth. And now, newly discovered articulated caudal vertebrae of the hadrosaurid *Brachylophosaurus* cf. *canadensis* provides evidence that another tyrannosaurid, probably *Daspletosaurus*, was an active predator as well.

The specimen (Fig. 15.1) consists of dorsals 15–18, sacrals 1–9, and caudals 1–74 with associated chevrons and ossified tendons, the ilia, proximal ends of the left and right pubes, iliac peduncle of the left and right ischia, and proximal ends of the left and right femora. Because the tail is complete to the terminal caudal, we believe the specimen was originally complete and that the missing parts were lost to erosion. The specimen shows trauma to the tail (Fig. 15.2) that we believe demonstrates a failed attack similar to the specimen described by Carpenter (2000).

Institutional Abbreviations DMNH, Denver Museum of Natural History (now Denver Museum of Nature and Science), Denver, Colorado; JRF, Judith River Foundation/ Judith River Dinosaur Institute, Malta, Montana.

15.2. Close-up of the bite-marked region in JRF 1002. A) Bitten off neural spine (arrow) at the base of the tail. B) Traumatic amputation of the neural spine of the second caudal (arrows). C) Detail of the traumatized region of the caudals. Note exostosic bone at the site of the traumatic amputation and on the neural spines of sacral neural spines 8 and 9 and caudal 1.

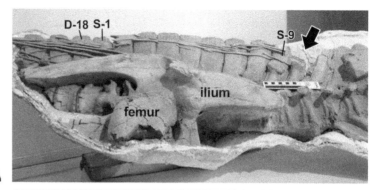

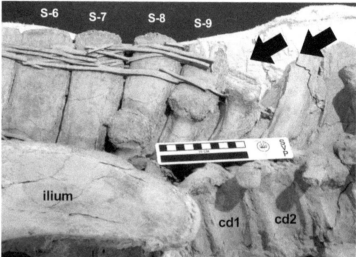

The specimen, JRF-1002, was found in the lower portions of the Judith River Formation in Phillips County, along the Little Cottonwood drainage in northeastern Montana. The specimen was found 138 feet above the marine Claggett Formation. The lithology of the site is composed of loosely consolidated, fine-grained sandstone containing some clay clasts and some laminates of carbonaceous plant material; the depositional environment is believed to be meandering river. The depositional environment and taphonomy is similar to that of other *Brachylophosaurus* specimens recovered from the same general area (see Murphy et al. 2006).

Description

The specimen, JRF-1002, has a total length of 354 cm, of which the sacrum is 46 cm and the tail 297 cm; the left ilium is 67 cm long. From these measurements, the individual is estimated to have been 4.35 m long and 1.4 m tall at the hips. The sutures between the centra and neural arches are still visible, indicating that the individual was not fully grown. The neural spine of the last sacral (S-8) and of caudals 1, 2, and 3 show trauma (Fig. 15.2) that we interpret to have been caused by a bite most likely inflected by a large tyrannosaurid. In addition, the ossified tendons are missing in this region but are otherwise in situ elsewhere on the specimen. The damaged sacral neural spine (S-8) shows localized exostosis on the lateral surface (Fig. 15.2C), as does the neural spine on caudal 1 (C-1). Caudal 2 is missing the distal third of the neural spine (Fig. 15.2C), and a sliver (Fig. 15.2B) is missing from the anterior distal portion of C-3. The exostosis of S-8 measures about 5.5 cm long and 4.5 cm tall, and that of C-1 measures 4.8 cm long and 3.8 cm tall. Both exostoses show a slight filigree pattern around the edges, which is hallmark of osteomyolitis, or bone infection (McWhinney et al. 2001). The exostosis is a reactive bone, probably caused by trauma to the periosteum and to the underlying bone. About 4.5 cm of caudal neural spine C-2 is missing, presumably bitten off, and the preserved portion of the spine also shows some filigreeing and some lateral lipping. This lipping has expanded the bone laterally, as is also seen in the bite zone of *Edmontosaurus* DMNH 1493 (Fig. 15.3; Fig. 14.13 this volume). A small sliver of bone, 2.8 cm tall and 0.5 cm long, is also missing along the distal edge of caudal neural spine 3. There is a slight spur of the bone where it was amputated. The exostosic sites occur just above the level of the postzygopophysis of S-8 and about mid-shaft on the neural spine of S-9. Together with the amputated neural spines, they form a very gentle arc that we interpret as delineating the left side of the attacker's mouth.

Discussion

The trauma suffered by JRF-1002 is most parsimoniously interpreted as a failed attack by a large predator when the remodeled bone and amputated neural spine are taken into account. This hadrosaur represents the second such specimen known, the previous being that of an *Edmontosaurus* (Carpenter 2000). In both cases the attack was directed to the tail.

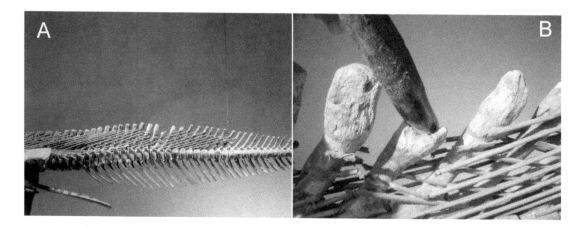

15.3. Tail of *Edmontosaurus annectens* (DMNH 1493) showing evidence of attack by *Tyrannosaurus rex* (A). Detail of injured neural spine showing how well a tooth of *Tyrannosaurus* fits (B). Note tooth puncture in adjacent neural spine.

Described by Carpenter (2000).

Based on the angle of the bites, the predator must have approached from the rear. We may infer then, that the victims were fleeing at the time of the attacks. Unfortunately, in neither instance can anything be said about the speeds of the prey or predators.

Why this particular individual of *Brachylophosaurus* was attacked is not known because too much of the skeleton is missing and key regions are not represented. In the case of the *Edmontosaurus*, the left ilium shows a well-developed callus around a fracture in which the preacetabular blade was displaced ventrally by actions of the *M. ilio-tibialis*. Most likely this individual limped and for that reason may have been singled out for attack. Whether some comparable feature was evident in the *Brachylophosaurus* is unknown. What we do know is that the attacker approached from the rear, although we have not determined whether from the right or left. In the case of the *Edmontosaurus*, the attack came from the right rear based on the angle of the tooth mark on the amputated caudal neural spine.

The *Edmontosaurus* attacker was most likely a *Tyrannosaurus* because of the broad tooth mark on the missing neural spine and because it was the only contemporaneous predator capable of reaching the 2.9 m height of the neural spines. The contemporaneous *Nanotyrannus* had narrow-bladed teeth and stood considerably less than 2.9 m tall. In the case of *Brachylophosaurus*, two equally large theropod candidates are known from the Judith River Formation: *Gorgosaurus* and *Daspletosaurus*. In the coeval Dinosaur Park Formation of southern Alberta (approximately 370 km to the northwest), *Gorgosaurus* specimens outnumber *Daspletosaurus* specimens two to one (Béland and Russell 1978:table 4 [*Gorgosaurus* as *Albertosaurus*]; Currie 2003:table 1). However, in the lower Judith River Formation in the vicinity of Malta, *Dapletosaurus* specimens are more common, at least as isolated teeth. This may be supported by an inventory of shed teeth referable to *Daspletosaurus* recovered from the lower Judith River Formation of Phillips County and its Canadian equivalent, the Oldman Formation. This inventory shows that 68 percent of tyrannosaurid teeth from the Judith River are referable to

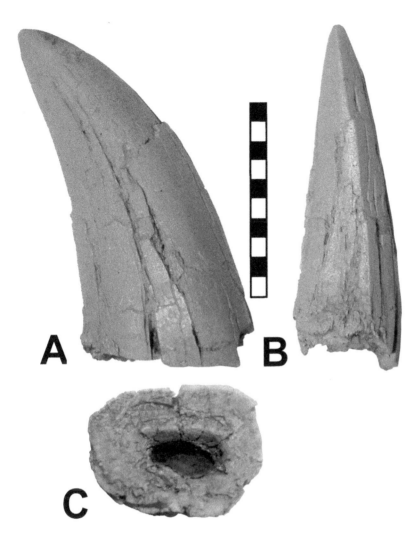

15.4. Tooth of *Daspletosaurus* (JRF-91-ASH) in multiple views showing its robust design (similar to *Tyrannosaurus*). The tooth is well designed for shearing bone. A) Tooth in lateral view. B) Tooth in posterior view. C) Tooth in ventral view showing the robustness of the crown. Scale in mm.

Daspletosaurus. Thus, comparing the ratios of *Gorgosaurus* and *Daspletosaurus* from the Dinosaur Park Formation and from the Judith River and Oldman formations suggests that there was some overlapping of paleobiogeographic ranges, with one or the other taxon dominant. A somewhat similar conclusion for these two tyrannosaurids was independently reached by Farlow and Pianka (2003) on ecological grounds.

Russell (1970) suggested a correspondence between predator and prey in the Dinosaur Park Formation (as the Oldman Formation) based on relative abundance. Thus, *Gorgosaurus* (as *Albertosaurus*) preyed on hadrosaurs, whereas *Daspletosaurus* preyed on ceratopsians. There is, however, no a priori reason to assume a correlation between the relative abundance of prey and their predators as long as prey outnumber predators. A more reliable method would be the discovery of an articulated skeleton of a predator in association with the skeleton of its prey in the manner of *Velociraptor* and *Protoceratops* (Carpenter 2000). Such an association was identified as a category 1 record of behavior by Carpenter et al. (2005). In contrast, the bite marks on the caudal vertebrae of the

15.5. Comparison of the skulls of (A) *Daspletosaurus* and (B) *Gorgosaurus.* Note the broad, rounded snout of *Daspletosaurus,* which is similar to that seen in *Tyrannosaurus.* Scale in cm. *(A) Modified from Russell (1970). (B) Adapted in part from Currie (2003).*

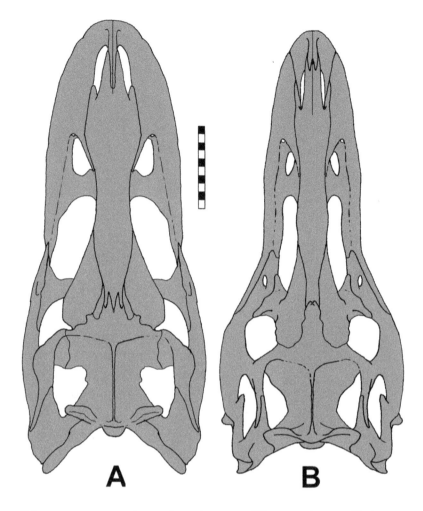

A   B

Edmontosaurus was deemed a category 4: "Trace evidence of behavior where the identity of the maker is inferred, but not established conclusively" (Carpenter et al. 2005:table 17.1). JRF-1002 is also considered an example of category 4 evidence of behavior. Although we cannot rule out that the attacker of the *Brachylosaurus* was *Gorgosaurus,* it seems more probable that the attacker was *Daspletosaurus* for the following reasons: (1) the teeth of *Daspletosaurus* resemble those of *Tyrannosaurus* and are well adapted for shearing bone (Fig. 15.4); (2) the mandibular force profile of *Daspletosaurus* is similar to that of *Tyrannosaurus,* suggesting a similar bite force (Therrien et al. 2005); (3) the bite pattern on the *Brachylophosaurus* shows only a single row of teeth, indicating a very broad mouth (the other row of teeth bit air); and (4) the skull of *Daspletosaurus* is broad, rather than narrow, as in *Gorgosaurus* (Fig. 15.5), and probably had a similar skull bending strength as *Tyrannosaurus* (Therrien et al. 2005).

The hypothesis that hadrosaurs formed a major component of the diet of *Daspletosaurus* is strengthened by acid-etched hadrosaur bones associated with a specimen of *Daspletosaurus.* These bones were interpreted by Varrichio (2001) to be stomach contents of the carnivore. The prey specimen is also of a sub-adult and was smaller, hence presumably younger, than JRF-1002. However, it is not known whether this hadrosaur was attacked or scavenged.

A growing body of evidence indicates that tyrannosaurid theropods were active hunters of hadrosaurs. Tanke (1989) has previously suggested that damaged neural spines in the caudal region of some hadrosaurs are indicative of "rough sex." However, in light of the specimens showing aborted attacks by a predator, perhaps these other specimens need to be reexamined. To date, evidence of failed attacks, as inferred from partial healing of injuries, includes two specimens bearing bite marks in the caudal regions and one specimen bearing bite marks on a rib.

Conclusion

We thank all those who were involved in the excavation of JRF-1002. We also thank the organizers of the tyrannosaurid conference held at the Burpee Museum of Natural History, where this specimen was first revealed. Thanks also to Michael Parrish and James Farlow for review comments.

Acknowledgments

Literature Cited

Béland, P., and D. A. Russell. 1978. Paleoecology of Dinosaur Provincial Park (Cretaceous), Alberta, interpreted from the distribution of articulated vertebrate remains. *Canadian Journal of Earth Sciences* 15:1012–1024.

Carpenter, K. 2000. Evidence of predatory behavior by carnivorous dinosaurs. *Gaia* 15:135–144.

Carpenter, K., F. Sanders, L. A. McWhinney, and L. Wood. 2005. Evidence for predator-prey relationships: examples for *Allosaurus* and *Stegosaurus;* pp. 325–50 in K. Carpenter (ed.), *The Carnivorous Dinosaurs.* Indiana University Press, Bloomington.

Currie, P. J. 2003. Cranial anatomy of tyrannosaurid dinosaurs from the Late Cretaceous of Alberta, Canada. *Acta Palaeontologica Polonica* 48:191–226.

Farlow, J. O., and E. R. Pianka. 2003. Body size overlap, habitat partitioning and living space requirements of terrestrial vertebrate predators: implications for the paleoecology of large theropod dinosaurs. *Historical Biology* 16:21–40.

Horner, J. R., and D. Lessem. 1993. *The Complete* T. rex. Simon & Schuster, New York.

McWhinney, L., B. Rothschild, and K. Carpenter, 2001. Posttraumatic chronic osteomyeltis in *Stegosaurus* dermal spikes; pp. 141–156 in Carpenter, K. (ed.), *The Armored Dinosaurs.* Indiana University Press, Bloomington.

Murphy, N. L., D. Trexler, and M. Thompson. 2006. "Leonardo," a mummified *Brachylophosaurus* (Ornithischia: Hadrosauridae) from the Judith River Formation of Montana; pp. 117–33 in K. Carpenter (ed.), *Horns and Beaks: Ceratopsian and Ornithopod Dinosaurs.* Indiana University Press, Bloomington.

Russell, D. A. 1970. Tyrannosaurs from the Late Cretaceous of Western Canada. *National Museum of Natural Sciences Publications in Paleontology* 1:1–34.

Tanke, D. 1989. Paleopathologies in Late Cretaceous hadrosaurs (Reptilia: Ornithischia) from Alberta, Canada. *Journal of Vertebrate Paleontology* 9(3):41A.

Therrien, F., D. M. Henderson, and C. B. Ruff. 2005. Bite me: biomechanical models of theropod mandibles and implications for feeding behavior; pp. 179–237 in K. Carpenter (ed.), *The Carnivorous Dinosaurs.* Indiana University Press, Bloomington.

Varrichio, D. J. 2001. Gut contents from a Cretaceous tyrannosaurid: implications for theropod digestive tracts. *Journal of Paleontology* 75:401–406.

Wegweiser, M., B. Breithaupt, and, R. Chapman. 2004. Attack behavior of tyrannosaurid dinosaur(s): Cretaceous crime scenes, really old evidence, and "smoking guns." *Journal of Vertebrate Paleontology* 24(3, suppl.):127A.

Index

This book was designed by Jamison Cockerham at Indiana University
Press, set in type by Jamie McKee at MacKey Composition, and
printed by Sheridan Books, Inc.

The fonts are Electra, designed by William A. Dwiggins in 1935,
Frutiger, designed by Adrian Frutiger in 1975, and Futura, designed
by Paul Renner in 1927. All were published by Adobe Systems
Incorporated.

Milton Keynes UK
Ingram Content Group UK Ltd.
UKHW022257250324
439855UK00003B/38